THE IMPACT OF
STELLA

Stars are mostly found in binary and mul
stars having companions; this fraction a
large proportion of the stars in these syst
of their components, leading to exotic ob
other chemically peculiar stars, but also
nebulae, supernovae and gamma-ray bu
critical role in the initial mass function, th
cantly affect the dynamical evolution of s
volume presents results from state-of-th
the impact of binaries on stellar evolution
as a bridge between observational and th
for researchers and advanced students of

GIACOMO BECCARI is a staff astrono
work is focused on the study of Blue St
winner of the Levi-Montalcini Prize and
(2014).

HENRI M. J. BOFFIN is a staff ast
Recently he has shown the importance
including discovering the binary star
interferometry to study mass transfer in

CAMBRIDGE ASTROPHYSICS SERIES

Series editors:

Andrew King, Douglas Lin, Stephen Maran, Jim Pringle, Martin Ward and Robert Kennicutt

Titles available in the series

27. *RR Lyrae Stars*
    by Horace A. Smith
28. *Cataclysmic Variable Stars*
    by Brian Warner
29. *The Magellanic Clouds*
    by Bengt E.Westerlund
30. *Globular Cluster Systems*
    by Keith M. Ashman and Stephen E. Zepf
33. *The Origin and Evolution of Planetary Nebulae*
    by Sun Kwok
34. *Solar and Stellar Magnetic Activity*
    by Carolus J. Schrijver and Cornelis Zwaan
35. *The Galaxies of the Local Group*
    by Sidney van den Bergh
36. *Stellar Rotation*
    by Jean-Louis Tassoul
37. *Extreme Ultraviolet Astronomy*
    by Martin A. Barstow and Jay B. Holberg
39. *Compact Stellar X-ray Sources*
    edited by Walter H. G. Lewin and Michiel van der Klis
40. *Evolutionary Processes in Binary and Multiple Stars*
    by Peter Eggleton
41. *The Physics of the Cosmic Microwave Background*
    by Pavel D. Naselsky, Dmitry I. Novikov and Igor D. Novikov
42. *Molecular Collisions in the Interstellar Medium, 2nd Edition*
    by David Flower
43. *Classical Novae, 2nd Edition*
    edited by M. F. Bode and A. Evans
44. *Ultraviolet and X-ray Spectroscopy of the Solar Atmosphere*
    by Kenneth J. H. Phillips, Uri Feldman and Enrico Landi
45. *From Luminous Hot Stars to Starburst Galaxies*
    by Peter S. Conti, Paul A. Crowther and Claus Leitherer
46. *Sunspots and Starspots*
    by John H. Thomas and Nigel O. Weiss
47. *Accretion Processes in Star Formation, 2nd Edition*
    by Lee Hartmann

48. *Pulsar Astronomy, 4th Edition*
    by Andrew Lyne and Francis Graham-Smith
49. *Astrophysical Jets and Beams*
    by Michael D. Smith
50. *Maser Sources in Astrophysics*
    by Malcolm Gray
51. *Gamma-ray Bursts*
    edited by Chryssa Kouveliotou, Ralph A. M. J. Wijers and Stan Woosley
52. *Physics and Chemistry of Circumstellar Dust Shells*
    by Hans-Peter Gail and Erwin Sedlmayr
53. *Cosmic Magnetic Fields*
    by Philipp P. Kronberg
54. *The Impact of Binary Stars on Stellar Evolution*
    by Giacomo Beccari and Henri M. J. Boffin

THE IMPACT OF BINARY STARS ON STELLAR EVOLUTION

Edited by

GIACOMO BECCARI
European Southern Observatory

HENRI M. J. BOFFIN
European Southern Observatory

CAMBRIDGE
UNIVERSITY PRESS

CAMBRIDGE
UNIVERSITY PRESS

University Printing House, Cambridge CB2 8BS, United Kingdom

One Liberty Plaza, 20th Floor, New York, NY 10006, USA

477 Williamstown Road, Port Melbourne, VIC 3207, Australia

314–321, 3rd Floor, Plot 3, Splendor Forum, Jasola District Centre, New Delhi – 110025, India

79 Anson Road, #06–04/06, Singapore 079906

Cambridge University Press is part of the University of Cambridge.

It furthers the University's mission by disseminating knowledge in the pursuit of education, learning, and research at the highest international levels of excellence.

www.cambridge.org
Information on this title: www.cambridge.org/9781108428583
DOI: 10.1017/9781108553070

First published 2019

Printed in the United Kingdom by TJ International Ltd. Padstow Cornwall

A catalogue record for this publication is available from the British Library.

Library of Congress Cataloging-in-Publication Data
Names: Beccari, Giacomo, editor. | Boffin, H. (Henri), 1965– editor.
Title: The impact of binary stars on stellar evolution / edited by Giacomo Beccari (European Southern Observatory, Garching), Henri M. J. Boffin (European Southern Observatory, Garching).
Description: Cambridge, United Kingdom ; New York, NY : Cambridge University Press, 2019. | Series: Cambridge astrophysics series ; 54
Identifiers: LCCN 2018042461 | ISBN 9781108428583 (hardback)
Subjects: LCSH: Stars–Evolution. | Double stars.
Classification: LCC QB806 .I567 2019 | DDC 523.8–dc23
LC record available at https://lccn.loc.gov/2018042461

ISBN 978-1-108-42858-3 Hardback

Contents

Contributors

Marc Audard
Department of Astronomy, University of Geneva, Chemin d'Ecogia, 1290 Versoix, Switzerland

Henri M. J. Boffin
European Southern Observatory, Karl-Schwarzschild-str. 2, 85748 Garching, Germany

Cathie J. Clarke
Institute of Astronomy, Madingley Road, Cambridge, CB3 0HA, United Kingdom

Sylvia Ekström
Department of Astronomy, University of Geneva, Maillettes 51 – Sauverny, 1290 Versoix, Switzerland

J. J. Eldridge
Department of Physics, University of Auckland, Private Bag 92019, Auckland, New Zealand

Dafydd W. Evans
Institute of Astronomy, University of Cambridge, Madingley Road, Cambridge, CB3 0HA, United Kingdom

Laurent Eyer
Department of Astronomy, University of Geneva, Chemin des Maillettes 51, 1290 Versoix, Switzerland

Francesco R. Ferraro
Department of Physics and Astronomy, Bologna University, Gobetti 93/2, I–40129 Bologna, Italy

Cyril Georgy
Department of Astronomy, University of Geneva, Maillettes 51 – Sauverny, 1290 Versoix, Switzerland

Ghina M. Halabi
Institute of Astronomy, Madingley Road, Cambridge, CB3 0HA, United Kingdom

Na'ama Hallakoun
School of Physics and Astronomy, Tel Aviv University, 6997801 Tel Aviv, Israel

Edward P. J. van den Heuvel
Anton Pannekoek Institute of Astronomy, University of Amsterdam, Science Park 904, 1098XH Amsterdam, The Netherlands

Simon Hodgkin
Institute of Astronomy, University of Cambridge, Madingley Road, Cambridge, CB3 0HA, United Kingdom

Berry Holl
Department of Astronomy, University of Geneva, Chemin d'Ecogia, 1290 Versoix, Switzerland

Robert G. Izzard
Astrophysics Research Group, Faculty of Engineering and Physical Sciences, University of Surrey, Guildford, Surrey, GU2 7XH, United Kingdom
Institute of Astronomy, Madingley Road, Cambridge, CB3 0HA, United Kingdom

Tereza Jerabkova
European Southern Observatory, Karl-Schwarzschild-str. 2, 85748 Garching, Germany

David Jones
Institute for Astrophysics in Canarias, E-38205 La Laguna, Tenerife, Spain

Pavel Kroupa
Helmholtz-Institut für Strahlen und Kernphysik, Universität Bonn
Astronomical Institute, Charles University in Prague, Czech Republic

Barbara Lanzoni
Department of Physics and Astronomy, Bologna University, Gobetti 93/2, I–40129 Bologna, Italy

Isabelle Lecoeur-Taibi
Department of Astronomy, University of Geneva, Chemin des Maillettes 51, 1290 Versoix, Switzerland

Emily M. Leiner
University of Wisconsin–Madison, 475 N. Charter St., Madison, WI 53706, USA

Sara Lucatello
National Institute of Astrophysics (INAF) – Astronomical Observatory in Padova, Vicolo dell'Osservatorio 5, 35122 Padova, Italy

Michela Mapelli
Institute for Astrophysics and Particle Physics, University of Innsbruck, Technikerstrasse 25/8, A–6020, Innsbruck, Austria
National Institute of Astrophysics (INAF) – Astronomical Observatory in Padova, Vicolo dell'Osservatorio 5, I–35122, Padova, Italy

Robert D. Mathieu
University of Wisconsin–Madison, 475 N. Charter St., Madison, WI 53706, USA

Andrea Mehner
European Southern Observatory, Alonso de Cordova 3107, Vitacura, Santiago de Chile, Chile

Maxwell Moe
Steward Observatory, University of Arizona, 933 N. Cherry Ave., Tucson, AZ 85721, USA

Nami Mowlavi
Department of Astronomy, University of Geneva, Chemin d'Ecogia, 1290 Versoix, Switzerland

Ulisse Munari
National Institute of Astrophysics (INAF), Astronomical Observatory of Padova, 36012 Asiago (VI), Italy

Gijs Nelemans
Department of Astrophysics/Institute for Mathematics, Astrophysics and Particle Physics (IMAPP), Radboud University, Nijmegen, The Netherlands
Institute of Astronomy, Catholic University (KU) Leuven, Leuven, Belgium

Ferdinando Patat
European Southern Observatory, Karl-Schwarzschild-str. 2, 85748 Garching, Germany

Dimitri Pourbaix
Institute of Astronomy and Astrophysics, Free University of Brussels (ULB), 1050, Brussels, Belgium

Andrej Prša
Villanova University, Department of Astrophysics and Planetary Sciences, 800 E Lancaster Avenue, Villanova, PA 19085, USA

Lorenzo Rimoldini
Department of Astronomy, University of Geneva, Chemin d'Ecogia, 1290 Versoix, Switzerland

Maurizio Salaris
Astrophysics Research Institute, Liverpool John Moores University, 146 Brownlow Hill, Liverpool, L3 5RF, United Kingdom

Hugues Sana
Institute of Astrophysics, Catholic University Leuven, Celestijnlaan 200D, 3001 Leuven, Belgium

George Seabroke
Mullard Space Science Laboratory, University College London, Dorking, Surrey, United Kingdom

Elizabeth R. Stanway
Department of Physics, University of Warwick, Gibbet Hill Road, Coventry, CV4 7AL, United Kingdom

Nial R. Tanvir
Department of Physics and Astronomy, University of Leicester, University Road, Leicester, LE1 7RH, United Kingdom

Hans van Winckel
Institute for Astronomy, Catholic University, Leuven, Celestijnlaan 200D, 3001 Leuven, Belgium

Lukasz Wyrzykowsi
Warsaw University Astronomical Observatory, Al. Ujazdowskie 4, PL-00-478 Warsaw, Poland

Foreword

Some 40 years ago Abt and Lewis (1976) discovered that more than half of all solarlike stars are members of binary systems. Inspection of their period distribution shows that many of them have orbital dimensions such that at some point in time they will interact with one another. A few years later the same was found to hold for the B2-B5 main-sequence stars (Abt and Lewis, 1978; Abt, 1979, 1983), and more recently Abt et al. (1990) found that on average one B2-B5 main-sequence star has 0.8 companions more massive than two solar masses, and 1.9 companions more massive than the Sun. Other studies have shown that these high percentages of potentially interacting binaries are found all along the main sequence. In recent years, Sana et al. (2012) discovered that binary evolution even completely dominates the evolution of the massive OB-type stars. It thus appears that a large fraction of all stars will during their lives interact with a companion star. It is for this reason, as Henri Boffin expresses in his introduction chapter (see Chapter 1), that the textbooks on stellar evolution will have to be rewritten, since so far, these concentrated mostly on the evolution of single stars. In view of the high incidence of binary interactions among all kinds of stars, it is not surprising that almost any kind of important or interesting class of stellar object in the sky has been found to be related to binary evolution, as is demonstrated by the following nonexhaustive list of examples, which also figure prominently in this book.

Novae are stars that suddenly appear in the sky and rise to a luminosity of between 10^4 and 10^6 times the luminosity of the Sun, and then fade away over several weeks to months. They are close binaries consisting of a white dwarf and a normal solarlike star. The white dwarf is the seat of the explosive behaviour. Also, dwarf novae, which flare up to much lower luminosities than novae, are binary systems of the same type.

The brightest X-ray sources in the sky are the X-ray binaries, which consist of a normal star and a neutron star or a black hole. The same is true for the microquasars, of which SS 433 and Cygnus X-3 are key examples. They show relativistic jet outflows, closely resembling those of the supermassive black holes in quasars and Active Galaxy Nuclei.

All three classes of Type I supernovae, which together form about half of all supernovae, are products of binary evolution. Their defining characteristic is the lack of hydrogen in their spectra. The Type Ia supernovae are thermonuclear explosions of carbon-oxygen white dwarfs, triggered by mass transfer from a companion star. Type Ib and Ic supernovae are core-collapse supernovae in which, before the explosion, the cores have been stripped of their hydrogen-rich envelopes by binary interaction.

Binary radio pulsars are products of evolution of binaries in which both stars have reached their final state; in most cases the companion of the pulsar is a neutron star or a white dwarf, and in a few cases a brown dwarf. Particularly the double neutron stars, of which we now know 20 cases in our Galaxy, are objects of outstanding (astro-)physical importance. Thanks to the fact that radio pulsars are extremely accurate clocks, close double neutron stars allow one to measure at least four General and Special Relativistic effects with extremely high precision, yielding highly precise mass measurements of neutron stars, with uncertainties below a fraction of a promille. This was first demonstrated for the Hulse–Taylor binary pulsar PSR B1913+16, discovered in 1974. It has a 7h45m orbital period and an orbital eccentricity of 0.617, induced by the second supernova explosion in the system. The highly accurate measurement of the rate of orbital decay of this system is, as Taylor and Weisberg (1989) showed, in exact agreement with the predictions of Einstein's General Theory of Relativity for orbital decay due to the emission of gravitational waves. This earned Joseph Taylor and Russell Hulse the 1993 Physics Nobel Prize. Of the presently known 20 double neutron stars, about half have orbital periods sufficiently short to show the same relativistic effects as the Hulse–Taylor system, and in some systems, these can be measured with even higher precision than in that system (Kramer and Stairs, 2008).

Another binary-related Physics Nobel Prize was the one of 2002 for Riccardo Giacconi, for discovering the first celestial X-ray sources, which are the *X-ray binaries*.

And a third binary-related Physics Nobel Prize was that of 2011, awarded to Saul Perlmutter, Brian Schmidt and Adam Riess for the discovery of Dark Energy, by using the 'standard-candle' properties of the *Type Ia supernovae*: exploding white dwarfs in binary systems.

The fourth binary-related Physics Nobel Prize is the one of 2017, awarded to the Laser Interferometer Gravitational-Wave Observatory (LIGO) pioneers for the detection of gravitational waves from merging *double black holes*. Apart from showing for the first time that double black-hole binaries exist in nature, the LIGO–Virgo detections of gravitational waves show, as had been predicted based on binary-star evolution, that the strongest gravitational wave sources in the Universe are the merger events of close double black holes and close double neutron stars.

Most of the 20 double neutron stars known in our Galaxy are in the galactic disk, and are clear products of binary evolution; two of them are found in globular clusters and are expected to have formed by gravitational capture processes in the dense cores of these clusters. The same is expected to be true for double black holes: the majority are expected to have formed by binary evolution, and a fraction may also have formed by gravitational capture processes in dense star clusters.

The first LIGO-observed merger event of a double neutron star GW170817 was accompanied by a *short Gamma-Ray Burst*. We therefore now know for sure that also these Gamma-Ray Bursts, which are among the most energetic explosions in the Universe, are due to a binary interaction, in this case of two neutron stars, as had been predicted by Eichler et al. (1989).

Apart from the aforementioned quite spectacular binary stellar phenomena, many other important and intriguing stellar phenomena appear to be due to binary interactions: *run-away OB stars, barium stars (giants as well as dwarfs), millisecond radio pulsars, Carbon Enhanced Metal Poor (CEMP) stars, blue stragglers in open as well as globular clusters, nuclei of planetary nebulae, symbiotic stars and supersoft X-ray sources*. All these objects are extensively discussed in this volume.

Even the *Luminous Blue Variables* (LBVs), whose properties, so far, had always been thought to be due to instabilities in the most massive single stars, appear now to be products of binary-star evolution (Smith and Tombleson, 2015). As pointed out in Chapter 11 in this volume, the LBVs can therefore no longer be considered as progenitors for the formation of Wolf–Rayet stars through single-star evolution. This implies that the only remaining channel to form *Wolf–Rayet stars* is through binary-star evolution.

It is amazing to see that now more Physics Nobel prizes have been awarded for binary-related physics than to single-star physics (the latter were the prize of 1974 for the discovery of radio pulsars; the prize of 1983 for the Chandrasekhar limit and for stellar nucleosynthesis; and the prize of 2002, which partially was for the detection of the neutrinos of Supernova 1987A).

In the July 2017 European Southern Observatory (ESO) workshop on 'The Impact of Binaries on Stellar Evolution', the world's leading experts in binary-star evolution and binary-related objects and phenomena were the invited speakers. They reviewed the present state of the art in their respective fields, and the many still unsolved problems. Most of these reviews are now collected in this book, which gives an excellent and most fascinating overview of all what we presently know about the role of binaries in stellar evolution.

It is clear from this book that we are still at the beginning of understanding how wide-ranging the effects of binary evolution are among the stars. Apart from being an excellent and up-to-date overview of the subject, this book therefore also is an inspiring starting point for future research.

Edward P. J. van den Heuvel

References

Abt, H. A., and Levy, S. G. 1976. Multiplicity among Solar-Type Stars. *ApJS*, **30**, 273–306.

Abt, H. A., and Levy, S. G. 1978. Binaries among B2-B5 IV, V Absorption and Emission stars. *ApJS*, **36**, 241–250.

Abt, H. A. 1979. The Frequencies of Binaries on the Main Sequence. *AJ*, **84**, 1591–1598.

Abt, H. A. 1983. Normal and Abnormal Binary Frequencies. *ARA&A*, **21**, 343–372.

Abt, H. A., Gomez, A. E., and Levy, S. G. 1990. The Frequency and Formation Mechanism of B2-B5 Main-Sequence Binaries. *ApJS*, **74**, 551–573.

1

The Zoo of Binary Stars

HENRI M. J. BOFFIN

1.1 Binaries Are Ubiquitous

Living on planet Earth in the Solar System, we are clearly in a very special place as, contrary to the majority of solarlike stars, the Sun has no stellar companion. This may be a good thing for us, but provides a severe bias in our understanding of binary stars in stellar evolution. Indeed, when looking away from the Sun, we can see that there are binaries everywhere!

Sirius, the brightest star in the night sky, is a binary, with a massive white dwarf companion in a 50-year orbit (Barstow et al., 2005). It was discovered, using astrometry, by Bessel (1844). Also, the closest star to our Sun, α Centauri, is a binary system (or even a triple, if Proxima Centauri is physically bound to α Cen AB). Distant 1.35 pc from the Sun, α Cen consists in a binary system with a 1.13 $M_\odot$ star orbiting a cooler 0.97 $M_\odot$ companion with a period of 79.91 years (Pourbaix and Boffin, 2016). As interesting, the closest brown dwarf to the Sun, Luhman 16AB, is also a binary system, composed of two brown dwarfs (Boffin et al., 2014b; Bedin et al., 2017) with a separation of about 3 au. A set of binary stars is also present around the visual pair Alcor and Mizar in the constellation of Ursa Major. If these two stars may not be physically linked, their aspect as visual binaries have had a profound cultural impact in Hindu and Japanese mythologies, and were also used as an eye test. However, more important for us, is that both Alcor and Mizar contain binaries. Alcor is thus a resolved binary, containing an A star with a low-mass companion, separated by 28 au (Mamajek et al., 2010), while Mizar is in fact a quadruple system: Mizar AB is a visual binary with a separation on sky of 14.4$''$, in which each component is itself a binary. Mizar A was apparently the first spectroscopic binary to be discovered, but was also resolved thanks to interferometry (Hummel et al., 1998), and contains two almost equal mass 2.4–2.5 $M_\odot$ stars with an orbital period of 20.5 days. Mizar B contains two stars in a 175.55-d orbit. Further out, and among younger stars, the Orion Trapezium stars also reveal a wealth of binaries and multiples. Two of the four Trapezium stars appear indeed binaries, while a third is a triple or even quadruple system. Thus the Trapezium is in fact composed of nine stars! Another famous binary star is Algol, the daemon. Now the prototype of a distinct type of variable stars, Algol, or β Persei, was found to be a variable star in 1667 by Geminiano Montanari. This eclipsing binary with a 2.86-d period is composed of a B8 V primary and a K2 IV secondary. Thus, the system presents the so-called Algol paradox as the most evolved of the two stars – the subgiant – is also the less massive of the two. Stellar evolution would have it the other way around, as a star evolves faster the more massive it is. The solution to this apparent paradox is mass transfer: the subgiant must have been the most massive of the two initially but then transferred most of its mass to its companion and the mass ratio was reversed.

A further dramatic example of a binary system is provided by η Carinae, which in the 1840s was the second brightest star in the night sky, and is still thought to be one of the most luminous stars in our Galaxy and perhaps also the next supernova that will explode in the Milky Way. The star itself is an extreme example of a Luminous Blue Variable (LBV) with an estimated mass of 120 $M_\odot$. The bright star, hidden in a thick nebula known as the Homunculus, belongs apparently to a binary system with a 30 $M_\odot$ companion in a 5.5-year orbit (Madura et al., 2012). At the other extreme, we have the close binary SDSS J010657.39-100003.3, which is a detached binary containing two white dwarfs (WDs) in a 39-minute orbit (Kilic et al., 2011) – due to gravitational wave radiation, it is thought that this system will merge within the next 37 Myr to become a subdwarf star. But if this wasn't extreme enough, there is also HM Cancri, which is another double system of white dwarfs, but interacting this time and with an orbital period of 5 minutes (Roelofs et al., 2010) and most likely the shortest period binary star known!

1.2 The Fraction of Binaries

The previous section showed how binaries are everywhere and come in a wide variety. However, are most stars indeed in binaries? Many works have been devoted to the study of the multiplicity of stars (e.g., Raghavan et al., 2010; Duchêne and Kraus, 2013; De Rosa et al., 2014), and they all reveal that the multiplicity of stars is a function of the mass of the primary[1] – the more massive stars being more often in binaries and multiple systems. The multiplicity of the most massive stars is well above 90%, while Raghavan et al. (2010) found that Sun-like stars have an overall observed fractions of single, double, triple and higher-order systems of 56%, 33%, 8% and 3%, respectively – thus, finding that, apparently, the majority of solartype stars are single. More recently, however, Fuhrmann et al. (2017) presented a multiplicity census for a volume-complete all-sky survey of 422 stars with distances less than 25 pc and found for the F- and G-type Population I stars that 58% are in binary or multiple systems, thus proving that the majority of solarlike stars are nonsingle. These authors also found a dependence on the mass of the primary. Moreover, these analyses also miss the important point that less massive stars will also be companions to more massive primaries. Boffin and Pourbaix (2018) found for example that also the

[1] The primary of a binary system is often defined as the more luminous component, with the companion being called the secondary. As the system evolves, a component that was initially the primary may become the secondary. Note that sometimes, primary is associated with the mass of the star – however, as there may be mass exchange in a binary system, this is not the best definition to use, as due to mass transfer, the less massive star may still be the more evolved, and thus the more luminous – e.g., in Algol.

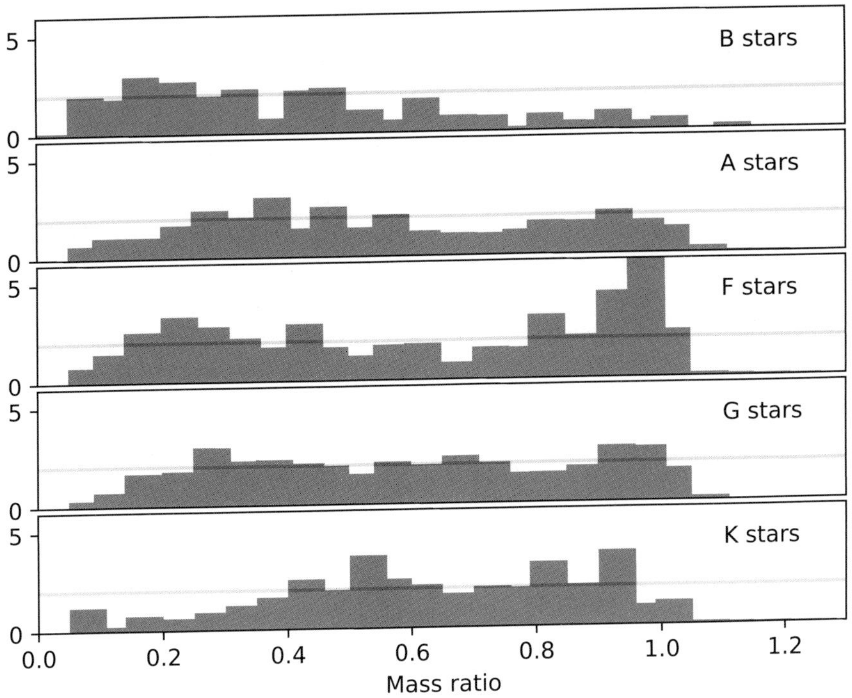

Figure 1.1 Mass-ratio distributions for the stars in the SB9 catalogue as a function of the spectral type of the primary. Adapted from Boffin and Pourbaix (2018).

mass ratio distribution (see Figure 1.1) depends on the primary mass: B stars will have more companions with smaller mass ratios than solarlike stars, and F stars seem to show an excess of twin systems, although overall, all stars seem to have a rather uniform mass-ratio distribution over a wide range of mass ratios. This means that there are many low-mass stars hidden in these binary systems and when accounted for, the multiplicity of G, K and M stars is increased, so that a majority of stars are in binary systems (see also Whitworth and Lomax, 2015).

The fraction of binaries is not the only important factor – one needs also to look at the distribution of separations or, alternatively, the distribution of orbital periods. Duchêne and Kraus (2013) show that this is also extremely dependent on the primary spectral type. Solarlike stars show a broad distribution centred around 30–50 au, while M stars are more centred around a few au, and OB stars are mostly seen in spectroscopic binaries with small separations and orbital periods below 100–1,000 days. This has also dramatic consequences on the fraction of stars that will interact during their life. This is particularly true for massive stars, as Sana et al. (2012) found that "more than 70% of all massive stars will exchange mass with a companion, leading to a binary merger in one-third of the cases," clearly

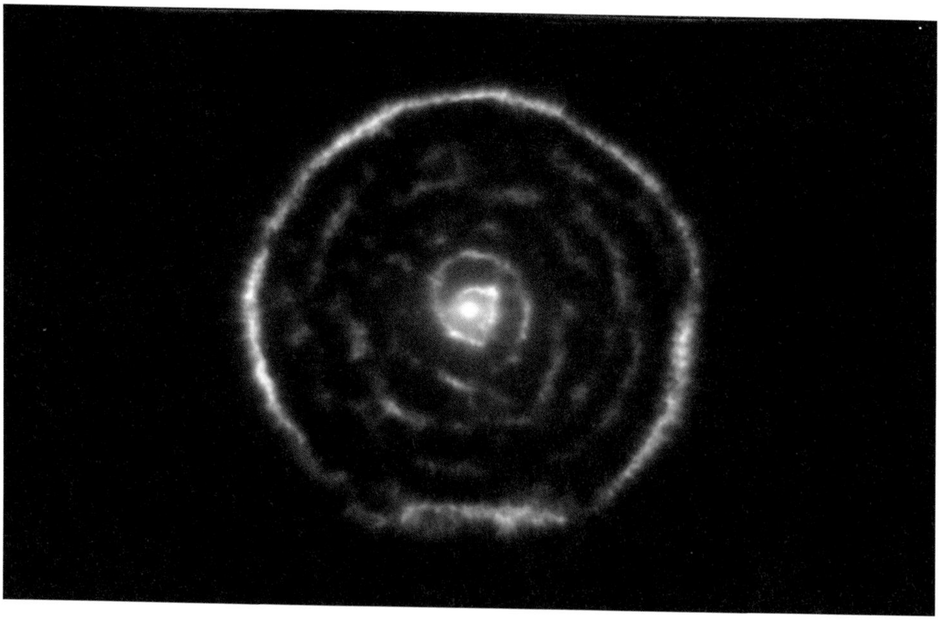

Figure 1.2 Observations using the Atacama Large Millimeter/submillimeter Array (ALMA) have revealed a spiral structure in the material around the AGB star R Sculptoris, probably caused by a hidden companion star. Credit: ALMA (ESO/NAOJ/NRAO)/M. Maercker et al. See Maercker et al. (2012).

indicating that binary interactions dominate the evolution of massive stars. For solarlike stars, the interaction will most likely happen when the star is evolving on the red giant and asymptotic giant branch (AGB), leading to many examples of interesting systems. It is not clear, however, what is the limit in orbital period for the interaction to be significant, as examples exist of systems with orbital periods as large as 350 years to show signs of interaction (see Figure 1.2; Maercker et al., 2012).

1.3 Interacting Binaries

Having seen that many stars will interact one way or another during their lives, when being part of a binary, one can now look at the kinds of interactions to expect. These come in four flavours:

- Tidal interaction
- Wind accretion
- Roche-lobe overflow
- Common-envelope evolution (CE)

Depending on the kind of interaction, as well as on the nature of the components of the binary system, the outcome will be a different type of outcome, many of which are discussed in this book. Here, we will quickly provide an overview of examples of mass transfer in binaries.

1.3.1 Wind Accretion

When the system is detached, i.e., none of the components are filling their Roche lobe, mass transfer can still happen though stellar wind. Massive stars as well as more evolved low- and intermediate-mass stars possess strong winds that have two effects: (i) they allow mass transfer between the two stars, and possibly, chemical pollution; and (ii) they induce angular momentum loss and thus a variation of the orbital elements. As shown by Theuns et al. (1996) and Nagae et al. (2004), the structure of the flow (see Figure 1.3) – and the resulting mass accretion rate – is extremely dependent on the ratio between the wind velocity and the orbital velocity, moving from an asymmetric Bondi–Hoyle kind of flow (as seen in, e.g., ζ Aur stars or massive OB stars) to a much complicated flow, where the wind is filling the Roche lobe of the primary, hence is now given the name wind Roche-lobe overflow. The latter is particularly relevant in systems where an AGB star transfers mass to a companion, as in this case, the wind velocity (5–10 km/s) is smaller or of the order of the orbital velocity (Theuns et al., 1996; Morris et al., 2006; Maercker et al., 2012). The angular momentum loss – and the subsequent evolution of the orbital elements – is also very dependent on the wind velocity and thus the respective effects of the orbital motion as shown by Hachisu et al. (1999) and Jahanara et al. (2005). Unfortunately, these effects have until now been ignored in binary evolution models, although it seems that this is starting to change (Saladino et al., 2018), which is most welcome given the current inability to provide a satisfactory answer to the origin of many systems that must have undergone such mass transfer. Indeed, the increased angular momentum loss that is seen in these simulations would imply that the orbit will shrink, instead of increase as

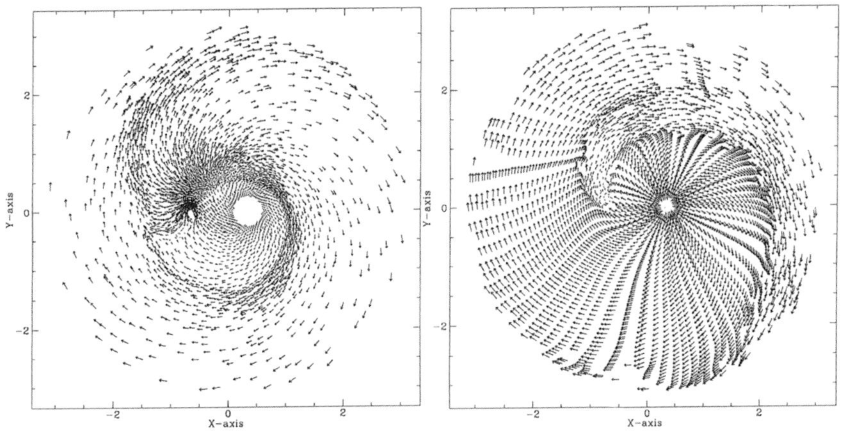

Figure 1.3 SPH simulation of mass transfer by wind in the case of an AGB star in a 3 au orbit with a companion, for two different values of the wind speed (Boffin, 1993, priv. comm.).

generally assumed, thereby providing a natural explanation for a whole class of post–mass transfer systems, including blue straggler stars, Barium stars, symbiotic stars, and post-AGB binaries. Barium stars (McClure and Woodsworth, 1990; Van der Swaelmen et al., 2017; Escorza et al., 2017) are chemically polluted (in carbon and s-process elements) stars, members of a binary system with a white dwarf companion. The material has been transferred to the barium star when the former primary, now a white dwarf, was an AGB star. Their orbital elements, similar to those seen for blue straggler stars in open clusters, for example (Geller et al., 2009), and especially their orbital periods and nonzero eccentricities, have until now resisted any tentative of explanations and therefore remain one of the most important problems in binary stellar evolution. In symbiotic stars, mass is also transferred between a red (or sometimes a yellow) giant and a hot companion that may be a WD or a main sequence star. Using interferometry, Boffin et al. (2014a) and Boffin et al. (2014c) showed that red giants in symbiotic systems are rather normal and the fact that they have larger mass-loss rates than single giants must be linked in some way to their binary nature, adding credence to the companion-reinforced attrition process mechanism (Tout and Eggleton, 1988) – a mechanism that is also not getting the attention it deserves in binary evolution models. A more detailed description of the wind mass transfer can be found in Boffin (2015).

1.3.2 Roche-Lobe Overflow

A star will fill its Roche lobe and transfer mass to its companion, if its radius is larger than the Roche-lobe radius, which in most cases can be approximated (Paczyński, 1971) by

$$R_L = a(0.38 + 0.2\log q), \tag{1.1}$$

where a is the semimajor axis and q is the mass ratio. Thus, the ability to fill the Roche lobe will depend mostly on the separation (or equivalently, the orbital period) and the evolutionary stage of the star. For a solarlike star on the main sequence, the separation will have to be of the order of a few solar radii and the orbital period of a few hours. This is, for example, the case of cataclysmic variables. When the star evolves and moves on to the red giant branch, systems with orbital periods between a few days (e.g., Algol) and a few years (i.e., symbiotic stars) are concerned. This range of possibilities of when the Roche-lobe overflow (RLOF) will occur is at the origin of a whole zoo of binary systems. When the star fills its Roche lobe, matter is transferred to its companion and because this matter has some angular momentum, a disc may form around the accretor – for short period systems, this will be the case when the accretion is a WD (cataclysmic variable), a neutron star (low-mass X-ray binaries) or a black hole (high-mass X-ray binary), while if the system is wide, this may happen when the accretion is a main sequence star, for example in symbiotic stars. On the other hand, for binary systems with short orbital periods, when the accretion happens onto a main sequence star, no disc can form, and there will be direct impact onto the star (as in Algols).

1.3.3 Common Envelope Evolution

In some cases, the RLOF will be dynamically unstable, leading to the formation of a common envelope (CE) around the two components of the binary system (Paczyński, 1976). This will happen when the mass-losing star is a red giant with a deep convective envelope. In this case, the mass transfer leads to a runaway process where the star expands when losing mass, thereby continuously overfilling its Roche lobe. The mass transfer happens on a dynamical timescale, and the companion is unable to process this and will expand as well, until a common envelope forms. Once inside the envelope, dynamical friction will make the two cores spiral towards each others, while the envelope will eventually be ejected. This scenario would explain the formation of very close binary systems containing a WD with orbital periods below 1d – such as cataclysmic variables. and progenitors of Type Ia supernovae. It could also explain mergers and red transients (Ivanova et al., 2013b). Yet, despite its importance, the CE is still one of the least understood phases in the evolution of compact binaries (Ivanova et al., 2013a), although here also progress is being made in simulating it (e.g., Iaconi et al., 2018).

The CE phase also seems to play a very important role in the formation and evolution of planetary nebulae (PNe; Jones and Boffin, 2017). Indeed, it is now known that 80% of all PNe are nonspherical and many present jetlike structures. Moreover, there is now overwhelming evidence for a large fraction of post-CE systems lying at the heart of PNe[2]. Compelling evidence has been found for a clear correlation between the morphology of the nebula and the presence of a binary, as well as between chemical abundances in the nebula or the central stars and the binarity. It is therefore likely that PNe are for a large part the outcome of binary interactions. This clearly indicates that textbooks need to be rewritten (see Figure 1.4)!

1.4 Massive Star Interactions and Exploding Events

As shown earlier, the large fraction of OB stars that reside in relatively short period binary systems leads to the staggering fact that 70% of them will interact one way or another during the course of their lives. This leads to a variety of phenomena, many of which are further described in this book. Chief among them is perhaps the existence of Luminous Blue variable stars (LBVs). These are among the most luminous stars in the Milky Way and as described at the beginning of this chapter, the most extreme case, η Car, was quite a sight in the nineteenth century. It is now more and more thought that binaries have to be at the origin of this class of star (Smith and Tombleson, 2015; Smith et al., 2018). The discovery, based on interferometry, that another of these stars, HR Car, is also a binary, with an orbital period of about five to six years (Boffin et al., 2016) lends further credit to the binary hypothesis, which is further discussed in Chapter 11.

2 An up-to-date list is available at http://drdjones.net/?q=bCSPN.

Figure 1.4 With so many examples of the importance of binarity in the evolution of stars, it is clear that textbooks need to be rewritten! Photo: H. M. J. Boffin.

The mass transfer in these binaries will most likely lead to the formation of different kinds of supernovae as expected from single star evolution. Similarly, it is now well established that short gamma-ray bursts are due to the merging of neutron stars in binary systems. But evidence is now mounting that even long gamma-ray bursts may have a binary origin. As another class of exploding events, it is also well established that the origin of Type Ia supernovae are due to binary interactions, although the real channel – if there is only one – is still unknown: is it the accretion of mass onto a WD, is it a merger of two WDs or is it even some other mechanism? All these are discussed further in this book.

Without any doubt, one of the most extraordinary observations of recent years is the discovery of gravitational waves coming from the merging of two black holes or two neutron stars. These, again, are the outcome of binary interactions, and as always, the recent observations have shed new light on the possible mechanisms at play. A chapter is also devoted to this in this book.

References

Barstow, M. A., Bond, H. E., Holberg, J. B., et al. 2005. Hubble Space Telescope spectroscopy of the Balmer lines in Sirius B*. *MNRAS*, **362**, 1134–1142.

Bedin, L. R., Pourbaix, D., Apai, D., et al. 2017. Hubble Space Telescope astrometry of the closest brown dwarf binary system – I. Overview and improved orbit. *MNRAS*, **470**, 1140–1155.

Bessel, F. W. 1844. On the variations of the proper motions of Procyon and Sirius. *MNRAS*, **6**, 136–141.

Boffin, H. M. J. 2015. Mass transfer by stellar wind. Page 153 of: Boffin, H. M. J., Carraro, G., and Beccari, G. (eds), *Ecology of Blue Straggler Stars*. Astrophysics and Space Science Library. Springer.

Boffin, H. M. J., and Pourbaix, D. 2018. The mass-ratio distribution of spectroscopic binaries along the main-sequence. Recio-Blanco, A., de Laverny, P., Brown, A. G. A., and Prusti, T. (eds), *IAU Symp.*, **330**, 330–340.

Boffin, H. M. J., Miszalski, B., Rauch, T., et al. 2012. An interacting binary system powers precessing outflows of an evolved star. *Science*, **338**, 773.

Boffin, H. M. J., Blind, N., Hillen, M., et al. 2014a. A PIONIER view on mass-transferring red giants. *The Messenger*, **156**, 35–37.

Boffin, H. M. J., Pourbaix, D., Mužić, K., et al. 2014b. Possible astrometric discovery of a substellar companion to the closest binary brown dwarf system WISE J104915.57-531906.1. *A&A*, **561**, L4.

Boffin, H. M. J., Hillen, M., Berger, J. P., et al. 2014c. Roche-lobe filling factor of mass-transferring red giants: the PIONIER view. *A&A*, **564**, A1.

Boffin, H. M. J., Rivinius, T., Mérand, A., et al. 2016. The luminous blue variable HR Carinae has a partner: discovery of a companion with the VLTI. *A&A*, **593**, A90.

De Rosa, R. J., Patience, J., Wilson, P. A., et al. 2014. The VAST Survey – III. The multiplicity of A-type stars within 75 pc. *MNRAS*, **437**, 1216–1240.

Duchêne, Gaspard, and Kraus, Adam. 2013. Stellar multiplicity. *Annual Review of Astronomy and Astrophysics*, **51**, 269–310.

Escorza, A., Boffin, H. M. J., Jorissen, A., et al. 2017. Hertzsprung–Russell diagram and mass distribution of barium stars. *A&A*, **608**, A100.

Fuhrmann, K., Chini, R., Kaderhandt, L., and Chen, Z. 2017. Multiplicity among solar-type stars. *ApJ*, **836**, 139.

Geller, A. M., Mathieu, R. D., Harris, H. C., and McClure, R. D. 2009. WIYN Open Cluster Study. XXXVI. Spectroscopic binary orbits in NGC 188. *AJ*, **137**, 3743–3760.

Hachisu, I., Kato, M., and Nomoto, K. 1999. A wide symbiotic channel to type IA supernovae. *ApJ*, **522**, 487–503.

Hummel, C. A., Mozurkewich, D., Armstrong, J. T., et al. 1998. Navy prototype optical interferometer observations of the double stars Mizar A and Matar. *AJ*, **116**, 2536–2548.

Iaconi, R., De Marco, O., Passy, J.-C., and Staff, J. 2018. The effect of binding energy and resolution in simulations of the common envelope binary interaction. *MNRAS*, **477**, 2349–2365.

Ivanova, N., Justham, S., Chen, X., et al. 2013a. Common envelope evolution: where we stand and how we can move forward. *A&A Rev.*, **21**, 59.

Ivanova, N., Justham, S., Avendano Nandez, J. L., and Lombardi, J. C. 2013b. Identification of the long-sought common-envelope events. *Science*, **339**, 433.

Jahanara, B., Mitsumoto, M., Oka, K., et al. 2005. Wind accretion in binary stars. II. Angular momentum loss. *A&A*, **441**, 589–595.

Jones, D., and Boffin, H. M. J. 2017. Binary stars as the key to understanding planetary nebulae. *Nature Astronomy*, **1**, 0117.

Kilic, M., Brown, W. R., Hermes, J. J., et al. 2011. SDSS J163030.58+423305.8: a 40-min orbital period detached white dwarf binary. *MNRAS*, **418**, L157–L161.

Madura, T. I., Gull, T. R., Owocki, S. P., 2012. Constraining the absolute orientation of η Carinae's binary orbit: a 3D dynamical model for the broad [Fe III] emission. *MNRAS*, **420**, 2064–2086.

Maercker, M., Mohamed, S., Vlemmings, W. H. T., et al. 2012. Unexpectedly large mass loss during the thermal pulse cycle of the red giant star R Sculptoris. *Nature*, **490**, 232–234.

Mamajek, E. E., Kenworthy, M. A., Hinz, P. M., and Meyer, M. R. 2010. Discovery of a faint companion to Alcor using MMT/AO 5 μm imaging. *AJ*, **139**, 919–925.

McClure, R. D., and Woodsworth, A. W. 1990. The binary nature of the barium and CH stars. III – Orbital parameters. *ApJ*, **352**, 709–723.

Morris, M., Sahai, R., Matthews, K., et al. 2006. A binary-induced pinwheel outflow from the extreme carbon star, AFGL 3068. Pages 469–470 of: Barlow, M. J., and Méndez, R. H. (eds), *Planetary Nebulae in our Galaxy and Beyond*. IAU Symposium, vol. 234.

Nagae, T., Oka, K., Matsuda, T., et al. 2004. Wind accretion in binary stars. I. Mass accretion ratio. *A&A*, **419**, 335–343.

Paczyński, B. 1971. Evolutionary processes in close binary systems. *ARA&A*, **9**, 183.

Paczyński, B. 1976. Common envelope binaries. Page 75 of: Eggleton, P., Mitton, S., and Whelan, J. (eds), *Structure and Evolution of Close Binary Systems*. IAU Symposium, vol. 73.

Pourbaix, D., and Boffin, H. M. J. 2016. Parallax and masses of α Centauri revisited. *A&A*, **586**, A90.

Raghavan, D., McAlister, H. A., Henry, T. J., et al. 2010. A survey of stellar families: multiplicity of solar-type stars. *ApJS*, **190**, 1–42.

Roelofs, G. H. A., Rau, A., Marsh, T. R., et al. 2010. Spectroscopic evidence for a 5.4 minute orbital period in HM Cancri. *ApJ L*, **711**, L138–L142.

Saladino, M. I., Pols, O. R., van der Helm, E., Pelupessy, I., and Portegies Zwart, S. 2018. Gone with the wind: the impact of wind mass transfer on the orbital evolution of AGB binary systems. *ArXiv e-prints*, 1805.03208.

Sana, H., de Mink, S. E., de Koter, A., et al. 2012. Binary interaction dominates the evolution of massive stars. *Science*, **337**, 444.

Smith, N., and Tombleson, R. 2015. Luminous blue variables are antisocial: their isolation implies that they are kicked mass gainers in binary evolution. *MNRAS*, **447**, 598–617.

Smith, N., Aghakhanloo, M., Murphy, J. W., et al. 2018. On the Gaia DR2 distances for galactic luminous blue variables. *ArXiv e-prints*, 1805.03298.

Theuns, T., Boffin, H. M. J., and Jorissen, A. 1996. Wind accretion in binary stars. II. Accretion rates. *MNRAS*, **280**, 1264–1276.

Tout, C. A., and Eggleton, P. P. 1988. The formation of Algols without catastrophes. *ApJ*, **334**, 357–361.

Van der Swaelmen, M., Boffin, H. M. J., Jorissen, A., and Van Eck, S. 2017. The mass-ratio and eccentricity distributions of barium and S stars, and red giants in open clusters. *A&A*, **597**, A68.

Whitworth, A. P., and Lomax, O. 2015. Are the majority of Sun-like stars single? *MNRAS*, **448**, 1761–1766.

2
Statistics of Binary and Multiple Stars

MAXWELL MOE

2.1 Observational Techniques

Radial velocity monitoring: The close binary fractions of M-type (Fischer and Marcy, 1992; Clark et al., 2012), solar-type (Duquennoy and Mayor, 1991; Latham et al., 2002; Halbwachs et al., 2003; Raghavan et al., 2010), B-type (Levato et al., 1987; Abt et al., 1990), and O-type (García and Mermilliod, 2001; Sana et al., 2012; Kobulnicky et al., 2014) main-sequence (MS) primaries are routinely measured with multi-epoch, high-resolution spectra. Even a few epochs distinguish radial velocity (RV) variables with $\Delta RV_{max} > 10$ km s^{-1} as close binaries (Chini et al., 2012; Gao et al., 2014; Badenes et al., 2017). Many spectroscopic binary surveys obtain $\gtrsim 20$ epochs in order to specify unique orbital periods P and eccentricities e. Surveys with increased cadence and sensitivity probe binaries across a larger parameter space, in some cases up to $P \sim 10$ years (Figure 2.1), and are therefore more complete (Duquennoy and Mayor, 1991; Raghavan et al., 2010). For double-lined spectroscopic binaries (SB2s), the measured ratios of velocity semiamplitudes directly provide the mass ratios $q = M_{comp}/M_1$. Meanwhile, SB1s have faint companions, typically low-mass MS secondaries ($q \lesssim 0.6$ for solar-type primaries or $q \lesssim 0.3$ for OB primaries). In some cases, SB1s contain compact remnant companions such as white dwarfs (WDs), neutron stars (NSs) or black holes (BHs). In the galactic field, $\approx$30% of solar-type and early-type SB1s have compact remnant companions (de Mink et al., 2014; Moe and Di Stefano, 2017; Murphy et al., 2018). Assuming random orientations, the intrinsic mass-ratio distributions of SB1s are reconstructed from the observed distributions of binary mass functions (Mazeh and Goldberg, 1992; Boffin et al., 1993).

Deblending composite spectra: For late-type binaries with $q > 0.7$ (or for early-type binaries with $q > 0.4$), absorption features from both components are distinguishable in a single-epoch spectrum (Chini et al., 2012; El-Badry et al., 2018). With improved spectral resolution and sensitivity, SB2s down to $q \approx 0.1$ are detected by cross-correlating the observed spectrum with atmospheric models (Gullikson et al., 2016). Binary companions that are hotter than their brighter primaries produce an ultraviolet (UV) excess, e.g., B-type MS companions to Cepheids (Evans et al., 2013). Conversely, companions that are substantially cooler than their primaries may exhibit a near-infrared excess, e.g., M-dwarf companions to WDs or central stars of planetary nebulae (Farihi et al., 2010; De Marco

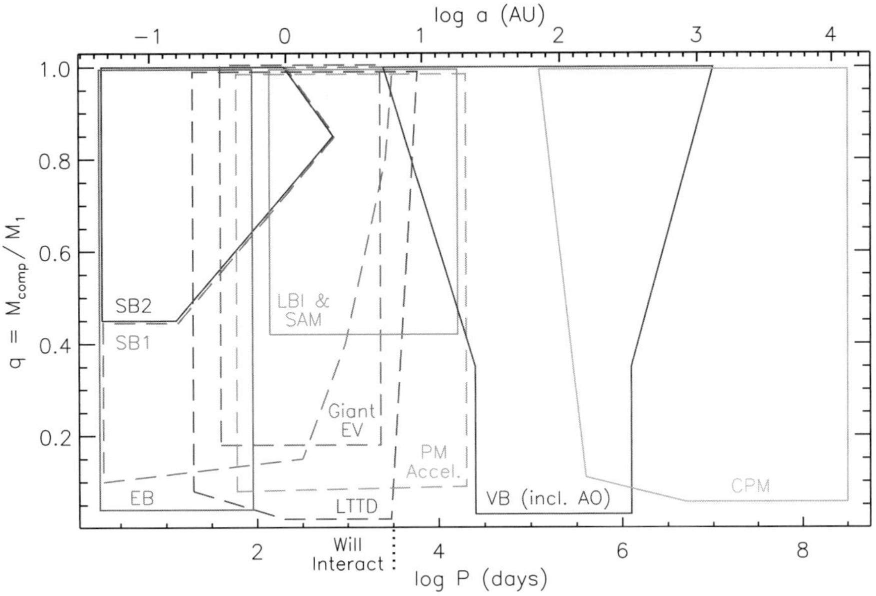

Figure 2.1 Various observational techniques discover binaries across specific intervals of periods and mass ratios. The exact regions of sensitivity depend on the survey and the properties of the binaries, e.g., their distances, spectral types, etc. (see text for details). Binaries in which the secondaries are discovered indirectly (dashed regions: SB1s, giant ellipsoidal variables, light-travel-time delay and proper-motion acceleration) may contain compact remnant companions such as WDs, NSs or BHs. Binaries with $\log P\,(\text{days}) \lesssim 3.5$ (left of dotted) will eventually interact via Roche-lobe overflow.

et al., 2013). B-type MS stars are expected to be X-ray quiet, and so B-type MS stars that coincide with X-ray sources likely contain young, active, late-type companions (Hubrig et al., 2001; Stelzer et al., 2003). While these techniques alone do not yield orbital separations (not shown in Figure 2.1), they nonetheless provide a census of spatially unresolved companions with single-epoch observations.

Eclipses and ellipsoidal variables: Wide-field time-domain photometric surveys have discovered a plethora of eclipsing binaries (EBs), including ≈2,000 EBs in the *Kepler* field (Kirk et al., 2016), ≈ 450,000 EBs towards the galactic bulge (Soszyński et al., 2016), and tens of thousands of early-type EBs in the Magellanic Clouds (Pawlak et al., 2016). *Gaia* and the Large Synoptic Survey Telescope (LSST) are expected to discover millions of EBs (Prša et al., 2011; Eyer et al., 2015). The component masses, ages and eccentricities of EBs with known distances can be estimated by fitting stellar models to the observed light curves (Moe and Di Stefano, 2015b). Due to geometrical selection effects, the probability of detecting EBs beyond $P \approx 100$ days becomes extremely small (Figure 2.1). Noneclipsing binaries that are nearly filling their Roche lobes produce small-amplitude periodic ellipsoidal variability (EV). Unlike EBs, EV is detected in binaries with giant primaries and small companions (Nicholls et al., 2010), which extends the parameter space up to $P \approx 2{,}000$ days (Figure 2.1).

Light-travel-time delay: The orbital motion of companions to pulsating stars, e.g., δ Scuti stars, induce small differences in subsequent pulsation timings due to light-travel-time delay (LTTD) effects (Murphy et al., 2018). Similarly, tertiary companions to EBs are discovered by measuring the changes in eclipse timings (Rappaport et al., 2013; Borkovits et al., 2016). These methods probe companions with periods of weeks to several years, depending on the duration of the survey (Figure 2.1). Moreover, companions with very small mass ratios $q \approx 0.02$ induce large enough changes in the pulsation or eclipse timings to be detectable with *Kepler* photometry.

Interferometry and sparse-aperture masking: Long-baseline interferometry (LBI) and sparse-aperture masking (SAM) identify companions with projected separations $\rho \sim$ 1 - 100 milli-arcseconds, filling in the period gap between spectroscopic RV monitoring and adaptive optic techniques (Raghavan et al., 2010; Rizzuto et al., 2013; Sana et al., 2014). The mass ratios derive from the observed brightness contrasts Δmag and empirical MS mass-luminosity relation. LBI and SAM detect binaries with $\Delta\text{mag} \lesssim 4$, and so can identify MS binaries with $q \gtrsim 0.4$, depending on the spectral type (Figure 2.1).

Visual binaries: Visual binaries (VBs) are spatially revolved with ground-based and space-based direct imaging, adaptive optics (AO), lucky imaging and speckle imaging (Duchêne et al., 2001; Shatsky and Tokovinin, 2002; Bergfors et al., 2010; Kraus et al., 2011; Peter et al., 2012; De Rosa et al., 2014; Sana et al., 2014; Aldoretta et al., 2015). The orbital separations are generally estimated from the projected offsets while the mass ratios are determined from the observed brightness contrasts. For $P \lesssim 100$ years, multi-epoch imaging reveals orbital motions that allow for more accurate determinations of the orbital elements. At small separations $\rho \lesssim 0.5''$, only binaries with small brightness contrasts, and therefore large mass ratios $q \gtrsim 0.3$, can be resolved (Figure 2.1). At larger separations $\rho \gtrsim 5''$, confusion with background/foreground stars becomes nonnegligible, especially when discriminating extreme mass-ratio binaries from optical doubles.

Proper motion: Common proper motion (CPM) is the only technique to confirm that very wide companions are gravitational bound (Lépine and Bongiorno, 2007; Chini et al., 2014). At closer separations, companions induce detectable accelerations in the astrometry of the primary stars (Kaplan and Makarov, 2003; Raghavan et al., 2010). Orbital elements can be estimated even if only a fraction of the orbits are observed, and so proper-motion (PM) acceleration can characterise binaries up to $P \sim 100$ years (Figure 2.1). *Gaia* is expected to discover millions of binaries through both CPM and PM acceleration (Eyer et al., 2015).

2.2 Statistics

2.2.1 *Binary Fraction and Multiplicity Frequency*

The bias-corrected binary and multiple star fractions as a function of spectral type were extensively reviewed by Duchêne and Kraus (2013) and Moe and Di Stefano (2017). In this section, I summarise two statistical quantities: the binary star fraction $F_{\text{bin}}(M_1)$, which is the fraction of MS primaries with at least one *stellar MS companion* with $q > 0.1$, and the multiplicity frequency $f_{\rm mult}(M_1)$, which is the average frequency of stellar MS companions with $q > 0.1$ per primary. After utilising several observational techniques and correcting for various selection effects, Duquennoy and Mayor (1991) measured the binary star fraction of field solar-type primaries to be $F_{\rm bin} = 0.57 \pm 0.05$. More recently, Raghavan et al. (2010) estimated the binary star fraction and multiplicity frequency of solar-type stars within < 25 pc to be $F_{\rm bin} = 0.46 \pm 0.02$ and $f_{\rm mult} = 0.59 \pm 0.05$, respectively. The difference of $\Delta F_{\rm bin} = 0.11$ between the two surveys is mainly because Duquennoy and Mayor (1991) added undetected WD companions during their corrections for incompleteness. Moe and Di Stefano (2017) found $11\% \pm 4\%$ of solar-type MS stars in the field have WD companions, and so the inclusion vs. exclusion of WD companions leads to the largest discrepancies in the observed vs. initial MS solar-type binary star fractions.

The binary fraction of field M-dwarfs is measurably smaller. For early-M primaries, the binary fraction is $F_{\rm bin} \approx 0.35$–0.40 and the multiplicity frequency is $F_{\rm mult} \approx 0.45$–$0.50$ (Fischer and Marcy, 1992; Bergfors et al., 2010; Janson et al., 2012; Ward-Duong et al., 2015). For late M-dwarfs, the binary star fraction is only $F_{\rm bin} \approx 0.20$ - 0.25 (Bouy et al., 2003; Basri and Reiners, 2006; Law et al., 2008). Systems with late M-dwarf primaries exhibit an extremely small triple star fraction, i.e., $f_{\rm mult}$ is only slightly larger than $F_{\rm bin}$ (Duchêne and Kraus, 2013). The mass-ratio distribution of binaries with late-M primaries are markedly weighted towards more equal masses (Bouy et al., 2003; Joergens, 2006; Basri and Reiners, 2006; Bergfors et al., 2010; Dieterich et al., 2012; Duchêne and Kraus, 2013). Hence, the fraction of late-M MS primaries ($M_1 \approx 0.12\,{\rm M}_\odot$) with late-M MS companions ($M_2 \approx 0.08$ - $0.12\,{\rm M}_\odot$) is $F_{\rm bin} \approx 0.20$ (Figure 2.2).

Despite the difficulty in identifying early-type binaries with intermediate periods (see Figure 2.1), the binary star fraction and multiplicity frequency of massive stars are definitively larger. The binary star fraction is $F_{\rm bin} = 0.69 \pm 0.07$ for A-type primaries (De Rosa et al., 2014), $F_{\rm bin} > 0.8$ for B-type primaries (Abt et al., 1990; Kouwenhoven et al., 2007; Kobulnicky and Fryer, 2007), and $F_{\rm bin} > 0.85$ for O-type primaries (Mason et al., 1998; Sana et al., 2014; Moe and Di Stefano, 2017). After correcting for selection effects, the multiplicity frequency is $f_{\rm mult} = 1.35 \pm 0.25$ for B-type primaries (Rizzuto et al., 2013), $f_{\rm mult} = 1.9 \pm 0.3$ for early-B primaries (Abt et al., 1990), and $f_{\rm mult} = 2.2 \pm 0.3$ for O-type primaries (Sana et al., 2012). The multiplicity frequency of OB primaries is close to $\approx$2, implying the majority of massive stars are in triples and higher-ordered multiples. While only $\approx$13% of solar-type MS primaries are in triples and higher-ordered multiples (Raghavan et al., 2010; Tokovinin, 2014), $\approx$60–80% of OB stars are in triples (Moe and Di Stefano, 2017). The multiplicity frequency $f_{\rm mult}$ and triple star fraction increase even more dramatically with respect to primary mass M_1 compared to the binary star fraction $F_{\rm bin}$ (see Figure 2.2).

2.2.2 *Period Distribution*

The binary period distributions provide insight into the fragmentation and migration processes of binary stars, and are used to determine the fraction of binaries that are close

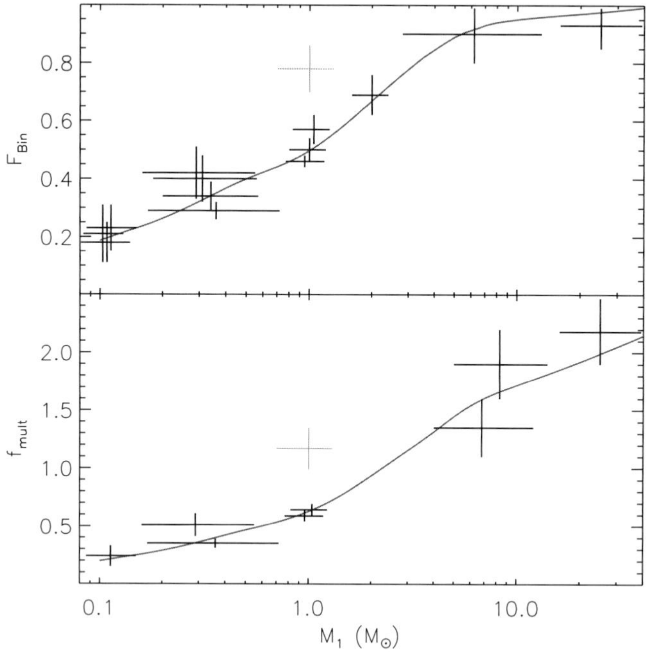

Figure 2.2 The binary star fraction (top) and multiplicity frequency (bottom) of MS stars as a function of primary mass (references for all data points provided in text). The curves derive from integrating the period distributions of inner binaries (top) and all companions (bottom) based on the analytic fit presented in Moe and Di Stefano (2017; see also Section 2.2.2). The binary fraction and multiplicity frequency of solar-type pre-MS stars (thin crosses) are larger than their MS counterparts in the field, but are still less than the observed values for OB stars.

enough to interact. In Figure 2.3, I show the frequency $f_{\log P}$ of MS companions per decade of orbital period as a function of M_1 and $\log P$ (adaptation from Moe and Di Stefano 2017). The companion frequency $f_{\log P}$ describes the period distribution of all companions, including those in binaries and triples. Nearly all companions with $P < 100$ days are members of inner binaries (Sana et al., 2014; Tokovinin, 2014). Meanwhile, at longer periods $\log P\,$(days) > 7, more than half of companions to solar-type MS primaries (Raghavan et al., 2010; Tokovinin, 2014) and nearly all companions to OB primaries (Sana et al., 2014; Moe and Di Stefano, 2017) are outer tertiaries in hierarchical triples. The resulting period distributions of only those companions that are members of inner binaries are shown as the thin lines in Figure 2.3.

Both Duquennoy and Mayor (1991) and Raghavan et al. (2010) demonstrated companions to field solar-type MS primaries follow a log-normal period distribution with a peak at $\log P\,$(days) $= 4.8 - 5.0$ ($a \approx 40\,$au), dispersion of $\sigma_{\log P} = 2.3$, and normalisation such that $f_{\rm mult} = \int f_{\log P}\, d\log P \approx 0.6$ (dotted line in Figure 2.3). The separations of early-M binaries also follow a log-normal distribution, but with a slightly smaller mean separation $a \approx 10$ au (Fischer and Marcy, 1992; Kraus et al., 2011; Janson et al., 2012; Duchêne and

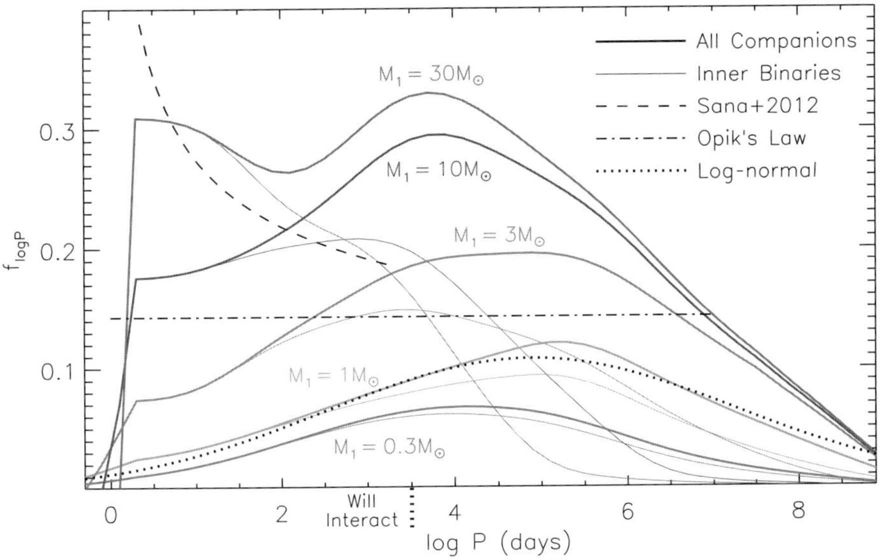

Figure 2.3 Plotted according to primary mass M_1, analytic fit to the measured frequency $f_{\log P}$ of all companions (thick) and inner binaries (thin) per decade of orbital period (Moe and Di Stefano, 2017). Integrating the thick and thin curves provide the multiplicity frequency $f_{\rm mult}(M_1)$ and binary star fraction $F_{\rm bin}(M_1)$, respectively. Field solar-type MS binaries follow a log-normal period distribution that peaks at $\log P\,(\text{days}) \approx 4.9$ (dotted; Duquennoy and Mayor, 1991; Raghavan et al., 2010). Companions to B-type MS primaries approximately obey Opik's law, i.e., a uniform distribution in $\log P$ (dash-dotted; Abt et al., 1990; Kobulnicky and Fryer, 2007; Kouwenhoven et al., 2007). Inner companions to O-type MS primaries are skewed significantly towards very short periods (dashed; Sana et al., 2012). Close binaries with $\log P\,(\text{days}) \lesssim 3.5$ (left of thick dotted line) will eventually interact via Roche-lobe overflow.

Kraus, 2013). The separation distribution of binaries with late-M primaries narrowly peaks across a = 1–10 au, exhibiting a dearth of systems beyond $a \gtrsim 10$ au (Bouy et al., 2003; Basri and Reiners, 2006; Duchêne and Kraus, 2013). The companion frequency $f_{\log P} \approx 0.05$ across intermediate separations $a \approx 1$ - 10 au is nearly constant between $M_1 = 0.1\,\mathrm{M}_\odot$ and $M_1 = 1\,\mathrm{M}_\odot$ (Murphy et al., 2018). The smaller M-dwarf binary fraction is therefore largely due to the dearth of wide companions.

For B-type MS primaries, the close binary fraction inferred from EBs and RV monitoring (Levato et al., 1987; Abt et al., 1990; Moe and Di Stefano, 2013) and the wide binary fraction measured with direct imaging and adaptive optics (Abt et al., 1990; Shatsky and Tokovinin, 2002) are both larger than the values for solar-type MS primaries. The similarity in the measurements of $f_{\log P} \approx 0.15$ at both close and wide separations led Kouwenhoven et al. (2007) and Kobulnicky and Fryer (2007) to infer the period distribution of B-type MS binaries was consistent with Opik's law, i.e., a uniform distribution with respect to $\log P$ (dash-dotted line in Figure 2.3). However, recent observations have filled in the gap at intermediate periods $\log P\,(\text{days})$ = 3–5, demonstrating early-type binaries actually peak with $f_{\log P} \approx 0.20$ - 0.25 at such intermediate separations (Rizzuto et al., 2013; Evans et al., 2015; Murphy et al., 2018).

The close binary fraction of O-type MS primaries is even larger (Chini et al., 2012; Sana et al., 2012; Kobulnicky et al., 2014). In particular, Sana et al. (2012) found 69% $\pm$ 9% of O-type MS primaries have companions with $P < 1{,}500$ days, and that the period distribution of inner companions is skewed towards very short periods $P < 20$ days (dashed line in Figure 2.3). Other observational techniques, including lucky imaging (Peter et al., 2012), speckle interferometry (Mason et al., 2009), direct imaging with *HST* (Aldoretta et al., 2015), LBI, SAM and AO (Sana et al., 2014), show the frequency of companions at intermediate and wide separations is larger for O-type MS primaries compared to intermediate-mass primaries. While inner companions to O-type primaries are weighted towards short periods as found in Sana et al. (2012), the distribution of all companions, including outer tertiaries, cover a much broader range of periods (see Figure 2.3).

The multiplicity statistics shown Figure 2.2 and the period distributions shown in Figure 2.3 are representative of the volume-averaged multiple star statistics of MS primaries. The majority of late-type MS stars are in the galactic field, while the majority of OB stars are in young clusters and associations. While the overall period distributions and binary fractions vary with age and environment (see Section 2.2.5), the observed differences in the period distributions below $\log P\,(\text{days}) < 5$ ($a \lesssim 100$ au) are almost entirely due to changes in primary mass (see discussion in Moe and Di Stefano 2017 and Section 2.3).

2.2.3 Mass-Ratio Distribution

Binary mass ratios are set by the processes of fragmentation and accretion, and so the measured mass-ratio distributions provide stringent tests for models of binary star formation. In addition, close binaries with large mass ratios $q \gtrsim 0.8$ are more likely to evolve through stable mass transfer, while close binaries with extreme mass ratios $q \lesssim 0.2$ will undergo common envelope evolution (Belczynski et al., 2008). Knowing the mass-ratio distribution is critical for determining the fraction of binaries that evolve through the various channels.

Most studies of binary stars fit a single power-law distribution $f_q \propto q^{\gamma}$ to the observed mass ratios (Shatsky and Tokovinin, 2002; Peter et al., 2012; Sana et al., 2012; Duchêne and Kraus, 2013; De Rosa et al., 2014). However, with large enough samples, it becomes evident that a single-parameter model cannot adequately fit the distribution across all mass ratios (Duquennoy and Mayor, 1991; Halbwachs et al., 2003; Gullikson et al., 2016; Murphy et al., 2018). Moe and Di Stefano (2017) therefore adopted a three-parameter model: a power-law slope γ_{smallq} across small mass ratios $q = 0.1$–0.3, a power-law slope γ_{largeq} across large mass ratios $q = 0.3$–1.0, and an excess fraction F_{twin} of twins with mass ratios near unity (adaptation shown in Figure 2.4).

For solar-type binaries, there is a modest excess twin fraction $F_{\text{twin}} \approx 0.20$ at short periods $P < 100$ days (Tokovinin, 2000; Halbwachs et al., 2003). Solar-type (Raghavan et al., 2010) and A-type (De Rosa et al., 2014) binaries with intermediate separations $a \approx 1$–100 au exhibit a smaller but statistically significant excess twin fraction $F_{\text{twin}} \approx 0.05$–$0.10$.

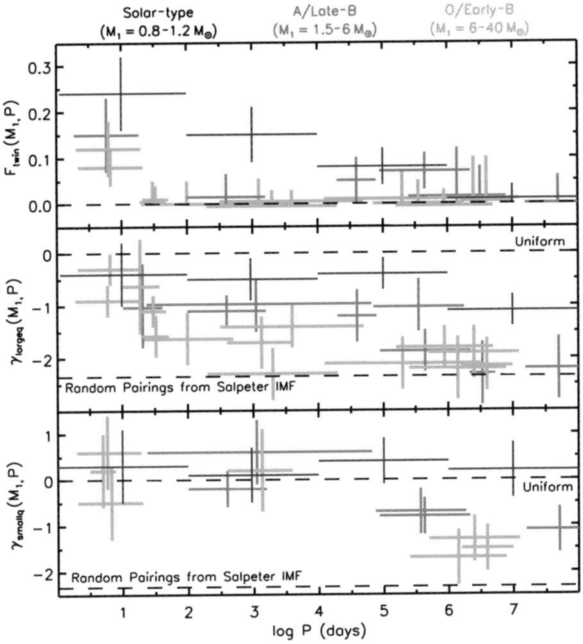

Figure 2.4 The mass-ratio distribution as parameterised by an excess twin fraction $F_{\rm twin}$ (top), power-law slope $\gamma_{\rm largeq}$ across large mass ratios q = 0.3–1.0 (middle) and power-law slope $\gamma_{\rm smallq}$ across small mass ratios q = 0.1–0.3 (bottom) as a function of period and greyed according to primary mass. Close binaries follow a uniform mass-ratio distribution ($\gamma_{\rm largeq}$ = $\gamma_{\rm smallq}$ = 0.0) with a small excess twin fraction $F_{\rm twin}$ = 0.1 - 0.2, while wider binaries become increasingly weighted towards smaller mass ratios, especially those with more massive primaries. Wide companions to early-type MS stars are significantly skewed towards extreme mass ratios, but their mass-ratio distribution is still mildly discrepant with random pairings drawn from a Salpeter initial mass function (IMF) ($\gamma_{\rm largeq}$ = $\gamma_{\rm smallq}$ = −2.35).

At wider separations $a > 200$ au, the excess twin fraction $F_{\rm twin} < 0.04$ is consistent with zero (Lépine and Bongiorno, 2007; Raghavan et al., 2010). For more massive primaries $M_1 \gtrsim 2\,{\rm M}_\odot$, the excess twin fraction $F_{\rm twin} \lesssim 0.05$ is negligible beyond $P > 20$ days (Abt et al., 1990; Shatsky and Tokovinin, 2002; Moe and Di Stefano, 2015b; Gullikson et al., 2016; Murphy et al., 2018). Only at very short periods $P < 20$ days do early-type MS binaries exhibit a small excess twin fraction $F_{\rm twin} \approx 0.10$ (Pinsonneault and Stanek, 2006; Moe and Di Stefano, 2013; Moe and Di Stefano, 2017).

Close companions to both solar-type and early-type primaries roughly follow a uniform mass-ratio distribution, i.e., $\gamma_{\rm largeq} \approx \gamma_{\rm smallq} \approx 0.0$ (Abt et al., 1990; Raghavan et al., 2010; Sana et al., 2012). The mass-ratio distribution of solar-type binaries with intermediate separations $a \approx 10$ au broadly peaks at $q \approx 0.3$ as found in Duquennoy and Mayor (1991), i.e., $\gamma_{\rm largeq} \approx -0.5$ and $\gamma_{\rm smallq} \approx 0.5$. At wider separations $a > 200$ au, the mass-ratio distribution of solar-type binaries becomes weighted towards smaller mass ratios q = 0.3

($\gamma_{\rm largeq} \approx -1.0$) and flattens below $q < 0.3$ ($\gamma_{\rm smallq} \approx 0.0$), but is still top heavy compared to random pairings drawn from the initial mass function (IMF; Lépine and Bongiorno, 2007; Moe and Di Stefano, 2017).

For early-type binaries, the power-law components $\gamma_{\rm largeq}$ and $\gamma_{\rm smallq}$ also decrease with increasing separation, but much more dramatically (see Figure 2.4). For A-type and late-B primaries, the fitted values are $\gamma_{\rm largeq} \approx -1.0$ and $\gamma_{\rm smallq} \approx 0.0$ at intermediate separations $a \approx 1$ - 100 au (Shatsky and Tokovinin, 2002; De Rosa et al., 2014; Gullikson et al., 2016; Murphy et al., 2018), which then decrease to $\gamma_{\rm largeq} \approx -2.0$ and $\gamma_{\rm smallq} \approx -1.0$ at very wide separations $a \gtrsim 500$ au (De Rosa et al., 2014). More massive binaries are even further skewed towards smaller mass ratios at intermediate and wide separations. For early-B and O-type primaries, the power-law slopes are $\gamma_{\rm largeq} \approx -1.5$ and $\gamma_{\rm smallq} \approx 0.0$ across $a \approx 1$ - 10 au (Abt et al., 1990; Rizzuto et al., 2013; Sana et al., 2014; Evans et al., 2015; Moe and Di Stefano, 2015b) and $\gamma_{\rm largeq} \approx -2.0$ and $\gamma_{\rm smallq} \approx -1.5$ at wide separations $a \gtrsim 100$ au (Abt et al., 1990; Peter et al., 2012; Sana et al., 2014). Even though wide companions to early-type MS stars are weighted significantly towards smaller mass ratios ($\gamma_{\rm largeq} \approx -2.0$), the mass-ratio distribution breaks below $q < 0.3$ ($\gamma_{\rm smallq} \approx -1.5$) and is therefore mildly discrepant with random pairings drawn from the IMF.

Although it is difficult to detect companions with $q < 0.1$, a bias-corrected census of extreme mass-ratio binaries is beginning to come to light. Close solar-type binaries with $P \lesssim 1{,}000$ days exhibit a dearth of $q = 0.02$–0.08 companions commonly known as the BD desert (Grether and Lineweaver, 2006). At wider separations $a > 10$ au, BD companions with $q < 0.1$ are as abundant as late-M companions with $q = 0.1$–0.2 (Kraus et al., 2011; Raghavan et al., 2010). For binaries with A/F primaries and intermediate periods $P \approx 100$–1,000 days, there is a deficit of extreme mass-ratio binaries $q < 0.1$ compared to systems with $q = 0.1$–0.2, but not a complete absence as observed for solar-type systems (Murphy et al., 2018). For more massive B-type MS primaries, the frequency of companions with $q = 0.05$–0.10 appears to be as plentiful as companions with $q = 0.10$–0.15, even at very short periods $P < 10$ days (Moe and Di Stefano, 2015a). As the mass-ratio distribution becomes weighted towards smaller values with increasing separation and primary mass, the fraction of binaries with extreme mass ratios $q < 0.1$ also increases.

M-dwarf binaries are weighted towards more equal masses compared to the mass-ratio distributions of solar-type MS binaries. For early M-dwarf binaries with intermediate separations $a \approx 10$ au, the mass-ratio distribution is nearly flat with a turnover in the substellar BD regime ($M_2 < 0.08\,{\rm M}_\odot$; Fischer and Marcy, 1992; Bergfors et al., 2010; Janson et al., 2012). As observed for more massive binaries, there may be an intrinsic trend whereby wider companions to M-dwarfs systematically favour smaller mass ratios (Duchêne and Kraus, 2013). For binaries with late M-dwarf primaries and intermediate separations $a \approx 1$ - 10 au, the mass-ratio distribution is weighted significantly towards $q \gtrsim 0.7$. (Bouy et al., 2003; Joergens, 2006; Basri and Reiners, 2006; Bergfors et al., 2010; Dieterich et al., 2012; Duchêne and Kraus, 2013). The mass-ratio distribution of late M-dwarf binaries can be modelled by either a large excess twin fraction $F_{\rm twin} \approx 0.5$ and/or a power-law slope $\gamma_{\rm largeq} \approx 2$ - 4 that is significantly skewed towards large mass ratios. As emphasised in Dieterich et al. (2012) and Duchêne and Kraus (2013), the dearth of BD companions to M-type primaries is not a selection bias but instead intrinsic to the population of M-dwarf binaries, similar to the BD desert observed for solar-type primaries.

2.2.4 Eccentricity Distribution

The eccentricity distribution and its variation with period, age and primary mass are used to constrain models of tidal evolution (Meibom and Mathieu, 2005; Belczynski et al., 2008; Moe and Di Stefano, 2015b). For solar-type binaries, the tidal circularisation period increases from $P_{\rm circ} \approx 6$ days during the pre-MS to $P_{\rm circ} \approx 12$ days for older populations in the field and halo (Meibom and Mathieu, 2005). The tidal circularisation period $P_{\rm circ} = 2$ days of early-type MS binaries is measurably smaller (Sana et al., 2012; Moe and Di Stefano, 2017), demonstrating tidal energy dissipation is substantially weaker for more massive stars with radiative envelopes (Zahn, 1977).

For periods beyond $P_{\rm circ}$, the eccentricities become increasingly weighted towards larger values. The eccentricity distribution is typically modelled with a power-law distribution $p_e \propto e^{\eta}$ up to a maximum value $e_{\rm max} \approx 0.8$ above which tidal interactions at periastron become important and/or the observational methods are incomplete (Harrington and Miranian, 1977; Moe and Di Stefano, 2017; Murphy et al., 2018). The observed eccentricity distributions are generally between a uniform distribution ($\eta = 0$) and a thermal distribution ($\eta = 1$; Ambartsumian, 1937; Kroupa, 2008), but there are slight variations with respect to primary mass and period. For late-type binaries with $M_1 < 2\,M_\odot$, the measured power-law slope increases from $\eta \approx -0.1$ at $\log P\,(\text{days}) = 2$ to $\eta \approx 0.5$ at $\log P\,(\text{days}) = 5$ (Meibom and Mathieu, 2005; Tokovinin and Kiyaeva, 2016; Moe and Di Stefano, 2017; Murphy et al., 2018). For young early-type MS binaries, observations indicate $\eta \approx 0.5$–0.8 across short and intermediate periods $1.5 < \log P\,(\text{days}) < 4.0$ (Moe and Di Stefano, 2015b; Moe and Di Stefano, 2017). Because binaries with intermediate periods are weighted towards modest eccentricities $\langle e \rangle \approx 0.5$, binaries with periods up to $\log P\,(\text{days}) \approx 3.5$ ($a \approx 10$ au, depending on M_1) will fill their Roche lobes at periastron as the primaries evolve into giants.

2.2.5 Variations with Age, Environment and Metallicity

There is clear observational evidence that the properties of wide binaries depend on age and environment. For example, the frequency of wide companions ($a \gtrsim 30$ au) to T Tauri stars is ≈ 2–3 times larger than that observed for field G-dwarfs (Ghez et al., 1993; Duchêne et al., 2007; Connelley et al., 2008; Tobin et al., 2016). In fact, the frequency of wide companions to pre-MS solar-type stars matches the frequency of wide companions to OB primaries (Shatsky and Tokovinin, 2002; Moe and Di Stefano, 2017). The wide binary fraction of early M-dwarfs in young moving groups may also show an excess relative to the field M-dwarf population (Shan et al., 2017). For both solar-type MS and pre-MS primaries in open clusters, the frequency and period distributions of wide companions are correlated with cluster density (Duchêne, 1999; Duchêne et al., 1999; Patience et al., 2002; Sollima et al., 2010; Kraus et al., 2011; Kounkel et al., 2016a). Runaway OB stars, i.e., massive stars with large peculiar velocities and/or not associated with open clusters, exhibit a substantially smaller wide binary fraction compared to OB stars embedded in clusters and associations (Mason et al., 2009). Although there are definitive variations in the wide binary fraction, the quantitative trends with primary mass, binary separation, age and cluster density have not yet been firmly established (Duchêne and Kraus, 2013).

Meanwhile, the properties of close binaries with $a \lesssim 1$ au appear to be relatively universal. The close binary fraction of solar-type primaries during the pre-MS phase (Mathieu, 1994; Melo, 2003; Kounkel et al., 2016b), in open clusters spanning a range of densities and ages (Geller and Mathieu, 2012; Leiner et al., 2015), in the galactic halo (Latham et al., 2002) and at low metallicity (Carney et al., 2005) are all consistent with the field population. The close binary fractions of B-type MS (Moe and Di Stefano, 2013) and O-type MS (Almeida et al., 2017) primaries in the low-metallicity Magellanic Clouds are similar to their counterparts in the Milky Way.

Considering the close binary fraction of T Tauri stars matches the close binary fraction of field G-dwarfs while the wide pre-MS binary fraction is comparatively larger, the overall separation distribution of solar-type pre-MS binaries is skewed towards larger values $a \approx 200$ - 5,000 au (Duchêne et al., 2007; Connelley et al., 2008; Tobin et al., 2016). By integrating $f_{\log P}$ across all orbital periods, the binary fraction and multiplicity frequency of the youngest pre-MS solar-type primaries is $F_{\rm bin} = 0.8 \pm 0.1$ and $f_{\rm mult} = 1.2 \pm 0.2$, respectively (Tobin et al., 2016; Moe and Di Stefano, 2017). These values are considerably larger than those measured for solar-type MS primaries in the field, but still smaller than those observed for more massive OB primaries (see Figure 2.2).

There are three notable exceptions suggesting the close binary fraction may depend on properties other than primary mass. First, the close binary fraction of OB stars in older and less dense associations are smaller than those observed in young, dense clusters (García and Mermilliod, 2001; Chini et al., 2012). Sana et al. (2012) and de Mink et al. (2014) argued this is a selection bias whereby OB stars in older environments are more likely to be affected by binary evolution. De Mink et al. 2014 estimated that ≈30% of massive stars are the products of binary evolution, and Moe and Di Stefano (2017) found that ≈20% of OB "primaries" in a volume-limited sample are actually the secondaries in which the true primaries have already evolved into compact remnants. About 10–20% of OB stars are runaway, and were either dynamically disrupted from their birth cluster or kicked from primaries that have since undergone supernova (Hoogerwerf et al., 2001). In either case, runaway OB stars were originally members of stable binaries (supernova scenario) or unstable triples/higher-ordered multiples (dynamical disruption scenario). The close binary fraction of O-type MS stars in young open clusters as measured by Sana et al. (2012) is therefore representative of the initial zero-age MS population.

Second, the close binary fraction of massive young stellar objects (mYSOs; $\tau \lesssim 1$ Myr) is substantially smaller than slightly older OB MS stars with ages $\tau \approx$ 2–3 Myr (Sana, 2017). The authors suggest this is observational evidence for orbital migration, whereby companions initially fragment in the disk at wider separations $a > 1$ au and subsequently migrate to very short periods after $\tau \gtrsim 1$ Myr. Another possible explanation involves a selection bias in which mYSOs that still exhibit signatures of protostellar disk emission after $\tau \approx 1$ Myr are less likely to have close binary companions. Indeed, for solar-type systems, the presence of close companions measurably truncates the disk lifetimes of pre-MS stars (Cieza et al., 2009; Kraus et al., 2012). Additional surveys of both mYSOs and normal OB MS stars in more clusters with different ages are needed to determine if the changes in the massive close binary fraction are due to orbital migration or the binary-induced truncation of disk lifetimes. In any case, at least some pre-MS companions to OB primaries migrate to very short periods $P < 10$ days within $\tau \lesssim 1$ Myr (Moe and Di Stefano, 2015a).

Finally, spectroscopic RV variability surveys of solar-type stars indicate the close binary fraction is a factor of $\approx$2–3 times larger at lower metallicity [Fe/H] $\lesssim -0.5$ (Gao et al., 2014; Badenes et al., 2017). Orbital periods of the RV variables have not yet been measured, but the observed factor of $\approx$2–3 variation cannot be easily explained by selection effects. It is nonetheless difficult to reconcile the recently observed metallicity trends with the earlier results of Latham et al. (2002) and Carney et al. (2005), who obtained enough epochs to measure orbital periods and found the close binary fraction does not vary by more than $\approx$40% across $-2.0 <$ [Fe/H] < 0.0. Additional epochs of the RV variables identified by Gao et al. (2014) and Badenes et al. (2017) are clearly needed to measure their orbital periods and firmly establish any metallicity trends in the close binary fraction.

2.3 Implications for Binary Star Formation and Evolution

Both observations of pre-MS binaries and theoretical simulations demonstrate stellar companions initially form via disk fragmentation on scales of 10s to 100s of au and via core fragmentation at several 100s to 1,000s of au (Bonnell and Bate, 1994; Bate et al., 1995; Tohline, 2002; Kratter et al., 2008; Kraus et al., 2011; Tobin et al., 2016). The observations discussed previously indicate the formation of wide binaries via core fragmentation is relatively insensitive to primary mass, and instead the wide binary fraction mainly depends on age and environment. Some theories suggest all stars are born in wide binaries, independent of primary mass, and that the mass-ratio distribution at birth is consistent with random pairings drawn from the IMF (Goodwin and Kroupa, 2005; Marks and Kroupa, 2011; Thies et al., 2015). Subsequent dynamical processing on timescales of $\tau \approx 1$ Myr preferentially disrupts wider binaries with lower binding energies, i.e., systems with smaller M_1 and q. The measured frequency of wide companions to solar-type pre-MS primaries matches the frequency of wide companions to OB stars (Ghez et al., 1993; Shatsky and Tokovinin, 2002; Duchêne et al., 2007; Connelley et al., 2008; Tobin et al., 2016; Moe and Di Stefano, 2017), which is consistent with this hypothesis. In addition, the mass ratios of wide binaries with younger OB primaries are closer to being randomly drawn from the IMF (albeit still mildly discrepant) compared to wide binaries with older solar-type MS stars in the field (Abt et al., 1990; Moe and Di Stefano, 2017; see Figure 2.4). However, the frequency of wide companions to very low-mass pre-MS stars and the mass-ratio distribution of wide pre-MS binaries have not yet been measured. Tokovinin (2017a) argued wide binaries derive from adjacent cores, *not* randomly paired cores, and so there should be at least some degree of correlation in the component masses even during the earliest pre-MS phase. Moreover, the properties of solar-type binaries with separations $a \approx 200$ au, which are more tightly gravitationally bound and less likely to be disrupted, vary more with cluster density and less so with age (Duchêne, 1999; Duchêne et al., 1999; Kraus et al., 2011; Kounkel et al., 2016a). It is therefore difficult to reconcile the observed trends of wide binaries under the assumption of a universal binary star population at birth.

At intermediate separations $a \approx 0.1$–100 au, the processes of fragmentation, protobinary accretion and migration within the disk are largely determined by the primary mass M_1. The binary fraction across intermediate separations is nearly constant below $M_1 \lesssim 1\ M_\odot$ but substantially increases above $M_1 \gtrsim 1\ M_\odot$ (Moe and Di Stefano, 2017; Murphy et al., 2018; see Figure 2.3). This is consistent with hydrodynamical simulations, which show the more massive disks around protostars above $\gtrsim 1\ M_\odot$ become increasingly more prone to disk fragmentation (Kratter and Matzner, 2006; Kratter et al., 2008). The observations suggest most stars with $M_1 \gtrsim 3\ M_\odot$ form at least one companion via disk fragmentation, and most stars with $M_1 \gtrsim 15\ M_\odot$ form two or more companions via disk fragmentation. For binaries with intermediate periods, the excess twin fraction is observed only in late-type binaries with $M_1 \lesssim 2\ M_\odot$, and the overall mass-ratio distribution becomes gradually weighted towards smaller values with increasing primary mass. During disk fragmentation, the ratio of fragment mass to final primary mass $q_{\rm frag} = M_{\rm frag}/M_1$ is also predicted to decrease with primary mass (Kratter and Matzner, 2006; Kratter et al., 2008). In addition, subsequent accretion of binary components embedded in disks tends to drive the mass ratios towards unity (Bate et al., 1995; Kroupa, 1995; Tokovinin, 2000; Bate, 2000). The disks of low-mass protostars are longer lived, and so their binary components are more likely to evolve towards larger mass ratios via shared accretion in the disk (Moe and Di Stefano, 2017). Models of disk fragmentation and subsequent accretion are therefore consistent with the observed trends with primary mass in which the binary fraction across intermediate periods increases and the mass ratios become weighted towards smaller values.

The measured inclination distributions of triples provide further evidence that wide companions form via core fragmentation while closer companions form via disk fragmentation. Only triples with wide tertiaries ($a_{\rm out} > 1{,}000$ au) exhibit random orientations (Sterzik and Tokovinin, 2002; Tokovinin, 2017b), indicating the processes of core fragmentation (wide outer tertiaries) and disk fragmentation (closer inner binaries) are relatively independent (see also Kraus et al., 2011). Meanwhile, almost all compact triples with $a_{\rm out} < 50$ au are in prograde configurations, a significant majority of which are nearly coplanar with small mutual inclinations $i < 30^\circ$ (Borkovits et al., 2016; Tokovinin, 2017b). Both companions in compact triples therefore derived from disk fragmentation.

Very close binaries with $a < 0.1$ au cannot have formed in situ, but instead initially fragmented within their natal disks on scales of 10s of au and subsequently migrated inwards (Bonnell and Bate, 1994; Bate et al., 2002). Nearly all very close binaries have outer tertiary companions (Tokovinin et al., 2006), suggesting dynamical interactions in triples hardened the inner binaries. Moe and Di Stefano (2017) showed the very close binary fraction is directly proportional to the overall triple star fraction, and argued the larger triple star fraction among massive systems causes the inner companions to be significantly skewed towards very short periods. One possible explanation for the preponderance of very close binaries with outer tertiaries is that Kozai–Lidov cycles in misaligned triples pumped the eccentricities of inner binaries, which then subsequently decayed towards shorter periods via tidal friction (Kiseleva et al., 1998; Fabrycky and Tremaine, 2007; Naoz and Fabrycky, 2014). However, the close binary fraction of T Tauri stars is consistent with the close binary fraction of solar-type stars in the field (Mathieu, 1994; Melo, 2003), demonstrating orbital migration occurred on rapid timescales $\tau \lesssim 5$ Myr. In addition, very close early-type and especially solar-type binaries with $P < 10$ days exhibit an excess twin fraction (Tokovinin, 2000; Halbwachs et al., 2003; Pinsonneault and Stanek, 2006; Moe and Di Stefano, 2017), indicating the components coevolved during the pre-MS phase via Roche-lobe overflow and/or shared accretion within the disk. Even considering very efficient tidal energy dissipation, population synthesis models cannot reproduce the observed population of close binaries by $\tau \approx 5$ Myr via secular evolution alone (Moe and Kratter, 2017). More than half of very close binaries in triples are in compact configurations with $a_{\rm out} < 50$ au (Tokovinin, 2014), the majority of which are nearly coplanar (as discussed previously in this section) and therefore could not have evolved via Kozai–Lidov cycles. Most close binaries with outer tertiaries therefore derived from the dynamical unfolding of

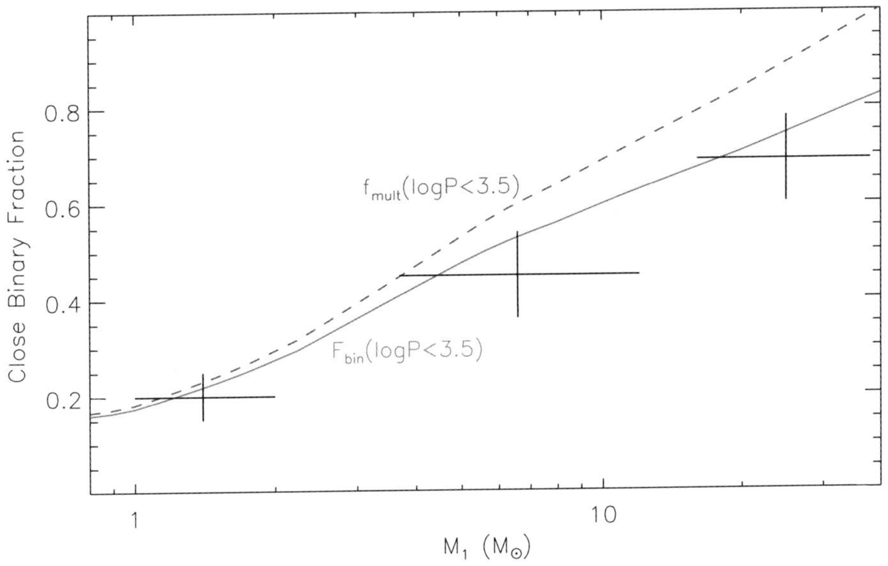

Figure 2.5 The fraction of primaries that have companions close enough to interact via Roche-lobe overflow (crosses). I also show the close binary fraction (solid line) and close multiplicity frequency (dashed line) calculated by integrating the period distributions of inner binaries and all companions, respectively, up to $\log P\,(\text{days}) = 3.5$. ($a \approx 10$ au).

unstable coplanar triples that fragmented within the disk coupled with significant energy dissipation within the disk (Bate et al., 2002; Bate, 2009; Moe and Kratter, 2017).

The close binary fraction increases dramatically with primary mass, implying post-MS binary interactions are more likely for more massive primaries (see Figure 2.5). About 15–25% of M_1 = 1 - 2 $M_\odot$ primaries will fill their Roche lobes at some point in their evolution (Raghavan et al., 2010; Geller and Mathieu, 2012). Late-B primaries with $M_1 \approx 3$ - 7 $M_\odot$ are the progenitors of Type Ia supernovae, and early-B primaries with $M_1 \approx 8$ - 16 $M_\odot$ produce the majority of core collapse supernovae. For such B-type MS primaries, $\approx$40–50% will interact with a companion (Abt et al., 1990; Kobulnicky and Fryer, 2007; Kouwenhoven et al., 2007). Finally, 69% $\pm$ 9% of O-type MS primaries have companions close enough to interact via Roche-lobe overflow (Sana et al., 2012). Moreover, about $\approx$10–20% of O-type MS primaries are in very compact triples ($a_{\rm out}$ < 10 au) in which even the outer tertiary is close enough to interact as the system evolves (Moe and Di Stefano, 2017). Indeed, binary (and triple) interactions dominate the evolution of the most massive stars.

References

Abt, H. A., Gomez, A. E., and Levy, S. G. 1990. The Frequency and Formation Mechanism of B2-B5 Main-Sequence Binaries. *ApJ S*, **74**, 551–573.

Aldoretta, E. J., Caballero-Nieves, S. M., Gies, D. R., et al. 2015. The Multiplicity of Massive Stars: A High Angular Resolution Survey with the Guidance Sensor. *AJ*, **149**, 26.

Almeida, L. A., Sana, H., Taylor, W., et al. 2017. The Tarantula Massive Binary Monitoring. I. Observational Campaign and OB-Type Spectroscopic Binaries. *A&A*, **598**, A84.

Ambartsumian, V. A. 1937. On the Statistics of Double Stars. *Astron. Zh.*, **14**, 207.

Badenes, C., Mazzola, C., Thompson, T. A., et al. 2017. Stellar Multiplicity Meets Stellar Evolution and Metallicity: The APOGEE View. *ArXiv e-prints*, **1711.00660**.

Basri, G., and Reiners, A. 2006. A Survey for Spectroscopic Binaries among Very Low Mass Stars. *AJ*, **132**, 663–675.

Bate, M. R. 2000. Predicting the Properties of Binary Stellar Systems: The Evolution of Accreting Protobinary Systems. *MNRAS*, **314**, 33–53.

Bate, M. R. 2009. Stellar, Brown Dwarf and Multiple Star Properties from Hydrodynamical Simulations of Star Cluster Formation. *MNRAS*, **392**, 590–616.

Bate, M. R., Bonnell, I. A., and Price, N. M. 1995. Modelling Accretion in Protobinary Systems. *MNRAS*, **277**, 362–376.

Bate, M. R., Bonnell, I. A., and Bromm, V. 2002. The Formation of Close Binary Systems by Dynamical Interactions and Orbital Decay. *MNRAS*, **336**, 705–713.

Belczynski, K., Kalogera, V., Rasio, F. A., et al. 2008. Compact Object Modeling with the StarTrack Population Synthesis Code. *ApJ S*, **174**, 223–260.

Bergfors, C., Brandner, W., Janson, M., et al. 2010. Lucky Imaging Survey for Southern M Dwarf Binaries. *A&A*, **520**, A54.

Boffin, H. M. J., Cerf, N., and Paulus, G. 1993. Statistical Analysis of a Sample of Spectroscopic Binaries Containing Late Type Giants. *A&A*, **271**, 125.

Bonnell, I. A., and Bate, M. R. 1994. The Formation of Close Binary Systems. *MNRAS*, **271**.

Borkovits, T., Hajdu, T., Sztakovics, J., et al. 2016. A Comprehensive Study of the Kepler Triples via Eclipse Timing. *MNRAS*, **455**, 4136–4165.

Bouy, H., Brandner, W., Martín, E. L., et al. 2003. Multiplicity of Nearby Free-Floating Ultracool Dwarfs: A Hubble Space Telescope WFPC2 Search for Companions. *AJ*, **126**, 1526–1554.

Carney, B. W., Aguilar, L. A., Latham, D. W., and Laird, J. B. 2005. A Survey of Proper-Motion Stars. XVII. A Deficiency of Binary Stars on Retrograde Galactic Orbits and the Possibility that ω Centauri Is Related to the Effect. *AJ*, **129**, 1886–1905.

Chini, R., Hoffmeister, V. H., Nasseri, A., Stahl, O., and Zinnecker, H. 2012. A Spectroscopic Survey on the Multiplicity of High-Mass Stars. *MNRAS*, **424**, 1925–1929.

Chini, R., Fuhrmann, K., Barr, A., et al. 2014. New Visual Companions of Solar-Type Stars within 25 pc. *MNRAS*, **437**, 879–886.

Cieza, L. A., Padgett, D. L., Allen, et al. 2009. Primordial Circumstellar Disks in Binary Systems: Evidence for Reduced Lifetimes. *ApJ L*, **696**, L84–L88.

Clark, B. M., Blake, C. H., and Knapp, G. R. 2012. The Close Binary Fraction of Dwarf M Stars. *ApJ*, **744**, 119.

Connelley, M. S., Reipurth, B., and Tokunaga, A. T. 2008. The Evolution of the Multiplicity of Embedded Protostars. II. Binary Separation Distribution and Analysis. *AJ*, **135**, 2526–2536.

De Marco, O., Passy, J.-C., Frew, D. J., Moe, M., and Jacoby, G. H. 2013. The Binary Fraction of Planetary Nebula Central Stars. I. A High-Precision, I-Band Excess Search. *MNRAS*, **428**, 2118–2140.

de Mink, S. E., Sana, H., Langer, N., Izzard, R. G., and Schneider, F. R. N. 2014. The Incidence of Stellar Mergers and Mass Gainers among Massive Stars. *ApJ*, **782**, 7.

De Rosa, R. J., Patience, J., Wilson, P. A., et al. 2014. The VAST Survey. III. The Multiplicity of A-Type Stars within 75 pc. *MNRAS*, **437**, 1216–1240.

Dieterich, S. B., Henry, T. J., Golimowski, D. A., Krist, J. E., and Tanner, A. M. 2012. The Solar Neighborhood. XXVIII. The Multiplicity Fraction of Nearby Stars from 5 to 70 AU and the Brown Dwarf Desert around M Dwarfs. *AJ*, **144**, 64.

Duchêne, G., Bontemps, S., Bouvier, J., et al. 2007. Multiple Protostellar Systems. II. A High Resolution Near-Infrared Imaging Survey in Nearby Star-Forming Regions. *A&A*, **476**, 229–242.

Duchêne, G. 1999. Binary Fraction in Low-Mass Star Forming Regions: A Reexamination of the Possible Excesses and Implications. *A&A*, **341**, 547–552.

Duchêne, G., and Kraus, A. 2013. Stellar Multiplicity. *ARA&A*, **51**, 269–310.

Duchêne, G., Bouvier, J., and Simon, T. 1999. Low-Mass Binaries in the Young Cluster IC 348: Implications for Binary Formation and Evolution. *A&A*, **343**, 831–840.

Duchêne, G., Simon, T., Eislöffel, J., and Bouvier, J. 2001. Visual Binaries among High-Mass Stars: An Adaptive Optics Survey of OB Stars in the NGC 6611 Cluster. *A&A*, **379**, 147–161.

Duquennoy, A., and Mayor, M. 1991. Multiplicity among Solar-Type Stars in the Solar Neighbourhood. II. Distribution of the Orbital Elements in an Unbiased Sample. *A&A*, **248**, 485–524.

El-Badry, K., Rix, H.-W., Ting, Y.-S., et al. 2018. Signatures of Unresolved Binaries in Stellar Spectra: Implications for Spectral Fitting. *MNRAS*, **473**, 5043–5049.

Evans, N. E., Bond, H. E., Schaefer, G. H., Mason, B. D., et al. 2013. Binary Cepheids: Separations and Mass Ratios in 5 $M_{\text{Ł}}$ Binaries. *AJ*, **146**, 93.

Evans, N. R., Berdnikov, L., Lauer, J., et al. 2015. Binary Properties from Cepheid Radial Velocities (CRaV). *AJ*, **150**, 13.

Eyer, L., Rimoldini, L., Holl, B., et al. 2015. The Gaia Mission, Binary Stars and Exoplanets. in: Rucinski, S. M., Torres, G., and Zejda, M. (eds), *Living Together: Planets, Host Stars and Binaries*. ASPC, **496**, 121.

Fabrycky, D., and Tremaine, S. 2007. Shrinking Binary and Planetary Orbits by Kozai Cycles with Tidal Friction. *ApJ*, **669**, 1298–1315.

Farihi, J., Hoard, D. W., and Wachter, S. 2010. White Dwarf–Red Dwarf Systems Resolved with the Hubble Space Telescope. II. Full Snapshot Survey Results. *ApJ S*, **190**, 275–296.

Fischer, D. A., and Marcy, G. W. 1992. Multiplicity among M Dwarfs. *ApJ*, **396**, 178–194.

Gao, S., Liu, C., Zhang, X., et al. 2014. The Binarity of Milky Way F,G,K Stars as a Function of Effective Temperature and Metallicity. *ApJ L*, **788**, L37.

García, B., and Mermilliod, J. C. 2001. High-Mass Binaries in the Very Young Open Cluster NGC 6231: Implication for Cluster and Star Formation. *A&A*, **368**, 122–136.

Geller, A. M., and Mathieu, R. D. 2012. WIYN Open Cluster Study. XLVIII. The Hard-Binary Population of NGC 188. *AJ*, **144**, 54.

Ghez, A. M., Neugebauer, G., and Matthews, K. 1993. The Multiplicity of T Tauri Stars in the Star Forming Regions Taurus-Auriga and Ophiuchus-Scorpius: A 2.2 Micron Speckle Imaging Survey. *AJ*, **106**, 2005–2023.

Goodwin, S. P., and Kroupa, P. 2005. Limits on the Primordial Stellar Multiplicity. *A&A*, **439**, 565–569.

Grether, D., and Lineweaver, C. H. 2006. How Dry Is the Brown Dwarf Desert? Quantifying the Relative Number of Planets, Brown Dwarfs, and Stellar Companions around Nearby Sun-Like Stars. *ApJ*, **640**, 1051–1062.

Gullikson, K., Kraus, A., and Dodson-Robinson, S. 2016. The Close Companion Mass-Ratio Distribution of Intermediate-Mass Stars. *AJ*, **152**, 40.

Halbwachs, J. L., Mayor, M., Udry, S., and Arenou, F. 2003. Multiplicity among Solar-Type Stars. III. Statistical Properties of the F7-K Binaries with Periods up to 10 Years. *A&A*, **397**, 159–175.

Harrington, R. S., and Miranian, M. 1977. Observational Selection in Binary-Star Eccentricities. *PASP*, **89**, 400.

Hoogerwerf, R., de Bruijne, J. H. J., and de Zeeuw, P. T. 2001. On the Origin of the O and B-Type Stars with High Velocities. II. Runaway Stars and Pulsars Ejected from the Nearby Young Stellar Groups. *A&A*, **365**, 49–77.

Hubrig, S., Le Mignant, D., North, P., and Krautter, J. 2001. Search for Low-Mass PMS Companions around X-ray Selected Late B Stars. *A&A*, **372**, 152–164.

Janson, M., Hormuth, F., Bergfors, C., et al. 2012. The AstraLux Large M-Dwarf Multiplicity Survey. *ApJ*, **754**, 44.

Joergens, V. 2006. Radial Velocity Survey for Planets and Brown Dwarf Companions to Very Young Brown Dwarfs and Very Low-Mass Stars in Chamaeleon I with UVES at the VLT. *A&A*, **446**, 1165–1176.

Kaplan, G. H., and Makarov, V. V. 2003. Astrometric Detection of Binary Companions and Planets: Acceleration of Proper Motion. *Astron. Nachr.*, **324**, 419–424.

Kirk, B., Conroy, K., Prša, A., et al. 2016. Kepler Eclipsing Binary Stars. VII. The Catalog of Eclipsing Binaries Found in the Entire Kepler Data Set. *AJ*, **151**, 68.

Kiseleva, L. G., Eggleton, P. P., and Mikkola, S. 1998. Tidal Friction in Triple Stars. *MNRAS*, **300**, 292–302.

Kobulnicky, H. A., and Fryer, C. L. 2007. A New Look at the Binary Characteristics of Massive Stars. *ApJ*, **670**, 747–765.

Kobulnicky, H. A., Kiminki, D. C., Lundquist, M. J., et al. 2014. Toward Complete Statistics of Massive Binary Stars: Penultimate Results from the Cygnus OB2 Radial Velocity Survey. *ApJ S*, **213**, 34.

Kounkel, M., Hartmann, L., Tobin, J. J., et al. 2016b. Spectroscopic Binaries in the Orion Nebula Cluster and NGC 2264. *ApJ*, **821**, 8.

Kounkel, M., Megeath, S. T., Poteet, C. A., Fischer, W. J., and Hartmann, L. 2016a. An HST Survey for 100-1000 au Companions around Young Stellar Objects in the Orion Molecular Clouds: Evidence for Environmentally Dependent Multiplicity. *ApJ*, **821**, 52.

Kouwenhoven, M. B. N., Brown, A. G. A., Portegies Zwart, S. F., and Kaper, L. 2007. The Primordial Binary Population. II. Recovering the Binary Population for Intermediate Mass Stars in Scorpius OB2. *A&A*, **474**, 77–104.

Kratter, K. M., and Matzner, C. D. 2006. Fragmentation of Massive Protostellar Discs. *MNRAS*, **373**, 1563–1576.

Kratter, K. M., Matzner, C. D., and Krumholz, M. R. 2008. Global Models for the Evolution of Embedded, Accreting Protostellar Disks. *ApJ*, **681**, 375–390.

Kraus, A. L., Ireland, M. J., Hillenbrand, L. A., and Martinache, F. 2012. The Role of Multiplicity in Disk Evolution and Planet Formation. *ApJ*, **745**, 19.

Kraus, A. L., Ireland, M. J., Martinache, F., and Hillenbrand, L. A. 2011. Mapping the Shores of the Brown Dwarf Desert. II. Multiple Star Formation in Taurus-Auriga. *ApJ*, **731**, 8.

Kroupa, P. 1995. The Dynamical Properties of Stellar Systems in the Galactic Disc. *MNRAS*, **277**.

Kroupa, P. 2008. Initial Conditions for Star Clusters. in: Aarseth, S. J., Tout, C. A., and Mardling, R. A. (eds), *The Cambridge N-Body Lectures*. Lecture Notes in Physics, Berlin Springer Verlag, **760**, 181.

Latham, D. W., Stefanik, R. P., Torres, G., et al. 2002. A Survey of Proper-Motion Stars. XVI. Orbital Solutions for 171 Single-Lined Spectroscopic Binaries. *AJ*, **124**, 1144–1161.

Law, N. M., Hodgkin, S. T., and Mackay, C. D. 2008. The LuckyCam Survey for Very Low Mass Binaries. II. 13 New M4.5–M6.0 Binaries. *MNRAS*, **384**, 150–160.

Leiner, E. M., Mathieu, R. D., Gosnell, N. M., and Geller, A. M. 2015. WIYN Open Cluster Study. LXVI. Spectroscopic Binary Orbits in the Young Open Cluster M35 (NGC 2168). *AJ*, **150**, 10.

Lépine, S., and Bongiorno, B. 2007. New Distant Companions to Known Nearby Stars. II. Faint Companions of Hipparcos Stars and the Frequency of Wide Binary Systems. *AJ*, **133**, 889–905.

Levato, H., Malaroda, S., Morrell, N., and Solivella, G. 1987. Stellar Multiplicity in the Scorpius-Centaurus Association. *ApJ S*, **64**, 487–503.

Marks, M., and Kroupa, P. 2011. Dynamical Population Synthesis: Constructing the Stellar Single and Binary Contents of Galactic Field populations. *MNRAS*, **417**, 1702–1714.

Mason, B. D., Gies, D. R., Hartkopf, W. I., et al. 1998. CCD Speckle Observations of Binary Stars. XIX. An Astrometric/Spectroscopic Survey of O Stars. *AJ*, **115**, 821.

Mason, B. D., Hartkopf, W. I., Gies, D. R., Henry, T. J., and Helsel, J. W. 2009. The High Angular Resolution Multiplicity of Massive Stars. *AJ*, **137**, 3358–3377.

Mathieu, R. D. 1994. Pre-Main-Sequence Binary Stars. *ARA&A*, **32**, 465–530.

3

Gaia and LSST: Their Importance in Binary Star Research

LAURENT EYER, NAMI MOWLAVI, ISABELLE LECOEUR-TAIBI, LORENZO RIMOLDINI, BERRY HOLL, MARC AUDARD, SIMON HODGKIN, DAFYDD W. EVANS, LUKASZ WYRZYKOWSI, GEORGE SEABROKE, ANDREJ PRŠA, AND DIMITRI POURBAIX

3.1 Introduction: The Data Revolution in Astronomy

Observational astronomy is benefitting from modern technological developments and is undergoing profound and deep changes. On one side, the precision/accuracy of the measurements (e.g., the astrometry accuracy, the photometric precision) is improved by several orders of magnitudes, and on the other side the number of sources and number of measurements per source is exploding. We are currently in an era of considerable data growth. Large surveys, such as Gaia and LSST, produce homogeneous data, i.e., the data are acquired with a single set of instruments, and the data need to be processed with software in automated ways. As a consequence, the data processing has become an enormous task, whose cost is of the same order of magnitude as the cost of the contribution of instruments/hardware for ground-based facilities, e.g., for LSST; see Ivezić et al. (2014). In space, the cost of processing is becoming a significant fraction of the hardware, launch and spacecraft operation budget. The Gaia webpage of ESA[1] states: 'The total cost of the mission, from the beginning of preliminary studies to the end of operations, is 740 million Euro. This does not include expenditures related to the DPAC[2] consortium, which are covered by the member states participating in the consortium, not ESA. These costs are currently estimated at around 250 million Euro.'

Besides Gaia and LSST, there are many other future surveys that will have a significant impact on Time Domain Astronomy: HATPI[3], the new generation of HAT (Bakos et al., 2004), TESS (Ricker et al., 2015), ZTF (Bellm, 2014), and Plato (Rauer et al., 2014), to name a few.

Nonsingle stars, binaries are extremely interesting objects and have an obvious importance in many domains of astronomy. Indeed, they are very common objects in the Universe, they allow to determine directly physical properties of stars such as mass and radius, they can be used as standard candles and they play important roles in star formation and dynamical processes in clusters. However, they add a layer of complexity to the data

[1] http://www.cosmos.esa.int/web/gaia.
[2] Data Processing and Analysis Consortium.
[3] https://hatpi.org/.

processing and require significant efforts to process and analyse. Such effort is undertaken in major surveys in order to achieve the detections of binaries, their characterisation, classification and the determination of their astrophysical parameters. From these data, one can then try to infer the properties of the underlying population of nonsingle stars, such as the multiplicity statistics, or the distribution of orbital elements; see, e.g., Duquennoy and Mayor (1991) and Raghavan et al. (2010).

3.2 Gaia and LSST

Eyer et al. (2012) gave some descriptions on Gaia and LSST and their impact on binaries. The article contained some comparisons at technical levels and also of precisions/accuracies. The corresponding diagrams were updated in Ivezić et al. (2014). It is often remarked that these two large-scale surveys complement each other very well.

3.2.1 The Gaia Mission

The Gaia mission is a cornerstone mission of the Horizon2000+ ESA program. It was selected in 2000, confirmed in 2002 (for a launch no later than 2012) and actually launched in December 2013. The spacecraft performs observations of all the objects brighter than a certain magnitude[4] limit $G = 20.7$, i.e., about 1.5 billion objects (mostly stars, but also quasars and solar system bodies). The observations consist of astrometric, photometric, spectrophotometric and spectroscopic (for the determination of radial velocities) data. The nominal mission lifetime is five years (each direction of the sky being scanned 70 times on average), from 2014 to 2019, with a possible extension of five additional years. The catalogue of the five-year nominal mission is planned for 2022. If the mission is extended by five years, the final Gaia catalogue is planned for 2027.

The Gaia scanning law: One Field-of-View transit measurement is composed of one measurement of the sky-mapper, followed by nine per-CCD measurements separated by 4.85 seconds each (astrometric field), by BP and RP spectrophotometric measurements, and by RVS (Radial Velocity Spectrometer) measurements. The subsequent Field-of-View transit is done 1h46m later in the majority of cases, possibly followed by another one 4h14m later. Thereafter, there are large gaps of typically three weeks. Statistics of the time sampling are given in Eyer et al. (2017).

Gaia data releases: Initially, five Gaia data releases were planned between 2014 and 2022 (Eyer et al., 2015). However, given the monumental work each data release represents, four releases are currently considered. They are described in the following paragraphs.

The first data release in September 2016 was based on the collection of 14 months of data from end of July 2014 to mid-September 2015. It consisted of (a) position and mean G-magnitude for 1.1 billion stars (Lindegren et al., 2016, and van Leeuwen et al.,

[4] The Gaia G-band magnitude is a broadband, white-light magnitude in the wavelength range 330–1050 nm defined by the telescope transmission and CCD quantum efficiency.

2017, respectively); (b) five-parameter astrometric solution (position, proper motion and parallax) for 2.1 million stars (Lindegren et al., 2016), a solution that makes use of the Hipparcos-Tycho astrometry and is called TGAS (Tycho-Gaia Astronometric Solution); (c) photometry (light curves) of 3,194 Cepheids and RR Lyrae stars in the direction of the South Ecliptic Pole (direction of the Large Magellanic Cloud) (Clementini et al., 2016; Eyer et al., 2017); (d) position and magnitude of 2,195 quasars (Mignard et al., 2016).

The second data release is planned for April 2018 and is based on 22 months of data. The foreseen products are (a) the five-parameter astrometric solution (position, proper motion and parallax) for more than 1.5 billion stars, (b) photometry (G band and integrated BP and RP bands), (c) astrophysical parameters for $G < 17$ mag, (d) median radial velocity for $G_{RVS} < 12$ mag, (e) epoch astrometry for > 10,000 asteroids and (f) a selected sample of variable stars (though no eclipsing binaries).

The third data release is planned for 2020 and would contain updated astrometry and photometry for all sources, source classification and astrophysical parameters, BP, RP and RVS spectra with astrophysical parameters, radial velocities for $G_{RVS} < 14$ mag, updated and extended solar system object catalogue and updated and extended variable star catalogue. In addition, the release would include nonsingle stars (astrometric binaries and eclipsing binaries).

The final release for the nominal mission is planned in 2022 and would contain the full astrometric, photometric, radial velocity catalogues, all variable stars and nonsingle stars and source classifications (probabilities), plus multiple astrophysical parameters (derived from BP, RP, RVS and astrometry) for stars, unresolved binaries, galaxies and quasars. A list of exoplanet candidates is also planned. All epoch and transit data for all sources, including all BP/RP/RVS spectra, would be made available.

Gaia Alerts: Gaia also provides a continuous alert stream (Hodgkin et al., 2013) under the responsibility of Cambridge University[5]. As of mid-November 2017, more than 3,800 alerts have been issued, at a constant rate of about six alerts/day for the last 18 months. Only about 17% of Gaia Alerts are robustly classified, via spectroscopic follow-up from ground-based observatories. The remaining 83% remain unclassified right now, but the Gaia Alerts team is working to automate the classification of all its remaining alerts (for the future, but also retrospectively) through machine learning approaches. Of the approximately 650 robustly classified alerts, the vast majority (about 80%) are supernovae, and half of the supernovae are of type Ia.

Gaia Alerts are not limited to the discovery of supernovae. There are suspected Cataclysmic Variables (> 400 alerts are predicted to be CVs), and 18 microlensing event candidates, including the exciting binary microlensing event Gaia16aye. The discovery of Gaia16aye triggered a huge, worldwide campaign of photometric follow-up, resulting in > 17,000 measurements from 33 telescopes lasting for over one year. There are also objects with abrupt declining brightness such as Young Stellar Objects (32 candidates) and R CrB stars (six candidates). Quasars are also detected in these alerts (46 candidates). No

[5] http://gsaweb.ast.cam.ac.uk/alerts/home.

doubt in the thousands of unclassified alerts, there remain many more exciting and unusual events yet to be uncovered and investigated. Gaia also alerted us of a very exciting AM CVn star, which is the first such eclipsing binary with a total eclipse (Campbell et al., 2015).

3.2.2 The Large Synoptic Survey Telescope (LSST)

LSST[6] is a project consisting of an 8-meter ground-based telescope, located in Chile. First light is foreseen in 2020 and regular observations will start in 2022. LSST will detect objects as faint as 24.5 mag (single epoch) to 27 mag (with co-added images). The number of observed objects by LSST is staggering, and is estimated to reach 20 billion stars and 20 billion galaxies. The observations will be done in the five Sloan photometric bands, plus a y band. The lifetime of the project is foreseen to be 10 years. During that period, half of the sky will have 825 visits.

The observing strategy of LSST is not definitively fixed and could evolve during the project. The current plan is that there will be two exposures of 15 seconds (or maybe one of 30 seconds). These two exposures are done to detect and remove cosmic rays. Then, three to five days later, the same field is observed, though very probably in a different filter.

3.2.3 LSST Data Stream

The impressive numbers given in the previous sections for Gaia seem small when compared to the ones of LSST. These are just stunning. The nightly alert system will consist of a stream of about 10 million time-domain events per night, detected and transmitted to event distribution networks within 60 seconds of observation. The early Data Releases will contain a catalogue of about 37 billion objects (20 billion galaxies, 17 billion stars), about 7 trillion single-epoch detections (sources), produced annually, accessible through online databases. There will also be deep co-added images.

Along with the data are 'Community Services'. They will consist of service and computing resources provided at the Data Access Centres to enable user-specified custom processing and analysis. Software and APIs (application program interfaces) enabling development of analysis codes are also provided.

3.3 The Detection of Nonsingle Stars

Nonsingle stars can have signatures in all astronomical fundamental technics (astrometry, photometry, spectroscopy). Therefore, all these techniques are contributing to their possible detection and characterisation. It is fundamental to understand the details of the selection effects so that real properties of samples of objects can be inferred. For example, the observed amplitude for the signal for a circular orbit is independent of the inclination i of

[6] For a recent description of LSST, please refer to Ivezić et al. (2017).

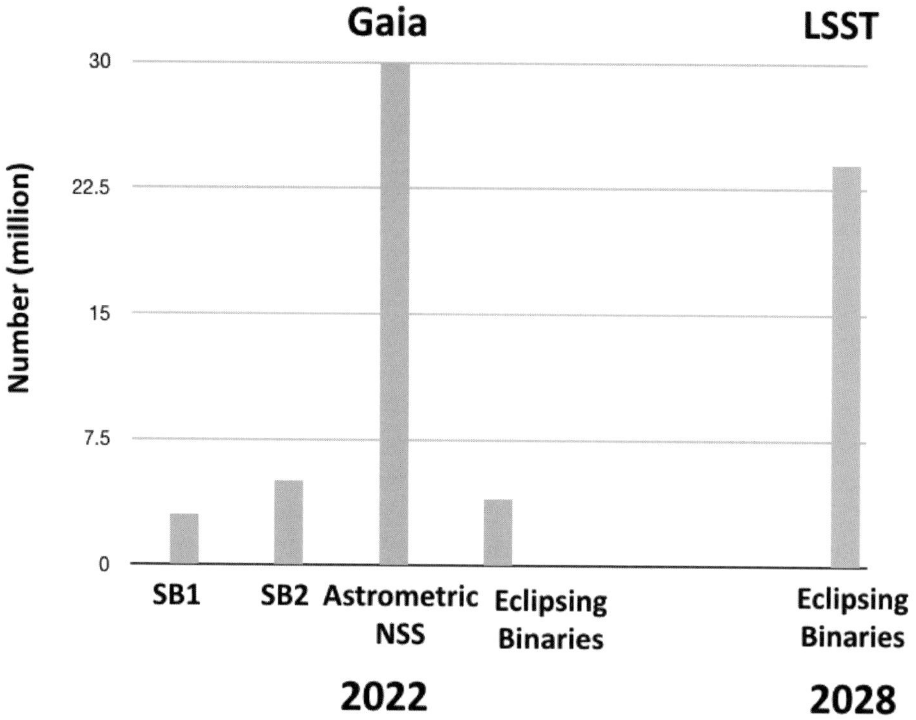

Figure 3.1 Histogram with the estimated number of nonsingle stars to be detected by Gaia and LSST. NSS: nonsingle star, SB1: single-lined spectroscopic binary, SB2: double-lined spectroscopic binary.

the orbit for the astrometry and is a function of $\sin(i)$ for the radial velocity. In the following subsections, we will go through several methods and give estimations (when possible) of the numbers of nonsingle stars detected by Gaia and LSST (see Figure 3.1).

3.3.1 Astrometric Detection of Binaries

The position as a function of time of a single star is described by the astrometric solution with five parameters (position, proper motion and parallax), while the astrometric solution of a binary system is fully described by 12 parameters (seven parameters are needed to describe a Keplerian orbit). At best, what can be derived from the orbital parameters are the period, the eccentricity, the inclination, the semimajor axes and the masses (Eyer et al., 2015, table 3). There are intermediate solutions that can be detected, for example a solution with acceleration terms for an orbital period (greatly) exceeding the observational baseline. There are also the Variability-Induced Movers (Wielen, 1996), where an unresolved double star can mimic an apparent astrometric motion if the variability of one component produces an apparent motion of the photocenter.[7]

[7] Such motion furthermore might be convolved with the true motion of the binary.

An other effect is that the positions of a star measured in distinct photometric bands exhibit some scatter around a central position, consistent with the astrometric precision. When two unresolved stars of very distinct colours are observed with the same distinct photometric bands, the photocentres get scattered around a line joining the two stars. The order of the photocentres further follow the order of the centre of bands. That is what is called a Colour-Induced Displacement double star. That technique was successfully applied with the Sloan Digital Sky Survey (SDSS) (Pourbaix et al., 2004, Pourbaix et al., 2016). Although filter-dependent astrometry was once considered in an early phase of the Gaia mission, it was discarded later on.

Predicted numbers of astrometric binaries from Gaia were presented in Eyer et al. (2013), and reach 30 million stars that are processed as nonsingle stars. There is no estimation of the number of astrometric binaries detected by LSST.

Comoving Pairs

Another astrometric method to detect binaries is by identifying comoving pairs. It is a way to detect wide binaries. The first Gaia data release was used by several authors to find such binaries, e.g., Andrews et al. (2017), Oh et al. (2017) and Oelkers et al. (2017). From these studies, the number of such detected comoving pairs ranges from 6,000 to 13,000. The improvement of the astrometric solution from the first Gaia data release to the second will be by several folds. These studies can be repeated, checks on the false detection rates can be estimated and understood and very large numbers of such pairs will be detected.

3.3.2 Spectroscopic Detection of Binaries

Spectroscopic binaries can be classified as single-lined (SB1), where several measurements are required to see the doppler shift of the lines, or double-lined (SB2), in which case the two binary components can be directly detected in the spectrum, even in a single measurement. In this latter case, several spectra should be taken to confirm that it is a bound system through the detection of a coherent motion of the two sets of lines.

The RVS on board of Gaia can detect both SB1 and SB2 binaries. It is expected that 150 million stars will have radial velocities. However, to detect binarity, the signal-to-noise level should be high enough. The estimated number of spectroscopic binaries is at the level of about 8 million, with about 5 million being SB2 (Eyer et al., 2013).

It is worth mentioning the Gaia-ESO survey, which consists of 300 nights of observations over five years at the VLT (UVES+ GIRAFFE). This survey is also detecting binaries: Merle et al. (2017) found 354 SBn candidates, where n is the number of components, and more precisely 342 SB2, 11 SB3 and even one SB4.

3.3.3 Photometric Detection of Binaries

There are several ways to detect binaries with photometry, such as through their position in the Hertzsprung–Russell diagram or in the colour–colour diagram. Furthermore, the photometric light curve can detect eclipses or ellipsoidal variations. There are also indirect ways, such as microlensing events or eclipse/transit timing.

Eclipsing Binaries

One very classical way to detect binaries is through the detection of the eclipses of the binary system when the orbital plane is favourably aligned with the observer. In recent years, the OGLE survey has provided the largest list to date with more than 450,000 eclipsing and ellipsoidal binary systems towards the Galactic Bulge (Soszyński et al., 2016) and more than 48,000 eclipsing binaries towards the Magellanic System (Pawlak et al., 2016).

From space, the *Kepler* mission has been a game-changer in the study of eclipsing binaries, with its unprecedented sampling and photometric precision. It allows an extremely detailed description of the light curves. One great surprise of the mission is that 18 % of eclipsing systems observed by *Kepler* are not strictly regular, showing changes in the eclipses, or even eclipse disappearance, such as in Guinan et al. (2012). The *Kepler* mission allowed us to estimate the fraction of eclipsing binaries that would be detected by Gaia or LSST. Surprisingly, both surveys would have a detection rate efficiency of about 70 % (Kochoska et al., 2017; Wells et al., 2017). If all the photometric bands of LSST are used, this detection efficiency could (or is expected to) reach 80 %.

The estimated number of eclipsing binaries to be detected by Gaia reaches 4 million (Eyer et al., 2013), though this estimate is quite uncertain and varies between 1/2 to 7 million. The expected number of eclipsing binaries to be detected by LSST reaches 24 million (Prša et al., 2011).

Detection of Special Binary Stars

Here are some additional ways to detect binaries with photometry. It should be noted, that in several of the following cases, the secondary or tertiary star is not necessarily in a stable (bound) orbit. Additional information is therefore required to determine the true nature of boundedness.

Colour–Magnitude diagram: In such a diagram a parallel sequence to the main sequence can be noted (see for example Gaia Collaboration et al., 2016a). It can be observed in clusters or in field stars when the parallax is available. The effect is stronger when the masses of the two stars are similar. When the masses differ, the position of the more massive star is moving towards the upper right in the Colour–Magnitude diagram due to its binarity, and therefore it appears redder.

Colour–Colour diagram: In such a diagram, binaries can be detected because of their location in zones where single stars are not present. Usually it concerns specific pairs; for example, Smolčić et al. (2004) detected with SDSS colours ($u - g$, $g - r$) a bridge from white dwarfs to M stars explained in terms of M dwarf and white dwarf pairs.

Microlensing: It was realised that binary systems and planetary systems can be detected through microlensing (Mao and Paczynski, 1991). The first system was detected with OGLE data (Udalski et al., 1994). Early warning systems were put in place to first detect mircolensing events and then also to detect any deviation from a usual microlensing event so that follow-up observations could be done. A detailed analysis of the light curve can in the best and rare cases give the mass, distance, period of rotation and inclination of the binary lens. More usually, the mass ratio and projected separation in units of Einstein radius can be determined (cf. Jaroszynski et al., 2006). We can note that Gaia has found a spectacular binary event, Gaia16aye[8] (Wyrzykowski et al., 2017), with multiple caustic crossings. The event is the first binary in the Disk, not in the Bulge. Such events show that one could detect binary black holes like the ones that merged in gravitational wave events.

Orbital variations: Several types of deviations from the strictly periodic light curves of eclipsing binaries can be attributed to a third body. Orbital variations have been studied theoretically by Fabrycky and Tremaine (2007). The variation of the eclipse timings in the *Kepler* data allowed the detection of 236 systems out of 1,279, where the timing deviation is compatible with the presence of a third body (Conroy et al., 2014).

Ellipsoidal variables: Flux variation as a function of phase does not necessarily imply eclipses; in general, the amount of flux received from a star is proportional to its cross-section. If, due to tidal and rotational distortion of the stellar shape, the cross-section varies, so will the flux. This effect is referred to as ellipsoidal variation, and in close circular orbits it amounts to $\sim$10% or less. However, the effect can be significantly augmented during periastron flybys of binaries in highly eccentric orbits. These objects feature unique light curves that demonstrate a strong periastron peak and frequent tidally induced pulsations. Because of their resemblance to a cardiogram, this type of objects are called heartbeat stars (Thompson et al., 2012). OGLE found large numbers of ellipsoidal binaries, and Pawlak et al. (2014) identified more than 2,000 such systems towards the Large Magellanic clouds.

Background EBs: When the primary object of interest is not an eclipsing binary, but its light is contaminated by a background eclipsing binary, its light curve will exhibit shallow eclipses. These objects are the main sources of exoplanetary false positives because eclipse depths are compatible with exoplanetary transits and their shapes are often indistinguishable because of the limited signal-to-noise ratio per data point. These are ubiquitous; for example, Abdul-Masih et al. (2016) report a discovery of 285 background EBs in the *Kepler* dataset.

Other effects: There are quite a few other second-order effects that, thanks to the exquisite precision of modern photometric surveys reaching $\sim$ppm accuracy, can be routinely detected. These include Doppler boosting (Bloemen et al., 2012), reflection effect (Derekas et al., 2015), accretion disks (Djurašević et al., 2010) and asteroseismic components (Huber, 2015).

3.4 The Classification of Binaries

In this section, we focus on the eclipsing binary classification. One of the great challenges of having large numbers of eclipsing binaries is their automated classification, with the

[8] See the Gaia Image of the Week of 27 October 2016: http://www.cosmos.esa.int/web/gaia/IoW_20161027.

twofold purposes of (a) studying known classes of binary systems and (b) identifying peculiar binary systems with properties deviating from the ones of known classes. Within the Gaia consortium, the detection and classification of eclipsing binaries are performed by a team based in Geneva (Switzerland) of the Coordination Unit 7 (CU7), while the modelling of the full binary system (orbital and stellar parameters) is performed by a team based in Brussels (Belgium) of the Coordination Unit 4 (CU4) (cf. Siopis and Sadowski, 2012). The full binary system solution is possible only for systems for which enough data, both photometric and spectroscopic, are available.

Four broad categories of binaries are usually considered: detached systems, semidetached systems where one of the stars fills its Roche lobe, contact (or overcontact) systems with potentially a common envelope and ellipsoidal systems. It is, however, rather difficult to provide a classification schema that directly matches these categories based on photometric measurements alone. The 'standard' classification into EA, EB and EW types described in the General Catalogue of Variable Stars,[9] for example, offers a convenient way for observers to classify eclipsing binary light curves, and explains its still widespread use. But it relies on visual, hence subjective, assessment of the light curve geometry and does not necessarily relate to the afore mentioned categories of binary systems. An attempt in this direction has been proposed by Pojmanski (2002) using Fourier series parameters.

A discrete separation between light curve morphological types is not very realistic, as there is a continuous progression of eclipsing binary properties across the borders from detached to overcontact systems. Matijevič et al. (2012) presented a procedure to classify the light curves with a single parameter that is a measure of the 'detachedness' of the system. The procedure uses locally linear embedding, a general nonlinear dimensionality reduction tool. They successfully apply the method to *Kepler* eclipsing binaries and show a good correlation of the single parameter with morphology type.

An additional step can be taken with an automated procedure to identify the main components in the light curves of eclipsing binaries. Mowlavi et al. (2017), for example, use Gaussian functions to model the light curve geometry of the eclipses, and sine functions to model the ellipsoidal variability due to tidal effects (the overall model is called the 2-Gaussian model). Applied to the OGLE-III set of eclipsing binaries in the Large Magellanic Cloud, they show how this characterisation of the light curve geometry allows the identification of systems that contain either additional physics (e.g., reflection effects, apsidal motions, discs) or additional variability intrinsic to one or both stars of the binary system. The procedure also provides a set of light curve geometry parameters that are favourable to the statistical analysis of populations of eclipsing binaries. The 2-Gaussian model is currently adopted by CU7 for the processing of Gaia eclipsing binary light curves; see Figure 3.2.

Additional modern classification techniques were proposed in the literature using, for example, a combination of the t-distributed stochastic neighbour embedding (t-SNE)

9 www.sai.msu.su/gcvs/gcvs.

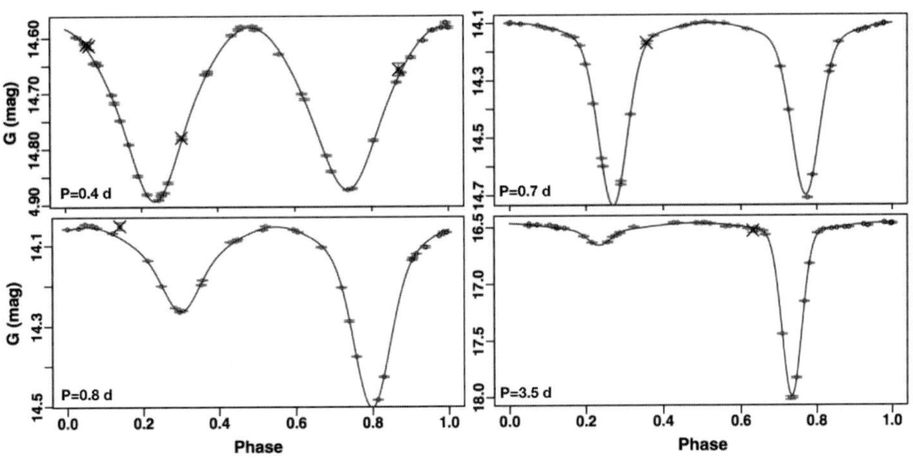

Figure 3.2 Four examples of eclipsing binary light curves of Gaia data fitted with the 2-Gaussian model.

and density-based spatial clustering of applications with noise (DBSCAN) algorithms (Kochoska et al., 2017) or machine learning algorithms (e.g., Süveges et al., 2017).

3.5 The Characterisation of Binaries

There are many characteristics that are of interests to binary systems. Examples include the binary fraction among different populations of stars as a function of mass, metallicity intervals, evolution status (dwarf/giant) or their underlying group (halo/disc stars, globular/open cluster, etc.). It is also interesting to determine the orbital properties, such as period, semimajor axis distribution, mass ratio, etc. Such an exploitation just started with the first data release of Gaia: Boffin and Pourbaix (2018) show the mass-ratio distribution of spectroscopic binaries along the main sequence.

The physical characteristics of binaries can also be compared to those obtained from stellar evolution and pulsation models.

There are fundamental warnings in such studies: there is a need to understand the selection effects of the different star samples, the statistical analysis should be very carefully done especially when using the parallax (there have been numerous misuses of parallaxes in the literature) and there are degeneracies that are sometimes difficult to lift (such as temperature versus extinction).

3.6 Conclusions

Binary stars will be one of the topics that is and will be extremely impacted by the Gaia and LSST surveys. The estimated numbers of new detected binaries presented in this chapter are going to reach in total many tens of millions. This data growth is exceptional. In this context, it is very relevant to get informed about the data products at each stage of the releases and to be prepared for their scientific exploitation. It is also important to understand some technical details of these surveys to determine their possible limitations, selection functions and statistical properties.

Acknowledgements

We are indebted to Professor Željko Ivezić (University of Washington), Project Scientist of LSST, who provided material for this contribution.

References

Abdul-Masih, M., Prša, A., Conroy, K., et al. 2016. Kepler Eclipsing Binary Stars. VIII. Identification of False Positive Eclipsing Binaries and Re-Extraction of New Light Curves. *AJ*, **151**, 101.

Andrews, J. J., Chanamé, J., and Agüeros, M. A. 2017. Wide Binaries in Tycho-Gaia: Search Method and the Distribution of Orbital Separations. *MNRAS*, **472**, 675.

Bakos, G., Noyes, R. W., Kovács, G., et al. 2004. Wide-Field Millimagnitude Photometry with the HAT: A Tool for Extrasolar Planet Detection. *PASP*, **116**, 266.

Bellm, E. 2014. The Zwicky Transient Facility. The Third Hot-Wiring the Transient Universe Workshop. Edited by P. R. Wozniak, 27.

Bloemen, S., Marsh, T. R., Degroote, P., et al. 2012. Mass Ratio from Doppler Beaming and Rmer Delay versus Ellipsoidal Modulation in the Kepler Data of KOI-74. *MNRAS*, **422**, 2600.

Boffin, H. M. J., and Pourbaix, D. 2018. The Mass-Ratio Distribution of Spectroscopic Binaries along the Main-Sequence. Recio-Blanco, A., de Laverny, P., Brown, A. G. A., and Prusti, T. (eds), *IAU Symp.*, **330**, 339.

Campbell, H. C., Marsh, T. R., Fraser, M., et al. 2015. Total Eclipse of the Heart: The AM CVn Gaia14aae/ASSASN-14cn. *MNRAS*, **452**, 1060.

Clementini, G., Ripepi, V., Leccia, S., et al. 2016. Gaia Data Release 1: The Cepheid and RR Lyrae Star Pipeline and Its Application to the South Ecliptic Pole Region. *A&A*, **595**, 133.

Conroy, K. E., Prša, A., Stassun, K. G., et al. 2014. Kepler Eclipsing Binary Stars. IV. Precise Eclipse Times for Close Binaries and Identification of Candidate Three-Body Systems. *AJ*, **147**, 45.

Derekas, A., Németh, P., Southworth, J., et al. 2015. A New sdO+dM Binary with Extreme Eclipses and Reflection Effect. *ApJ*, **808**, 179.

Djurašević, G., Latković, O., Vince, I., and Cséki, A. 2010. Accretion Disc in the Eclipsing Binary AU Mon. *MNRAS*, **409**, 329.

Duquennoy, A., and Mayor, M. 1991. Multiplicity among Solar-Type Stars in the Solar Neighbourhood. II. Distribution of the Orbital Elements in an Unbiased Sample. *A&A*, **248**, 485.

Eyer, L., Dubath, P., Mowlavi, N., et al. 2012. The Impact of Gaia and LSST on Binaries and Exoplanets. From Interacting Binaries to Exoplanets: Essential Modeling Tools, Proceedings of the International Astronomical Union. Edited by M. T. Richards and I. Hubeny. *IAU Symp.*, **282**, 33.

Eyer, L., Holl, B., Pourbaix, D., et al. 2013. The Gaia Mission. *Central European Astrophysical Bulletin*, **37**, 115.

Eyer, L., Mowlavi, N., Evans, D. W., et al. 2017. Gaia Data Release 1: The Variability Processing & Analysis and Its Application to the South Ecliptic Pole Region. *ArXiv e-prints*, 1702.03295.

Eyer, L., Rimoldini, L., Holl, B., et al. 2015. The Gaia Mission, Binary Stars and Exoplanets. Living Together: Planets, Host Stars and Binaries. Edited by S. M. Rucinski, G. Torres and M. Zejda. *ASPC*, **496**, 121.

Fabrycky, D., and Tremaine, S. 2007. Shrinking Binary and Planetary Orbits by Kozai Cycles with Tidal Friction. *ApJ*, **669**, 1298.

Gaia Collaboration, Brown, A. G. A., Vallenari, A., et al. 2016a. Gaia Data Release 1. Summary of the Astrometric, Photometric, and Survey Properties. *A&A*, **595**, 2.

Gaia Collaboration, Prusti, T., de Bruijne, J. H. J., et al. 2016b. The Gaia Mission. *A&A*, **595**, 1.

Gaia Collaboration, van Leeuwen, F., Vallenari, A., et al. 2017. Gaia Data Release 1. Open Cluster Astrometry: Performance, Limitations, and Future Prospects. *A&A*, **601**, 19.

Guinan, E., Bonaro, M., Engle, S., and Prša, A. 2012. Eclipsing Binaries That Don't Eclipse Anymore: The Strange Case of the Once (and Future?) Eclipsing Binary QX Cassiopeiae. *JAAVSO*, **40**, 417.

Hodgkin, S. T., Wyrzykowski, L., Blagorodnova, N., and Koposov, S. 2013. Transient Astronomy with the Gaia Satellite. *Phil. Trans. Royal Soc. of London Series A*, **371**, 20120239.

Huber, D. 2015. Asteroseismology of Eclipsing Binary Stars. in Giants of Eclipse: The ζ Aurigae Stars and Other Binary Systems, *ASSL*, **408**, 169.

Ivezić, Ž., Kahn, S. M., and Eliason, P. 2014. The Gaia-LSST Synergy. *EAS Publ. Ser.*, **67**, 211.

Ivezić, Ž., Connolly, A. J., and Jurić, M. 2017. Everything We'd Like to Do with LSST Data, but We Don't Know (Yet) How: Astroinformatics. *IAU Symp.*, **325**, 93.

Jaroszynski, M., Skowron, J., Udalski, A., et al. 2006. Binary Lenses in OGLE-III EWS Database: Season 2004. *Acta Astron.*, **56**, 307.

Kochoska, A., Mowlavi, N., Prša, A., et al. 2017. Gaia Eclipsing Binary and Multiple Systems: A Study of Detectability and Classification of Eclipsing Binaries with Gaia. *A&A*, **602**, 110.

Lindegren, L., Lammers, U., Bastian, U., et al. 2016. Gaia Data Release 1. Astrometry: One Billionj Positions, Two Million Proper Motions and Parallaxes. *A&A*, **595**, 4.

Mao, S., and Paczynski, B. 1991. Gravitational Microlensing by Double Stars and Planetary Systems. *ApJ*, **374**, L37.

Matijevič, G., Prša, A., Orosz, J. A., et al. 2012. Kepler Eclipsing Binary Stars. III. Classification of Kepler Eclipsing Binary Light Curves with Locally Linear Embedding. *AJ*, **143**, 123.

Merle, T., Van Eck, S., Jorissen, A., et al. 2017. The Gaia-ESO Survey: Double, Triple and Quadruple-Line Spectroscopic Binary Candidates. *A&A*, **608**, 34.

Mignard, F., Klioner, S., Lindegren, L., et al. 2016. Gaia Data Release 1. Reference Frame and Optical Properties of ICRF Sources. *A&A*, **595**, 5.

Mowlavi, N., Lecoeur-Taïbi, I., Holl, B., et al. 2017. Gaia Eclipsing Binary and Multiple Systems. Two-Gaussian Models Applied to OGLE-III Eclipsing Binary Light Curves in the Large Magellanic Cloud. *A&A*, **606**, 92.

Oelkers, R. J., Stassun, K. G., and Dhital, S. 2017. Gaia Assorted Mass Binaries Long Excluded from SLoWPoKES (GAMBLES): Identifying Ultra-Wide Binary Pairs with Components of Diverse Mass. *AJ*, **153**, 259.

Oh, S., Price-Whelan, A. M., Hogg, D. W., Morton, T. D., and Spergel, D. N. 2017. Comoving Stars in Gaia DR1: An Abundance of Very Wide Separation Comoving Pairs. *AJ*, **153**, 257.

Pawlak, M., Soszyński, I., Pietrukowicz, P., et al. 2014. Period-Luminosity Relations for Ellipsoidal Binary Stars in the OGLE-III Fields of the Large Magellanic Cloud. *Acta Astron.*, **64**, 293.

Pawlak, M., Soszyński, I., Udalski, A., et al. 2016. The OGLE Collection of Variable Stars: Eclipsing Binaries in the Magellanic System. *Acta Astron.*, **66**, 421.

Pojmanski, G. 2002. The All Sky Automated Survey. Catalog of Variable Stars. I. 0 h - 6 hQuarter of the Southern Hemisphere. *Acta Astron.*, **52**, 397.

Pourbaix, D., Ivezić, Ž., Knapp, G. R., Gunn, J. E., and Lupton, R. H. 2004. Color-Induced Displacement Double Stars in SDSS. *A&A*, **423**, 755.

Pourbaix, D., Knapp, G. R., Gunn, J. E., et al. 2016. Robust Detection of CID Double Stars in SDSS. *A&A*, **591**, 96.

Prša, A., Pepper, J., and Stassun, K. G. 2011. Expected Large Synoptic Survey Telescope (LSST) Yield of Eclipsing Binary Stars. *AJ*, **142**, 52.

Raghavan, D., McAlister, H. A., Henry, T. J., et al. 2010. A Survey of Stellar Families: Multiplicity of Solar-Type Stars. *ApJ S*, **190**, 1.

Rauer, H., Catala, C., Aerts, C., et al. 2014. The PLATO 2.0 Mission. *Experimental Astronomy*, **38**, 249.

Ricker, G. R., Winn, J. N., Vanderspek, R., et al. 2015. Transiting Exoplanet Survey Satellite (TESS). *Journal of Astronomical Telescopes, Instruments, and Systems*, **1**, 014003.

Siopis, C., and Sadowski, G. 2012. Towards an Automated Processing of Gaia Eclipsing Binaries. In Proceedings of the Workshop 'Orbital Couples: Pas de Deux in the Solar System and the Milky Way'. Edited by F. Arenou, D. Hestroffer, 59.

Smolčić, V., Ivezić, Ž., Knapp, G. R., et al. 2004. In A Second Stellar Color Locus: A Bridge from White Dwarfs to M Stars. *ApJ L*, **615**, L141.

Soszyński, I., Pawlak, M., Pietrukowicz, P., et al. 2016. The OGLE Collection of Variable Stars: Over 450 000 Eclipsing and Ellipsoidal Binary Systems Toward the Galactic Bulge. *Acta Astron.*, **66**, 405.

Süveges, M., Barblan, F., Lecoeur-Taïbi, I., et al. 2017. Gaia Eclipsing Binary and Multiple Systems: Supervised Classification and Self-Organizing Maps. *A&A*, **603**, 117.

Thompson, S. E., Everett, M., Mullally, F., et al. 2012. A Class of Eccentric Binaries with Dynamic Tidal Distortions Discovered with Kepler. *ApJ*, **753**, 86.

Udalski, A., Szymanski, M., Mao, S., et al. 1994. The Optical Gravitational Lensing Experiment: OGLE No. 7: Binary Microlens or a New Unusual Variable? *ApJ*, **436**, L103.

van Leeuwen, F., Evans, D. W., De Angeli, F., et al. 2017. Gaia Data Release 1. The Photometric Data. *A&A*, **599**, 32.

Wells, M., Prša, A., Jones, L., and Yoachim, P. 2017. Initial Estimates on the Performance of the LSST on the Detection of Eclipsing Binaries. *PASP*, **129**, 065003.

Wielen, R. 1996. Searching for VIMs: An Astrometric Method to Detect the Binary Nature of Double Stars with a Variable Component. *A&A*, **314**, 679.

Wyrzykowski, L., Mroz, P., Rybicki, K., et al. 2017. Gaia16aye Binary Microlensing Event Is Rising for the 5th time. *Astronomer's Telegram*, **10341**.

4

Population Synthesis of Binary Stars

ROBERT G. IZZARD[1] AND GHINA M. HALABI[2]

4.1 Introduction

The evolution of binary stars is often quite different from that of their solitary cousins (De Marco and Izzard, 2017). Many types of stars *only* form in binaries, and their properties allow investigation into astrophysics that is impossible to probe in single stars. Good examples include type Ia supernovae; merging neutron stars (kilonovae) and black holes; the most massive stars, e.g., blue stragglers; many massive Wolf–Rayet stars; X-ray binaries; long and short gamma-ray bursts; barium/CH/CEMP-*s* stars, Algols with peculiar surface chemistry for their evolutionary state; W UMa contact binaries; low-mass helium stars and related sdB and sdO stars; sequence D variable stars; and thermonuclear novae. Other processes in which binaries are implicated include the formation of asymmetric planetary nebulae, many post-(asymptotic-)giant-branch stars with circumbinary discs and overmassive stars in the Galactic thick disc. Triple and higher-multiple systems are often hierarchical, e.g., triples are effectively a long-period outer binary system containing a short-period inner binary.

Many aspects of binary-star physics remain highly uncertain. Mass transfer, accretion and loss, and associated angular momentum redistribution, e.g., common envelope evolution, dominates our ignorance of binary-star evolution. One way to improve our knowledge is to apply statistical techniques to compare our stellar models to the stars we observe. This process is called stellar population synthesis. In this brief review, we consider the challenge and opportunity that big data from modern astronomy bring, the basic technique of binary-star population synthesis and the associated parameter space problem. As an example of binary population synthesis, we consider how the chemical ejecta from stars depend on model input parameters. We also show an example of how modelling metal-poor stars led to better astrophysical understanding. Finally, we report on recent population synthesis work, in particular with respect to predicting and understanding the discovery of gravitational waves from stellar and black-hole mergers.

[1] Supported by the Science and Technology Facilities Council (STFC) Rutherford fellowship ST/L003910/1 and Churchill College, Cambridge.
[2] Supported by STFC Rutherford grant ST/M003892/1.

4.2 Big Data and Big Challenges

The age of big data is very much upon us. From commerce to industry to pure blue-skies research, the volume of data continues to increase endlessly. Astronomy is no exception. The Gaia satellite is currently monitoring more than a billion stars with trillions of observations. Transient surveys, such as Pan-STARRS, Skymapper and the upcoming LSST, monitor billions of objects in the night sky to detect stellar explosions. Such large numbers are difficult to comprehend on a human level. Even the human population of Earth, currently in excess of 7×10^9, is difficult to imagine. To modern, massively parallel computers, the challenge is rather less daunting. Commercial companies, especially in the financial sector, process such data daily. There are thus natural links to astrophysics, where the challenge is to understand the stars and galaxies we see, both individually and as a collective. This is where population synthesis modelling is such a useful tool. The general idea is to simulate a population of stars on a computer, and compare simulated observables to those we see. By varying the input parameters of the population model to match the stars that are seen, we can pin down the underlying astrophysics. Often, new physics can be ruled out because the numbers fail to match, and uncertain physics can be far better constrained.

Population synthesis is particularly useful because astrophysics is mostly not an experimental science. We want to answer fundamental questions, such as how do stars function, how did they get there and what are they going to do in the future. But we cannot – yet! – experiment on stars. This is analogous to a dendrologist examining the trees of a forest.[3] Some are small, some are large, some are green and some yellow, some are in flower and some are dead. From this snapshot in time the dendrologist must determine the entire biology and evolutionary history of trees: a difficult task! Astronomers have been trying to do this for millennia with the stars in the sky. Although we cannot experiment on stars directly, through population synthesis we can experiment inside our computers and statistically compare to our forest of stars.

First, we make a set of models of stars to attempt to match those under investigation. Quantitative computer models of stars have been constructed for at least fifty years. The current state of the art is to combine one-dimensional stellar structure equations with parametrised models of internal mixing and rotational deformation. Despite these simplifications, such computer codes take hours to compute the evolution of a single star from birth to death, even when details such as thermal pulses and supernovae are ignored. Assuming the evolution can be calculated, the second step is to turn such models into a stellar population. For this, we need a description of the population's initial parameters such as an initial mass function. Next, we must convert the models into simulated observations. This is rarely a trivial task. Stellar models predict physical parameters such as luminosity, effective temperature and chemical abundances. Observations are usually of stellar apparent magnitudes, colours and spectral line widths, so conversion is required. Survey selection biases must be taken into account when making the simulated population. If such biases

[3] Many thanks to Dr. Karl Kruszelnicki for the excellent analogy.

are simple – say, a luminosity cutoff – then this is relatively easy. In general, biases are poorly known and sometimes impossible to quantify. Finally, our population model should be compared to the observations in a quantitative way to tell us which model is best and as much as possible about the underlying astrophysics. Each of the aforementioned problems is its own challenge. Population synthesis thus pushes the limits of both theoretical and computational astrophysics.

4.2.1 The Single- and Binary-Star Parameter Spaces

The stellar parameter space is so large that simulation of every star in the sky, in detail, is impossible. We can, however, use our astrophysical knowledge to define a few key initial parameters that describe the evolution of a star. Usually these are, in order of importance, stellar mass M, metallicity[4] Z and rotation rate $v_{\rm rot}$. In the crudest approximation, the rotation rate and metallicity are fixed, usually $Z = 0.02$ ('solar' metallicity) and $v_{\rm rot} = 0$. This can be justified because most stars have a metallicity roughly equal to that of the Sun, and most stars rotate slowly enough that their structure is not affected substantially. It is thus possible to make a model population of N stars by varying only their mass M. If each star takes $\Delta t = 1\,\mathrm{h}$ to run on a modern CPU and we sample M between, say, $0.8\,\mathrm{M}_\odot$ and $100\,\mathrm{M}_\odot$, we may need $N = 100$ stars and the simulation takes $N \times \Delta t = 100$ hours. This is certainly tractable.

Binary stars extend the parameter space by introducing a secondary star of mass M_2, an orbital semimajor axis a (or period P, linked to M_1 and M_2 by Kepler's law), an orbital eccentricity e. Stellar spins are usually assumed to align with the orbit (Hut, 1981) which is often assumed circular, i.e., $e = 0$ (Hurley et al., 2002). However, even with $e = 0$ and aligned inclinations, the single-star parameter space of dimension 1 (M) becomes a binary-star parameter space of dimension 3 (M_1, M_2 and a), hence we have N^3 stellar models to run. If $N = 100$, we require 10^6 stellar models, which take about $2 \times 10^6\,\mathrm{h}$ of CPU time. This is possible on a modern computing cluster, and indeed has been done (albeit at lower resolution, e.g., Eldridge et al., 2017), but we are at the current computational limit and hence exploration of the model parameter space is restricted.

Single-star models contain many parameters that can vary, such as the initial abundance mixture; the initial angular momentum profile; the stellar mass loss prescription; angular momentum loss prescriptions (e.g., magnetic braking); uncertain nuclear reaction rates; models of internal mixing, e.g., convective mixing model (such as mixing length theory); rotational mixing efficiencies; and magnetic field evolution. In binary stars, we must include the strength of tidal interaction; mass and angular momentum transfer efficiencies; mass and angular momentum accretion efficiencies; an algorithm to describe when mass transfer is stable; companion-induced enhanced mass loss; gravitational radiation angular

[4] Metallicity, Z, is usually defined in stellar evolution as the mass content of a star that is not hydrogen or helium. Observational astronomy usually defines metallicity as $[\mathrm{Fe/H}] = \log_{10}\left(N_{\rm Fe}/N_{\rm Fe\odot}\right) - \log_{10}\left(N_{\rm H}/N_{\rm H\odot}\right)$, where N_x are the abundances, by number of particles, of species x. Conversion between the two can be done e.g. assuming a solar-scaled Z.

momentum loss; nova explosions and interaction with the ejected material; discs and associated jets; winds and orbital interactions; and supernova kicks. We also require a model for common envelope evolution and to allow for the possibility that stars merge. Triple and higher stellar multiples also exist. To be stable, these are hierarchical, so a triple can be modelled approximately as a close binary inside a wide binary (Toonen et al., 2016): it is still, approximately, binary-star evolution.

Finally, when making a population, each model parameter is either fixed or has an initial distribution. The best-known example of such is the initial mass function, $\xi(M)$, where M is the initial stellar mass. This is often described as a set of power laws in M (e.g., Kroupa et al., 1993; Kroupa, 2001). However, even this formalism is subject to uncertainty, particularly in massive stars ($M \gtrsim 10\,\mathrm{M}_{\odot}$). In binaries, the secondary mass, orbital separation (or period) and eccentricity all have their own input distributions, and a mass-dependent binary fraction is often employed (Raghavan et al., 2010; Sana et al., 2012; Duchêne and Kraus, 2013; Fuhrmann et al., 2017; Moe and Di Stefano, 2017).

So, in general, modelling involves a large number of input parameters N_{param}. In our *binary_c* single- and binary-star population synthesis code there are currently 238 model parameters for each binary star, not including the many parameters that are compile-time constants and parameters that form the initial distributions. Thus $N_{\mathrm{param}} \approx 1{,}000$ is not unreasonable.

Which parameters are important depends very much on which stars are being modelled and compared to. For example, changing mass loss rates on the red giant branch will not affect main sequence lifetimes, so if one is measuring the time spent on the main sequence this parameter is unimportant. However, if one is measuring the number of main sequence stars that have accreted material from chemically peculiar red giants, then the giant's mass loss rate is very important. There is currently no easy way to determine a priori which parameters are important to make which stars: the experience of the astrophysicist is crucial in deciding. This task is, however, vital. If N_{param} can be reduced from 1,000 to just a few, the problem is greatly reduced in complexity and may become tractable even if the model run time, Δt, is long. In general, however, N_{param} is so large that it is necessary to reduce Δt by many orders of magnitude, as described next.

4.2.2 Detailed and Synthetic Stellar Models

Full stellar evolution models of some stars, such as those in the thermally pulsing asymptotic giant branch (TPAGB) or those undergoing rapid mass transfer, are far more costly to create than the $\Delta t \sim 1\,\mathrm{h}$ suggested previously. Δt in these cases could be days, weeks or longer. It is impossible to make population models containing such stars covering any significant parameter space at reasonable resolution simply because of time constraints. Students have PhDs to finish and even tenured astrophysicists retire eventually!

One solution is to make a grid of models. The grid can be interpolated (e.g., De Donder and Vanbeveren, 2002; Brott et al., 2011; Kruckow et al., 2018) or fitted to functions (e.g., *SSE* and *BSE* of Eggleton et al., 1989; Tout et al., 1997; Hurley et al., 2000, 2002) to reduce the run time – often to less than 1 s. With $\Delta t = 1$ s, a typical binary grid takes about 10^6 s, or about one day on a 16-core desktop personal computer. Stellar codes that employ such techniques are referred to as *synthetic* stellar evolution codes, as opposed to their *detailed* full stellar evolution cousins. In the case of fitted formulae, internal stellar structure is sacrificed to gain a factor of at least 10^6 in run time. Grid lookup codes can retain internal structure, which is useful when calculating, e.g., the stellar binding energy, at the cost of computational storage space.

Our *binary_c* code (Izzard et al., 2004a, 2006, 2009, 2018) uses the a *C* version of the *BSE* library for most of its stellar structure calculations and nucleosynthesis from various sources, including Karakas et al. (2002) and Karakas (2010) during the TPAGB, and supernova yields from massive stars (Woosley and Weaver, 1995; Chieffi and Limongi, 2004). It runs at least 10^6 times faster than the full stellar evolution codes on which it is based, but still provides useful estimates of stellar luminosity, mass, core mass, radius and core radius and chemistry as a function of time for the entire evolution of the star. Binary-star interaction is included mostly according to Hurley et al. (2002) but with improvements to mass transfer by Roche-lobe overflow (Claeys et al., 2014), Wind-Roche-Lobe-overflow (Abate et al., 2013, 2015), tides (Siess et al., 2013), rejuvenation (de Mink et al., 2013; Schneider et al., 2014a) and supernovae (Boubert et al., 2017a,b). Binary-star nucleosynthesis includes accretion and thermohaline mixing (Stancliffe et al., 2007; Izzard et al., 2018) as well as explosions such as thermonuclear novae (José and Hernanz, 1998) and Ia supernovae (Claeys et al., 2014). Modern detailed binary star codes, including *MESA* (Paxton et al., 2011, 2015), implement similar binary interaction physics to *BSE* and *binary_c*, and take far longer to run. If one is interested in the effects of binaries on a stellar population, rather than the precise details of stellar structure, it makes no sense to throw endless CPU cycles at the problem.

Synthetic codes have another major advantage over detailed codes, that of stability. While detailed stellar evolution codes habitually stop because of physical or numerical problems (they 'crash'), synthetic codes rarely suffer such problems. Thus, if a task does not require full, detailed stellar evolutionary calculations, a synthetic code is often a good choice for both speed and reliability. Often the observations to which the models are being compared have uncertainties that do not justify more accurate, and costly, modelling.

4.3 Stellar Accountancy

Our aim is to model a population of stars that matches the stars we see now. These stars were born in the past, at rates given by a star formation rate $S(t)$, where t is time since the Big Bang. Stars in our model are labelled i, and each has a probability of existing, ψ_i, normalised such that $\sum_i \psi_i = 1$. Our model then predicts the number of stars of type j that exist now, when $t = t_{\max}$, is,

$$N_j = \sum_{t=0}^{t_{\max}} S(t, \mathbf{Z}) \sum_i \psi_i (M, \ldots, \mathbf{Z}) \sum_{\tau=0}^{t_{\max}-\tau} \delta_j (t - \tau, \mathbf{X}, \mathbf{Z})\, \delta\tau \,, \tag{4.1}$$

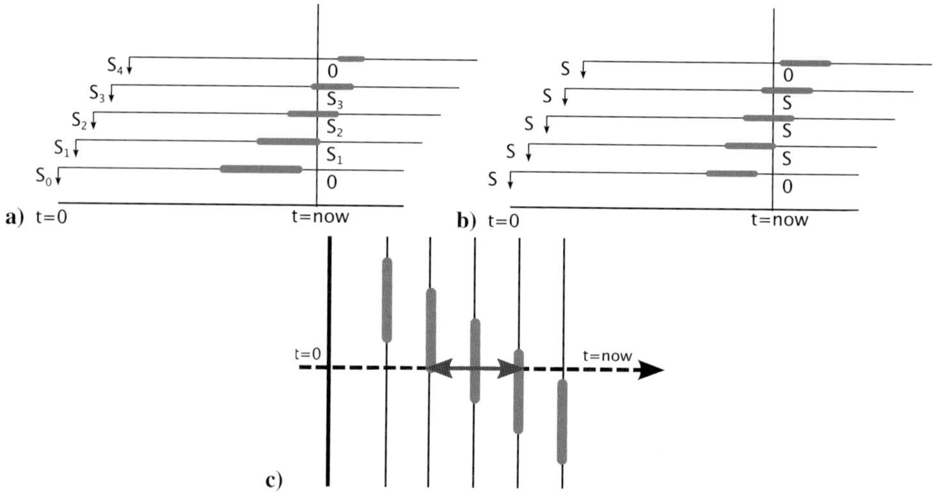

Figure 4.1 Graphical representation of population synthesis modelling. Time runs left to right from the Big Bang ($t = 0$) until now in finite time steps, Δt. Stars formed at 0 to $4\Delta t$ are shown, with corresponding star formation rates S_0, S_1, ..., but in a realistic simulation there may be many thousands of time steps. Stars evolve from left to right along horizontal lines, which, when they are in the phase of interest that is to be counted, are thick (equivalent to $\delta_j = 1$ in Equation (4.3), $\delta_j = 0$ otherwise). Stars are counted when the phase of interest intersects the vertical line which represents now. **(a)** In an ideal population model, at each time step the star formation rate varies ($S_0 \neq S_1 \neq S_2 \ldots$) and the properties of the stars also vary (indicated by the change in length of time spent in the phase of interest). **(b)** In a simplified model, the stars all have the same properties (e.g., the same metallicity) and hence length of time in the phase of interest. The star formation rate is constant, i.e., $S = S_0 = S_1 = S_2 \ldots$ **(c)** Instead of performing an expensive convolution over many models, the problem is rotated by 90° and population statistics are calculated from a single model run represented by the horizontal dashed line. The region denoted by the arrow is the phase of interest and the evolutionary model set has to be calculated only once, so the calculation is relatively cheap.

where τ is the age of each star, $\delta\tau$ is the time step of the model, and δ_j is a function that is 1 if the star is of the type j and 0 otherwise. The vector **Z** represents model input parameters, such as metallicity, and input distributions that can change with the time of birth of the stars. The vector **X** denotes the many parameters[5] that affect stellar evolution, such as metallicity, rotation rate, mass-loss prescription and mixing-length parameter. The function δ_j factors in stellar evolution, so is a function of **X**, **Z** and τ. The probability, ψ_i, is a function of at least mass, perhaps also metallicity (hence **Z**) and, for binary stars, the companion mass and orbital properties also. If the function ψ_i has n parameters, the sum $\sum_i$ is implied over all of them. The large number of elements of this sum illustrates the parameter space problem we face even if **X** and **Z** are fixed.

[5] Some parameters exist in both **X** and **Z** in our rather crude definition.

In its most general form, Equation (4.1) is a nasty convolution problem of the type faced by galactic chemical evolution models (Gibson et al., 2003). The problem can be simplified by fixing the stellar physics (i.e. $\mathbf{X} = \text{constant}$ and $\mathbf{Z} = \text{constant}$) and the input distributions and assuming a constant star formation rate, e.g.,[6] $S = 1$. The sum reduces to the following, as shown in Figure 4.1:

$$N_j = \sum_i \psi_i(M,\ldots) \sum_{\tau=0}^{t_{\max}} \delta_j(\tau)\,\delta\tau\,. \tag{4.2}$$

The parameter space wrapped into ψ_j is still potentially very large, but the number of models to be calculated is greatly reduced because $\mathbf{X}$ and $\mathbf{Z}$ are fixed. We can also reduce the size of the sum over time by fixing the time step, $\delta\tau$, and allowing δ_j to take values between 0 and 1 corresponding to partial occupancy in phase j during each time step. This is useful when calculating tables of model results with a fixed time resolution, e.g., for later use in a galactic chemical evolution context. The number of stars in a galaxy can be predicted by then scaling S appropriately.

All the preceding is for just one type of star, denoted by j. In reality, there are many different types of star and hence many different N_j. Our binary-star population synthesis code *binary_c* has more than 200 such types of star, both single and binary. It is also possible to count stars in bins, so that individual N_j can be combined to create distributions. Differential measurements, such as number ratios, are more useful than simple predictions of the number of different types of stars in a population. One would hope that errors in approximations, such as our assumption that the star formation rate is constant, at least partly cancel out.

Further, the definition can be extended to count, say, stellar mass ejecta as we show in the Section 4.4. We set $\delta_j = \dot{M}_j(\tau)$, the rate of mass lost from a system as isotope j, as a function of stellar age, τ. The normalised mass ejected as isotope j from a population of single stars is defined as,

$$y_{j,\mathrm{sin}} = \frac{\sum_{i=M} \psi_i \sum_{\tau=0}^{t_{\max}} \dot{M}_j(\tau)\,\delta\tau}{\sum_{i=M} \psi_i M}\,, \tag{4.3}$$

and from a population of binary stars as,

$$y_{j,\mathrm{bin}} = \frac{\sum_{i=M_1,M_2,a} \psi_i \sum_{\tau=0}^{t_{\max}} \dot{M}_j(\tau)\,\delta\tau}{\sum_{i=M_1,M_2,a} \psi_i (M_1 + M_2)}\,, \tag{4.4}$$

where M is the initial mass of a single star and M_1, M_2 and a are the initial primary mass, initial secondary mass and initial separation of a binary. The sums in the denominators are the total mass formed in stars in each population. This must be included to compare single to binary stars because binary systems have initially more mass, on average, than single stars. Note that y_j are dimensionless.

[6] Formally, $S = T^{-1}$ where T is the unit of time in which $\delta\tau$ is measured. A time-dependent star formation rate can also be implemented as $S(\tau)$, provided $\mathbf{X}$ is fixed, but we limit ourselves to the simpler case of constant star formation rate.

4.4 Slow and Fast Parameters

Equation (4.2) contains two main sets of parameters. The *fast* parameters are those that affect the input distributions, ψ_i. These include the slope and shape of the initial mass function, the secondary mass (or mass ratio) distribution and the initial orbital period distribution. These are called *fast* because changing ψ_i does not require a recalculation of the stellar evolution model (which is embedded in δ_j). Indeed, at any time step in a stellar population calculation, as many different ψ_i as desired can be calculated.

The *slow* parameters, on the other hand, require a complete recalculation of the stellar structure and evolution at every time step. These are the $\mathbf{X}$ of Equation (4.3). The calculation process is thus, in terms of computer speed, slow. Such parameters include the metallicity, mixing parameters, mass loss prescriptions and nuclear reaction rates. When stellar model grids are calculated, most of the slow parameters are fixed. Only one or a few parameters are then varied, such as the metallicity. Many of the constant parameters may not matter greatly to stellar evolution, but this often depends on the type of star under consideration, so the precise definition of our j and hence δ_j, and this dependence is often not known *a priori*.

In the following, we demonstrate the dependence of a stellar population statistic, the amount of mass ejected from a stellar population as a particular isotope, on various model input parameters[7] using Equations 4.3 and 4.4. The fast parameters affect the initial distributions, ψ_i, while the slow parameters affect the $\dot{M}_j(\tau)$, i.e., when and how much mass is ejected as isotope j as a function of time. The initial parameters, i, are sampled on logarithmic grids, with single-star masses in the range $0.1 \leq M/\mathrm{M}_\odot \leq 80$ and $n_M = 1{,}000$ stars, binary component masses M_1 and M_2 are treated identically to M but with $0.1 \leq M_2 \leq M_1$, $n_{M_1} = 100$ and $n_{M_2} = 100$, and separations in the range $3 \leq a/\mathrm{R}_\odot \leq 10^4$, also with $n_a = 100$ (so the total number of binary systems in each grid is $n_{M_1} \times n_{M_2} \times n_a = 10^6$). At this resolution, errors in the relative mass ejected caused by a limited resolution and time step are at most 2%.

4.4.1 Fast Parameters

In our example, we have three distributions that define the fast parameters. These are the single or primary star initial mass distribution, the initial secondary mass distribution and the initial separation distribution. The single or primary star mass distribution is also known as the initial mass function. Our standard model is the Kroupa et al. (1993) initial mass function, and we consider two alternatives from Chabrier (2003) and Salpeter (1955). Applying each gives us a range of uncertainty in stellar ejecta, y_j, as shown in Table 4.1.

We apply a similar process to the binary-star primary mass function, which is assumed to be identical to that of single stars. The results are similar to those in single stars, but

[7] These models are from an old version of *binary_c* first presented in the first author's PhD dissertation but demonstrate the principle under consideration accurately enough.

Table 4.1 *Masses of various isotopes ejected from a stellar population relative to the mass born into stars when reasonably varying* fast *parameters, i.e., the initial mass function in single stars, and in binary stars the primary star mass function, secondary star mass distribution and initial orbital separation within reasonable limits. These parameters are* fast *because they require only one run of the stellar evolution models.*

	Single stars	Binary stars		
Isotope	Vary M	Vary M_1	Vary M_2	Vary a
$^{1}H \times 10$	$1.9-3.7$	$1.9-3.6$	$1.7-2.2$	$1.6-2.6$
$^{12}C \times 10^3$	$4.6-15$	$5.7-18$	$3.8-6.1$	$4.0-6.2$
$^{14}N \times 10^4$	$9.6-24$	$8.1-19$	$7.6-9.8$	$7.9-9.5$
$^{16}O \times 10^3$	$8.4-30$	$9.6-32$	$7.2-10$	$7.1-11.5$
$^{56}Fe \times 10^3$	$1.0-3.0$	$2.5-5.4$	$1.3-2.7$	$0.9-3.4$
$^{65}Cu \times 10^6$	$10-32$	$8.8-28$	$7.0-9.2$	$8.0-12$
$Ba \times 10^9$	$8.1-16$	$6.5-13$	$6.0-7.8$	$6.2-9.4$

the impact of binary stars is seen in the reduced nitrogen and barium ejecta, because mass transfer truncates the evolution of thermally pulsing asymptotic giant branch stars, and in an increase in iron because of type Ia supernovae.

Our default distribution of secondary star masses is assumed to be flat in $q = M_2/M_1$, i.e., $\psi_q \propto q^0$, between $q = q_{\min} = 0.1\,M_\odot/M_1$ and 1. We also consider distributions with $\psi_q \propto q^{0.5}$ (cf. Garmany et al., 1980) and $\psi_q \propto q^{-1.5}$, for illustration. In general, the changes to the mass ejecta, y_j, are smaller than changes caused by variation in the initial or primary star mass distribution.

Finally, we consider the distribution of binary orbits. This is set either by a distribution of orbital semimajor axes, a, or orbital periods, P, where both are related by Kepler's law (we assume circular orbits). Our default distribution is $\psi_a \propto a^{-1}$, equivalent to Öpik (1924). We also consider $\psi_a \propto a^1$, $\psi_a \propto a^{-2}$ and the combination of $\psi_a \propto a^{-0.7}$ and $10 \leq a/R_\odot \leq 1.3 \times 10^3$ to match Goldberg et al. (2003). Table 4.1 shows that, as with the secondary star mass distribution, reasonable changes in the separation distribution matter less than reasonable changes in the initial/primary star mass distribution.

Please note that the numbers calculated here are only a demonstration. A proper exploration of this problem would consider the more recent distributions of, e.g., Moe and Di Stefano (2017) with their error bars. Such an exploration is unfortunately beyond the scope of this work.

4.4.2 Slow Parameters

There are many *slow* model parameters that alter the chemical ejecta from a population of stars. We limit our investigation to a population of 100% binary stars with fixed initial distributions of masses and orbital separations. Primary masses M_1 are distributed according to Kroupa et al. (1993) between 0.1 and $80\,M_\odot$, secondary mass ratios q follow a flat distribution between $0.1\,M_\odot/M_1$ and 1, and separations, a, have a constant distribution in $\log a$ between 3 and $10^4\,R_\odot$.

The parameters we vary are metallicity in the range $10^{-4} \leq Z \leq 0.02$ (default is 0.02), binary eccentricity $0 \leq e \leq 1$ (default is 0), red giant branch mass loss parameter $0.25 \leq \eta \leq 0.75$ (default is 0.5 using the Reimers, 1975, formalism as defined in Hurley et al., 2000's eq. [106]), ^{13}C pocket efficiency $0.01 \leq \xi_{13} \leq 2.0$ (default is 1, Izzard et al., 2006), wind loss rate during a Wolf–Rayet phase multiplier $0.1 \leq f_{WR} \leq 10$ (default is 1, cf. Hurley et al., 2000's section 7.1), an Eddington limit for accretion multiplied by 1 or 10^6 (default is 10^6, i.e., no limit, Hurley et al., 2002's eq. [67]), a supernova kick velocity dispersion $0 \leq \sigma_{SN}/\mathrm{km\,s^{-1}} \leq 400$ (default is $190\,\mathrm{km\,s^{-1}}$, Hansen and Phinney, 1997), companion reinforced attrition process (Tout and Eggleton, 1988) parameter $0 \leq B \leq 10^4$ (default is 0, i.e., no reinforcement), common envelope parameter $\alpha_{CE} = 3$ (Hurley et al., 2002) and third dredge up parameters $-0.1 \leq \Delta M_{c,\min} \leq 0$ (default $-0.07\,M_\odot$) and $0 \leq \lambda_{\min} \leq 1$ (default $0.8 - 37.5Z$, Izzard et al., 2004a; Izzard and Tout, 2004).

The normalised ejected mass of each isotope, y_j, is expanded as a series in the slow parameters labelled k with values x_k,

$$y_j(x_k) = y_{j0} + \frac{\partial y_j}{\partial x_k}(x_k - x_{k0}) + \cdots \approx y_{j0} + \left(1 + T_{jk}\frac{\delta x_k}{\Delta x_k}\right), \tag{4.5}$$

where x_{k0} are the default values that one may consider best estimates, y_{j0} are the ejecta corresponding to the default parameter set x_{k0}, $\delta x_k = x_k - x_{k0}$, and x_k varies from $x_{k,\min}$ to $x_{k,\min} + \Delta x_k$. The matrix of derivatives,

$$T_{jk} = \frac{\partial\left(y_j/y_{j0}\right)}{\partial\left(x_k/\Delta x_k\right)}, \tag{4.6}$$

provides a relatively unbiased comparison of parameters because it measures the rate of change of dimensionless and normalised numbers, y_j/y_{j0} and $x_k/\Delta x_k$. A larger absolute value of T_{jk} thus reflects greater sensitivity of amount of isotope j ejected to the parameter k. One drawback of this approach is that it only considers the linear terms in the expansion: it is possible that, at second or higher order, the effects of changing two parameters cancel each other out. Such considerations are beyond the scope of this work.

Table 4.2 shows how the T_{jk} vary with each model parameter. Absolute values above 0.1 are marked in bold. These are the parameters to which the chemical ejecta are most sensitive. As an example, the ejecta of most isotopes vary little with the ^{13}C s-process pocket efficiency. The exception is barium, an s-process element, which is very sensitive to the efficiency of s-processing, as one would expect. While this is a trivial example, it is perhaps less obvious that the ejection of ^{16}O depends 10 times more weakly on the common envelope parameter, α_{CE}, than does ^{14}N.

Table 4.2 *Ejecta dependence on various model parameters as measured by T_{jk}, the magnitude of which is an attempt to measure the dependence on each model parameter k of the relative mass ejected from a population of stars as isotope j (Equation 4.6, symbols are defined in the main text). $x^{(y)}$ means $x \times 10^y$. The numbers highlighted in bold have an absolute value exceeding* 0.1 *and are the most sensitive to their parameter.*

	Model parameter									
Isotope	Z	ξ_{C13}	f_{WR}	α_{CE}	$\Delta M_{c,min}$	e	f_{Edd}	η_{GB}	λ_{min}	σ_{SN}
^{1}H	**$-1.9^{(-1)}$**	$1.2^{(-4)}$	$-3.6^{(-3)}$	$-1.6^{(-2)}$	$5.2^{(-3)}$	$4.5^{(-2)}$	$-3.5^{(-4)}$	$8.2^{(-3)}$	$-9.9^{(-3)}$	$-5.3^{(-3)}$
^{12}C	$6.0^{(-2)}$	$-1.8^{(-3)}$	**$1.0^{(-1)}$**	**$1.1^{(-1)}$**	$-7.1^{(-2)}$	**$-1.6^{(-1)}$**	$-1.8^{(-2)}$	$-5.0^{(-2)}$	**$1.6^{(-1)}$**	$4.6^{(-2)}$
^{14}N	**$-6.5^{(-1)}$**	$-1.3^{(-4)}$	$-5.9^{(-3)}$	**$-4.1^{(-1)}$**	$-3.8^{(-3)}$	$9.4^{(-2)}$	$-2.7^{(-5)}$	$-1.9^{(-2)}$	$1.8^{(-2)}$	$2.7^{(-2)}$
^{16}O	$8.8^{(-2)}$	$-2.6^{(-3)}$	**$-4.9^{(-1)}$**	$-2.3^{(-2)}$	$2.0^{(-3)}$	$-5.3^{(-2)}$	$1.2^{(-3)}$	$-1.0^{(-2)}$	$-4.0^{(-4)}$	$1.0^{(-1)}$
^{56}Fe	**$9.6^{(-1)}$**	$-3.9^{(-4)}$	**$-1.8^{(-1)}$**	**$2.3^{(-1)}$**	$1.8^{(-3)}$	**$-2.5^{(-1)}$**	$5.8^{(-2)}$	$-9.3^{(-3)}$	$-2.5^{(-3)}$	$2.5^{(-2)}$
^{65}Cu	**$-7.2^{(-1)}$**	$-5.8^{(-4)}$	$-7.5^{(-2)}$	**$-3.1^{(-1)}$**	$1.0^{(-3)}$	$7.8^{(-2)}$	$1.1^{(-3)}$	$-2.0^{(-2)}$	$1.4^{(-3)}$	$7.0^{(-2)}$
Ba	**-2.2**	**1.3**	$2.5^{(-2)}$	**$-2.6^{(-1)}$**	$-9.1^{(-2)}$	**$1.1^{(-1)}$**	$-2.2^{(-3)}$	$-7.1^{(-2)}$	**$2.0^{(-1)}$**	$3.1^{(-3)}$

4.5 Matching Models to Observations, and Models to Models

The general problem of matching model results, with their distributions of input parameters, to many observations is one that is, in every sense of the word, nontrivial. The size of the parameter space is daunting. Efforts like those mentioned in Section 4.4.2 reduce the number of parameters to only those that matter for a given observational data set of a particular type of star. Perhaps the greatest problem is in how to match all stars from as many data sets as possible simultaneously, thus providing the best possible constraints. The solution to this problem is beyond our current reach.

Frameworks to match a grid of models to an individual observed star are relatively well developed. A good example is the Bayesian algorithm *BONNSAI* (Schneider et al., 2014b, 2017). Given, say, the effective temperature, luminosity and surface abundances of the star, all the models in a particular grid are compared to it to determine – in an automated way – which fits best. Bayesian techniques also naturally give posterior distributions of parameters, hence associated error bars. *BONNSAI* also tests that model resolution is sufficient to believe the best fit. Other related techniques such as Markov-chain Monte Carlo (MCMC) methods are starting to be employed by the community, particularly in the hunt for the progenitors of gravitational wave sources (e.g., Mandel and O'Shaughnessy, 2010). The next big step will be to use such techniques in general, e.g., to match the data of Gaia to whole population models.

When matching a large number of stars to population models, the selection effects of the survey in question should also be taken into account. All surveys, even those that are 'complete', have some kind of selection bias. In many cases, a simple cut in, say, magnitude or surface gravity, which effectively selects stars in a particular phase of evolution, is sufficient and easy to fit into a Bayesian analysis (e.g., Casagrande et al., 2014; Izzard et al., 2018). Modern surveys often have a well-described 'selection function' that describes in a probabilistic way the various biases inherent to their sample, e.g., the Gaia-ESO survey

(Stonkutė et al., 2016). However, older data often have poorly defined selection effects and, in some cases, data are selected 'by eye' which renders its modelling essentially impossible.

Finally, our computation models must be tested and verified. One way to do this is to compare the models of one group against those of another to test for consistency. In the case that both groups use identical input physics, the results should be the same. The *POPCORN* collaboration tried to do this, pitting *binary_c* against the *SeBa*, *Startrack* and *Brussels* population synthesis codes (Toonen et al., 2014, and references therein). The four codes were, for the most part, consistent, which at least suggests we are getting the numerics right. Whether the physics they employed is true to life is another question that *POPCORN* did not try to answer.

4.6 Headline News in Population Synthesis

There are many fields in which population synthesis has shown itself to be useful. The colours of galaxies are now accepted to be influenced by binary stars, for example the ultraviolet (UV) emission of elliptical galaxies is influenced by subdwarf-O/B stars, which can only form in binaries (Han et al., 2007). Massive blue straggler stars form when binaries transfer mass or merge and are common enough to be measurable in the present-day mass function of young stellar clusters (Schneider et al., 2014a). Similarly, type II supernovae can be delayed in binary systems, and predictions have recently been made that analyse the multi-dimensional parameter space of such objects in great detail (Zapartas et al., 2017). The rates of more exotic channels, such as long and short gamma-ray bursts, can also only be predicted by population synthesis (Izzard et al., 2004b; Szécsi, 2017). Binary stars affect stellar evolution such that the chemical yields of elements are altered, e.g., through type Ia supernova and nova channels. Population synthesis is required to estimate rates and yields of such events (e.g., Claeys et al., 2014) and the effect of mass transfer on chemistry more generally (De Donder and Vanbeveren, 2002; Izzard et al., 2018). Mass transfer also makes chemically peculiar stars. The number and properties of these – such as orbital periods and eccentricities – can be used to better understand stellar interior mixing, mass transfer, tides and interactions with circumstellar and circumbinary discs (Jorissen et al., 2009; Izzard et al., 2010; Dermine et al., 2013; and Figure 4.2).

We have also recently improved our *binary_c* models to combine them with a model of the Galaxy and Magellanic Clouds to predict the number of hypervelocity stars that originate from the Large Magellanic Cloud (Boubert et al., 2017b). These young stars are moving so fast that they are not bound to our Galaxy. One way they can be made is in binary systems in which one star explodes as a supernova leaving the companion to escape. We predicted large numbers of such stars, their positions on the sky and their stellar properties. Our predictions will soon be tested by Gaia.

The recent advances in gravitational wave physics have certainly put population synthesis in the headlines. Merging stellar-mass black holes and neutron stars are all the rage, and the predictions of population synthesis models of such objects can be put to the test (e.g., Belczynski et al., 2016b, and many other works). The systematic uncertainties on such

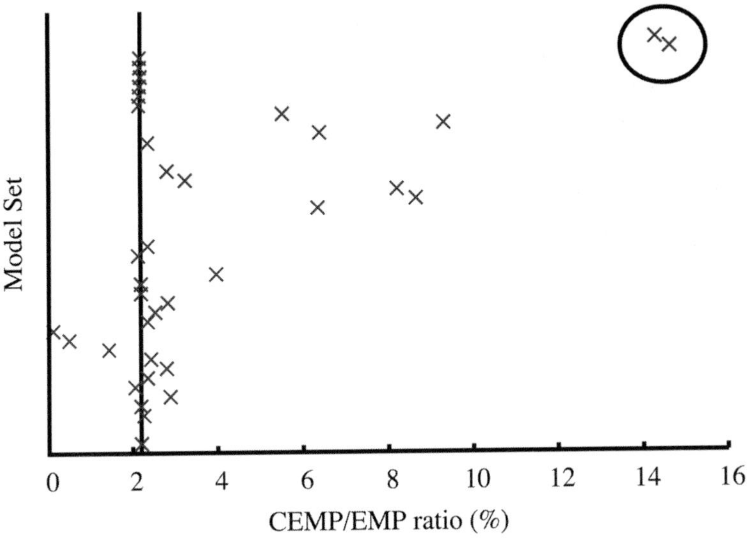

Figure 4.2 The ratio of carbon-enhanced extremely metal-poor (CEMP) stars to all extremely metal-poor stars (EMP) in the galactic halo as predicted by our population nucleosynthesis code, *binary_c*, as a function different input physics, here labelled 'model sets'. Full details of the model sets and general problem are in Izzard et al. (2009). Briefly, CEMP stars are red giants that formed by accretion from a carbon-star companion in the past (Abate et al., 2015, shows our latest models). Each model set predicts a CEMP/EMP ratio that should to be compared to the observed ratio in the galactic halo, which is 20%. Most of our model sets predict a 2% CEMP/EMP ratio, which is wrong by an order of magnitude. Only two model sets approach the observed 20%. These both require all the following. (1) Third dredge up must be efficient in low-mass stars, right down to $0.9\,M_\odot$, at low metallicity (here $Z = 10^{-4}$). (2) Third dredge up must be efficient in stars with small envelope masses (the minimum envelope mass for third dredge up, $M_{\mathrm{env,min}}$, is zero here). (3) A small amount of accretion occurs during any common envelope phase ($\Delta M_{\mathrm{CE}} = 0.05\,M_\odot$). (4) Thermohaline mixing in accreting stars is suppressed. At the time of writing the original paper (2009), it was considered impossible for $0.9\,M_\odot$ stars to undergo third dredge up and hence become carbon stars. More recent works have shown that not only is this possible, it is quite likely (e.g., Karakas, 2010). The population synthesis modelling results encouraged detailed stellar evolutionists to examine stars of both low mass and low metallicity in far more detail, resulting in a far better understanding of the problem (e.g., Fishlock et al., 2014; Matrozis and Stancliffe, 2016, 2017).

rate estimates are large, often many orders of magnitude. These realistically reflect our ignorance of the physical processes involved, especially binary problems such as common envelope evolution, and their sensitivity to the many model parameters that are often known only approximately.

4.7 *Not* Any Colour You Like ...

A persistent claim we have heard over the years is that, with population synthesis, you can get whatever you like out of the models. Indeed, with so many parameters to play with, this may seem at first glance to be true. An important counterargument, however, is that number statistics can be predicted by population synthesis *with quantified uncertainties*. It is these uncertainties that are large, and perhaps this is where the 'whatever you like' myth originates. In such cases, it is important to identify the parameters in the model that lead to the large uncertainties and pin these down, either by observation, experiment or development of a better theory. By some this is called progress and, with more and better data, it is simply no longer possible to get any result one would like. Also, ignorance of the effect of uncertain parameters, be they initial distributions of stars or the choice of input physics, does not mean the error on such model predictions is zero. Population synthesis at least allows us to quantify some of these systematic errors and remains a very powerful tool to investigate the evolution of both single and binary stars in the twenty-first century.

References

Abate, C., Pols, O. R., Izzard, R. G., Mohamed, S. S., and de Mink, S. E. 2013. Wind Roche-lobe overflow: application to carbon-enhanced metal-poor stars. *A&A*, **552**, A26.

Abate, C., Pols, O. R., Stancliffe, R. J., et al. 2015. Modelling the observed properties of carbon-enhanced metal-poor stars using binary population synthesis. *A&A*, **581**, A62.

Belczynski, K., Holz, D. E., Bulik, T., and O'Shaughnessy, R. 2016. The first gravitational-wave source from the isolated evolution of two stars in the 40–100 solar mass range. *Nature*, **534**, 512–515.

Boubert, D., Erkal, D., Evans, N. W., and Izzard, R. G. 2017b. Hypervelocity runaways from the Large Magellanic Cloud. *MNRAS*, **469**, 2151–2162.

Boubert, D., Fraser, M., Evans, N. W., Green, D. A., and Izzard, R. G. 2017a. Binary companions of nearby supernova remnants found with Gaia. *A&A*, **606**, A14.

Brott, I., Evans, C. J., Hunter, I., et al. 2011. Rotating massive main-sequence stars. II. Simulating a population of LMC early B-type stars as a test of rotational mixing. *A&A*, **530**, A116.

Casagrande, L., Silva Aguirre, V., Stello, D., et al. 2014. Strömgren Survey for Asteroseismology and Galactic Archaeology: let the SAGA begin. *ApJ*, **787**, 110.

Chabrier, G. 2003. Galactic stellar and substellar initial mass function. *PASP*, **115**, 763–795.

Chieffi, A., and Limongi, M. 2004. Explosive yields of massive stars from $Z = 0$ to $Z = Z_{\odot}$. *ApJ*, **608**, 405–410.

Claeys, J. S. W., Pols, O. R., Izzard, R. G., Vink, J., and Verbunt, F. W. M. 2014. Theoretical uncertainties of the Type Ia supernova rate. *A&A*, **563**, A83.

De Donder, E., and Vanbeveren, D. 2002. The chemical evolution of the solar neighbourhood: the effect of binaries. *New Astronomy*, **7**, 55–84.

De Marco, O., and Izzard, R. G. 2017. Dawes Review 6: the impact of companions on stellar evolution. *PASA*, **34**, e001.

de Mink, S. E., Langer, N., Izzard, R. G., Sana, H., and de Koter, A. 2013. The rotation rates of massive stars: the role of binary interaction through tides, mass transfer, and mergers. *ApJ*, **764**, 166.

Dermine, T., Izzard, R. G., Jorissen, A., and Van Winckel, H. 2013. Eccentricity-pumping in post-AGB stars with circumbinary discs. *A&A*, **551**, A50.

Duchêne, G., and Kraus, A. 2013. Stellar multiplicity. *ARA&A*, **51**, 269–310.
Eggleton, P. P., Fitchett, M. J., and Tout, C. A. 1989. The distribution of visual binaries with two bright components. *ApJ*, **347**, 998–1011.
Eldridge, J. J., Stanway, E. R., Xiao, L., et al. 2017. Binary Population and Spectral Synthesis Version 2.1: construction, observational verification and new results. *PASA*, **34**, e058.
Fishlock, C. K., Karakas, A. I., and Stancliffe, R. J. 2014. The effect of including molecular opacities of variable composition on the evolution of intermediate-mass AGB stars. *MNRAS*, **438**, 1741–1750.
Fuhrmann, K., Chini, R., Kaderhandt, L., and Chen, Z. 2017. Multiplicity among solar-type Stars. *ApJ*, **836**, 139.
Garmany, C. D., Conti, P. S., and Massey, P. 1980. Spectroscopic studies of O type stars. IX. Binary frequency. *ApJ*, **242**, 1063–1076.
Gibson, B. K., Fenner, Y., Renda, A., Kawata, D., and Lee, H.-c. 2003. Galactic chemical evolution. *PASA*, **20**, 401–415.
Goldberg, D., Mazeh, T., and Latham, D. W. 2003. On the mass-ratio distribution of spectroscopic binaries. *ApJ*, **591**, 397–405.
Han, Z., Podsiadlowski, P., and Lynas-Gray, A. E. 2007. A binary model for the UV-upturn of elliptical galaxies. *MNRAS*, **380**, 1098–1118.
Hansen, B. M. S., and Phinney, E. S. 1997. The pulsar kick velocity distribution. *MNRAS*, **291**, 569.
Hurley, J. R., Pols, O. R., and Tout, C. A. 2000. Comprehensive analytic formulae for stellar evolution as a function of mass and metallicity. *MNRAS*, **315**, 543–569.
Hurley, J. R., Tout, C. A., and Pols, O. R. 2002. Evolution of binary stars and the effect of tides on binary populations. *MNRAS*, **329**, 897–928.
Hut, P. 1981. Tidal evolution in close binary systems. *A&A*, **99**, 126–140.
Izzard, R. G., Dermine, T., and Church, R. P. 2010. White-dwarf kicks and implications for barium stars. *A&A*, **523**, A10.
Izzard, R. G., Dray, L. M., Karakas, A. I., Lugaro, M., and Tout, C. A. 2006. Population nucleosynthesis in single and binary stars. I. Model. *A&A*, **460**, 565–572.
Izzard, R. G., Glebbeek, E., Stancliffe, R. J., and Pols, O. R. 2009. Population synthesis of binary carbon-enhanced metal-poor stars. *A&A*, **508**, 1359–1374.
Izzard, R. G., Preece, H., Jofre, P., et al. 2018. Binary stars in the Galactic thick disc. *MNRAS*, **473**, 2984–2999.
Izzard, R. G., Ramirez-Ruiz, E., and Tout, C. A. 2004b. Formation rates of core-collapse supernovae and gamma-ray bursts. *MNRAS*, **348**, 1215–1228.
Izzard, R. G., and Tout, C. A. 2004. A binary origin for low-luminosity carbon stars. *MNRAS*, **350**, L1–4.
Izzard, R. G., Tout, C. A., Karakas, A. I., and Pols, O. R. 2004a. A new synthetic model for asymptotic giant branch stars. *MNRAS*, **350**, 407–426.
Jorissen, A., Frankowski, A., Famaey, B., and van Eck, S. 2009. Spectroscopic binaries among Hipparcos M giants. III. The eccentricity – period diagram and mass-transfer signatures. *A&A*, **498**, 489–500.
José, J., and Hernanz, M. 1998. Nucleosynthesis in classical novae: CO versus ONe white dwarfs. *ApJ*, **494**, 680.
Karakas, A. I. 2010. Updated stellar yields from asymptotic giant branch models. *MNRAS*, **403**, 1413–1425.
Karakas, A. I., Lattanzio, J. C., and Pols, O. R. 2002. Parameterising the third dredge-up in asymptotic giant branch stars. *PASA*, **19**, 515–526.

Age estimates of young-intermediate age clusters are obviously affected by the amount of CBM included in the models: the larger the CBM region, the older the age estimate of a given cluster. It has also been shown that the diffusive approach to mixing in the CBM region provides different results in terms of evolutionary times compared to instantaneous mixing, because of a slower addition of extra fuel from the overshooting region when this scheme is implemented (Ventura and Castellani, 2005).

5.2.2 Core He-Burning Mixing

Core-He burning mixing in low- and intermediate-mass stellar models is also complicated (see, e.g., Castellani et al., 1971; Dorman and Rood, 1993, and references therein). He burning is efficient in an almost pure He convective core, and the opacity is dominated by electron scattering, but an important contribution comes also from *free–free* absorption. The transformation of He to C due to nuclear burning increases the *free–free* opacity, and within the convective core the radiative temperature gradient $\nabla_{\rm rad}$ increases, developing a discontinuity at the inner convective boundary. This situation is unphysical, because at the inner side of the boundary the convective flux is larger than zero, while it is zero at the outer side of the boundary. According to Gabriel et al. (2014) the presence of this discontinuity is due to an incorrect application of the Schwarzschild criterion, for a proper implementation requires always equality of radiative and adiabatic temperature gradients ($\nabla_{\rm rad} = \nabla_{\rm ad}$) at the convective side of the boundary. In practical terms, at every computational timestep, one should place the boundary of the fully mixed convective region at the layer where $\nabla_{\rm rad} = \nabla_{\rm ad}$. After this early phase, when the extension of the convective core follows the increase of $\nabla_{\rm rad}$, with decreasing He the radiative gradient profile starts to show a local minimum (profile 3 in Figure 5.1), as a consequence of the progressive outward shift of the convective boundary. The presence of this minimum depends on the complex behaviour of the opacity, pressure, temperature and local energy flux. In this situation, outward mixing to eliminate the $\nabla_{\rm rad}$ discontinuity will induce a general decrease of the radiative gradient in the whole convective core (due to an average increase of He and consequent decrease of C after mixing – see profile 4 in Figure 5.1). The radiative gradient will eventually decrease and become equal to $\nabla_{\rm ad}$ at the location of the minimum of $\nabla_{\rm rad}$ (profile 5 in Figure 5.1).

A full instantaneous mixing between the external convective shell and the formally radiative surroundings would decrease the minimum of ∇_{rad} below $\nabla_{\rm ad}$, causing an inconsistency in the application of the convection criterion. A possible solution is to impose *by hand* a partial mixing in the formally convective shell (see, e.g., Castellani et al., 1985, for an example of implementation), such that the final chemical composition – shown in Figure 5.2 – satisfies the condition $\nabla_{\rm rad} = \nabla_{\rm ad}$ (profile 6 in Figure 5.1) in the shell – called now 'semiconvective' region. The mass location of the minimum of the radiative gradient moves outwards with time, because of the evolution of the chemical abundances caused by nuclear burning. As a final result, the C-enriched region increases its size outwards.

The effects of semiconvection on the evolution of these models are the following: the evolutionary tracks perform more extended loops in the HRD, and the central He-burning

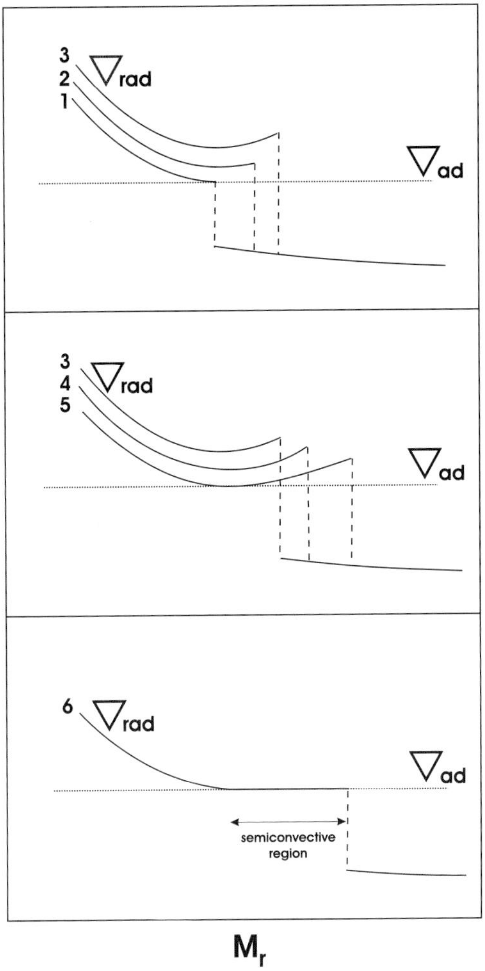

Figure 5.1 Sketch of the time evolution of the ∇_{rad} profile near the boundary of the convective core during the central He-burning stage. Increasing numbers denote a sequence of increasing time (see text for details).

phase lasts longer because of a larger amount of fuel to burn. Alternative options found in the literature envisage overshooting from the formal convective boundaries, like the prescription employed by Sweigart (1990), who lets the formal boundary propagate outwards with velocity

$$v_c = F_{ov} \frac{\nabla_{rad} - \nabla_{ad}}{\mu_i - \mu_e} \tag{5.2}$$

where μ_i and μ_e are the molecular weights at the inner and outer side of the formal convective boundary. The free parameter F_{ov} is set to 10^{-5}. This overshooting implementation roughly reproduces the semiconvective chemical stratification.

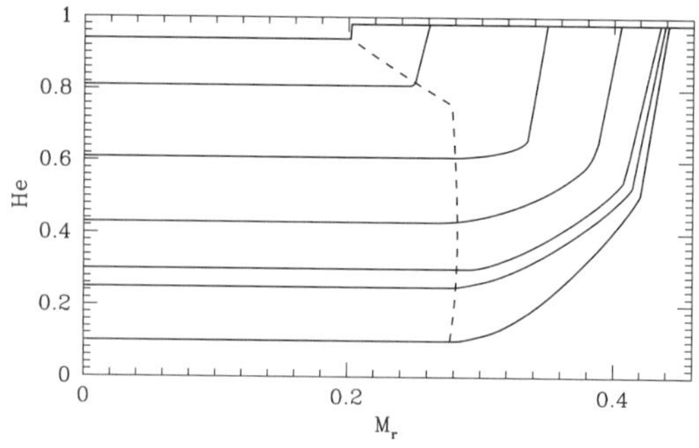

Figure 5.2 Helium abundance profile inside the convective core and the semiconvective zone at various levels of central He depletion, for a low-mass core He-burning model. The dashed line marks the location of the fully mixed core boundary.

5.3 Element Transport in Radiative Layers

In addition to convection, additional element transport mechanisms are efficient in stars. Their effect is negligible within convective regions, because of the much shorter convective timescales, but they are very important within regions in radiative equilibrium.

5.3.1 Thermohaline Mixing

In convectively stable layers where $\nabla_\mu < 0$, an instability is expected to set in, usually called 'thermohaline' instability. It is controlled by the heat leakage of the displaced element, and is observed for example when a layer of hot saltwater lies over a layer of cold freshwater. In recent years, the effect of thermohaline mixing in stellar evolution models has been widely explored, for example in connection with low-mass RGB evolution.

When the convective envelope deepens after the TO, the surface chemical composition is altered due to the dredge-up of H-burning processed matter – the so-called *first dredge-up*. The first dredge-up is completed when the convective region reaches its maximum extension, approaching closely the H-burning shell around the electron degenerate He core. From this moment on, the receding convective boundary is not expected to modify further the surface chemical abundances along the RGB. Spectroscopic observations of metal-poor Galactic halo stars (Gratton et al., 2000; Angelou et al., 2015) provide, however, clear evidence for an additional mixing process, occurring when RGB stars reach the luminosity of the RGB bump, that causes a sudden drop of the isotopic ratio $^{12}C/^{13}C$, a decrease of Li and C and an increase of N.

In low-mass stars, the main H-burning mechanism during the MS is the *p-p chain*, that is efficient also in stellar layers far from the star centre. As a consequence, ^3He accumulates in a broad zone outside the main energy production region. During the first dredge-up, this ^3He is mixed within the convective envelope, with the consequence that during the following RGB evolution, the layers above the H-discontinuity left over by the receding convective envelope at its maximum extension will have a uniform ^3He abundance, larger than the initial one.

In the outer wing of the advancing H-burning shell, there is a narrow region where ^{3}He is processed through the reaction ^{3}He(^{3}He, 2p)^{4}He. This reaction decreases the mean molecular weight, leading to a small local decrease of μ when moving from the surface towards the centre of the star. As long as the H-burning shell advances through layers below the H-discontinuity – when the star evolves before the RGB bump – this effect is negligible because the shell is moving in a region with a large positive gradient, due to the H profile left over at the end of the MS. However, when the H-burning shell enters the region of uniform H-abundance above the discontinuity, the local inversion of the μ profile, of the order of one part in 10^4, becomes important (Eggleton et al., 2006). This situation corresponds to the conditions for thermohaline mixing (Charbonnel and Zahn, 2007).

Thermohaline mixing is usually included in stellar evolution codes as a diffusive process that works in the direction to erase the molecular weight inversion, with diffusion coefficient

$$D_{\rm th} = C_{\rm th}\,\frac{K}{c_p\rho}\left(\frac{\phi}{\delta}\right)\frac{\nabla_\mu}{(\nabla_{\rm rad}-\nabla_{\rm ad})} \tag{5.3}$$

where $C_{\rm th} = (8/3)\pi^2\alpha^2$, α being a free parameter related to the aspect ratio (length/width) of the mixing elements.

Evolutionary calculations show that thermohaline mixing extends between the outer wing of the H-burning shell and the inner boundary of the convective envelope. Depending on the mixing efficiency (hence the choice of $C_{\rm th}$), a significant amount of nuclear processed matter in the hotter layers of the H-burning shell can be dredged up to the surface during the remaining RGB evolution, helping to explain the spectroscopic data.

Comparisons with observations require $C_{\rm th} \sim 100-300$ from Li observations in Galactic globular clusters, and $C_{\rm th} \sim 1000$ for C in globulars, which are clearly mutually inconsistent. On the other hand, models by Charbonnel and Lagarde (2010) are able to match the ^{12}C/^{13}C isotopic ratio, C and N abundances in halo RGB field stars with $C_{\rm th} \sim 1{,}000$; the same choice of $C_{\rm th}$ allows also to match measurements of carbon isotopic ratio and [N/C] ratio in open clusters, and Li abundances in field disk stars. Results from 3D hydrodynamics simulations (rescaled to match stellar conditions) add an additional source of uncertainty, for they predict an efficiency equivalent to just $C_{\rm th} \approx 10$ (Traxler et al., 2011).

A recent study by Lattanzio et al. (2015) has shown that the surface chemical abundances predicted by stellar models accounting for thermohaline mixing depend on numerical assumptions, such as spatial and time resolution adopted in the computations. As a consequence, the predicted surface chemical abundances should be treated with caution until a firmer assessment on how to include thermohaline mixing in model computations is achieved.

5.3.2 Atomic Diffusion

Microscopic effects related to collisions among the gas particles induce a *slow* element transport within radiative regions (Burgers, 1969). Individual ions are forced to move under the influence of pressure as well as temperature gradients, which both tend to displace the

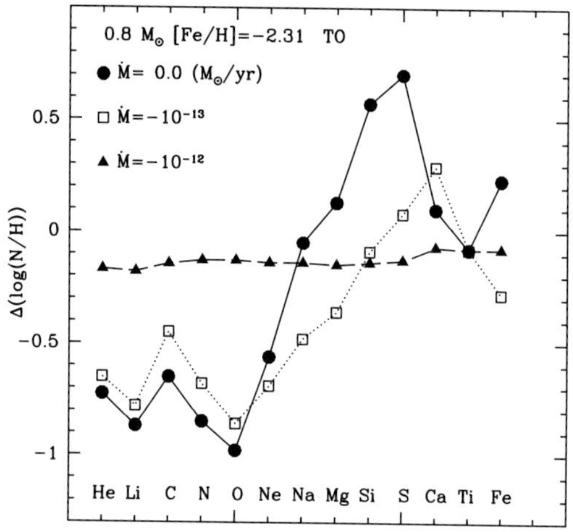

Figure 5.3 Difference between the TO surface abundances of the most relevant elements and their initial values for a low-mass metal-poor model (see labels). The element abundances are given as logarithms of their number fraction to hydrogen (Vick et al., 2013). These abundance variations are shown for models calculated with no mass loss, and two different mass loss rates during the MS evolution.

heavier elements towards the centre of the star, and of concentration gradients that oppose the aforementioned processes. Radiation – which doesn't have a major effect in the Sun (Turcotte et al., 1998) – pushes the ions towards the surface whenever the acceleration imparted to an individual ion species is larger than the local gravitational acceleration. Figure 5.3 shows the pattern of variations of TO surface abundances, for a typical low-mass stellar model at the TO of a metal poor globular cluster. These models have thin convective envelopes that maximise the effect of diffusion.

Elements like He, Li, C, N and O sink below the convective layers, other elements such as Mg, Si, S and Fe are pushed into the convective envelope by the radiation pressure and their surface abundances increase during the MS. The size of these variations decreases in models with larger convective envelopes. After the first dredge-up is completed, these surface abundance variations basically disappear. The abundances of metals (other than CNO) in the He-core are only a few 0.01 dex more abundant than at the start of the evolution, while they (and He) are less abundant than originally by about the same amount around the core.

Regarding the following horizontal branch (HB) phase, at T_{eff} above ~6,000 K the surface chemical composition of the models begins to be altered by diffusion; the selective effect increases during the HB evolution and is more pronounced for lower HB masses that evolve at higher T_{eff}. The abundance of He decreases while overall the surface metal content increases. Figure 5.4 displays the predicted behaviour of the surface Fe abundance in metal-poor HB models.

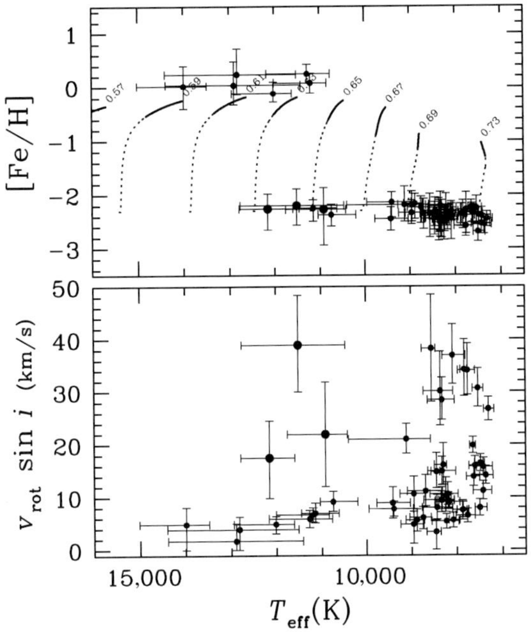

Figure 5.4 The top panel displays the surface [Fe/H] values measured in a sample of stars in the metal-poor Galactic globular clusters M 15, M 68 and M 92 as a function of their T_{eff}. The evolution (dotted segments for an HB age between 0 and 10 Myr, solid segments for ages between 30 and 60 Myr) of the surface [Fe/H] for HB models with the labelled masses including diffusion is also displayed (Michaud et al., 2011). The bottom panel displays the projected rotation velocities for the cluster stars in the top panel. The three objects with T_{eff} between 11,000 and 13,000 K and the highest rotational velocities display much lower surface [Fe/H] compared to the the other stars in the same temperature range (see the text for details).

These surface abundance changes are again erased by the deepening envelope convection for all HB masses that evolve to the following Asymptotic Giant Branch (AGB) phase. The subsequent hot white dwarf phase sees diffusion altering again in a major way the surface abundances. Due to the high surface gravity, gravitational settling of metals in the doubly layered H and He (DA type) or He-dominated (non-DA type) envelopes is very fast, with timescales of the order of 10^6 yr (Koester, 2009).

Helioseismic observations tell us that diffusion is efficient in the Sun (Bahcall et al., 1995; Villante et al., 2014), and the very efficient diffusion of metals from the white dwarf (WD) envelopes is confirmed by observations, whereby for the vast majority of hot WDs the chemical composition of the surface is either pure H or pure He. However, existing spectroscopic measurements of surface element abundances (e.g., Fe) in a sample of Galactic globulars partially contradict this picture (e.g., Korn et al., 2007; Mucciarelli et al., 2011; Gruyters et al., 2016). In a star cluster, the initial mass of stars evolving along post-MS stages is essentially constant, and measurements of chemical abundances at the

TO and the base of the RGB should display a difference, due to the effect of diffusion on evolutionary tracks, that is maximised at the TO and essentially disappears along the RGB. For example, in the case of NGC 6397 with [Fe/H]~ -2.1 along the RGB, the TO Fe abundance should be $\sim$0.3 dex lower, and the Ca abundance $\sim$0.1 dex larger than on the RGB according to models calculated with diffusion; observations show instead TO Fe and Ca abundances $\sim$0.15 and $\sim$ 0.1 dex lower than on the RGB, respectively (Korn et al., 2006). This points to a reduction of the effect of diffusion on the surface abundances by some competing mechanism. Nothing can be said about the efficiency of diffusion in the inner layers. The same qualitative result is found in the more metal-rich clusters NGC 6752 and M 4, and also on the HB of a number of clusters, as shown by Figure 5.4.

For HB stars hosted by three metal-poor globulars, the surface Fe abundance increases abruptly when T_{eff} goes above $\sim$11000 K, while theory predicts Fe enhancements (compared to the initial value [Fe/H]=$-$2.3 dex typical of these clusters) at much lower temperatures. Also comparisons of models in the mass range between $\sim$1.5 and $\sim 3.0 M_{\odot}$ calculated including diffusion with surface abundances measured in A and F stars, show that diffusion must be moderated by some competing mechanism (Deal et al., 2016). For example, in samples of A and F stars belonging to the Hyades, Pleiades and Coma Berenices clusters, observed abundances of elements such as Na, Fe and Ni display a general pattern and star-to-star scatters that point to diffusion being moderated to different degrees by some competing process (Gebran et al., 2010).

Rotational mixings (see Section 5.3.3) can counteract the effect of diffusion, and this has been invoked to explain the *delayed* (in terms of T_{eff}) onset of diffusion in hot HB stars (Quievy et al., 2009). This conclusion is related to the distribution of observed projected rotational velocities for the hot HB stars of Figure 5.4, as shown in the lower panel of the same figure. All observed stars with T_{eff} above $\sim$11,000 K that show spectroscopically the signature of diffusion, are very slow rotators, whereas at lower T_{eff} rotation rates are on average higher. In support of this idea, Figure 5.4 shows three stars with $T_{eff} >$11,000 K and large projected rotation velocities with no enhancement of surface Fe, at odds with the other slow-rotating objects at the same temperatures. The link between (partial) inhibition of diffusion and rotation is, however, probably not justified for A and F stars, which often show slow rotation velocities and surface abundances clearly affected by a reduced efficiency of diffusion compared to the model predictions.

Another process that can moderate diffusion in the stellar outer layers is mass loss (see Vick et al., 2013, and references therein). Assuming that mass loss (ML) is spherical and unseparated (chemical composition of the wind equal to the photospheric composition), its effect is to remove the outer layers of a model. Simple mass conservation constraints, coupled to the fact that the structure of the star is unaffected by the amount of mass lost between two consecutive computational timesteps, translates into the appearance of an outward-flowing interior velocity v_w due to the wind. Figure 5.3 displays the effect of ML on the surface abundances of a typical model at the TO of a metal-poor globular cluster. For elements affected mainly by gravitational settling, the effect of ML is to increase their surface abundance compared to the pure diffusion case. In the case of elements supported by radiative levitation in some layers below the convective envelope, the effect of ML on the surface abundances is due to the interplay between v_w and the depth at which the radiative acceleration $g_{\rm rad}$ is larger than the local gravitational acceleration g. For an element such as Mg, an ML rate $\dot{M} = -10^{-13} M_\odot/yr$ transforms the surface overabundance of about 0.10 dex into an underabundance of about 0.4 dex. A further increase of $\dot{M}$ reduces the underabundance to about just 0.1 dex. In the case of Ca, the overabundance caused by diffusion is instead increased by an ML rate $\dot{M} = -10^{-13} M_\odot/{\rm yr}$ and then transformed into an underabundance when $\dot{M} = -10^{-12} M_\odot/{\rm yr}$.

ML rates of the order of $\approx 10^{-12} M_\odot/{\rm yr}$ are required to explain the surface abundances in NGC 6397 (Vick et al., 2013). These rates are much larger than the solar current mass loss rate $\dot{M} = -2 \times 10^{-14} M_\odot/{\rm yr}$. Rates of the order of $\approx 10^{-13} M_\odot/{\rm yr}$ are invoked to explain at least some of the surface abundance patterns seen in A and F stars (Vick et al., 2010).

Another point to consider is the following. The modifications of the chemical abundance profiles caused by diffusion lead to variations of the local mean molecular weight. Helium always sinks, introducing a stabilising contribution to the local μ-gradient (increasing downwards). On the other hand, metals supported by radiative levitation lead to a destabilising μ-gradient (increasing outwards). If this second effect prevails, these layers are subject to thermohaline mixing that will tend to erase the abundance changes, hence to moderate the effect of diffusion. Published calculations of models for A and F stars including both diffusion and thermohaline mixing show that thermohaline mixing (with diffusion coefficient by Brown et al., 2013) plays an important role in moderating the effect of diffusion on the surface abundances, although the surface abundances of the models do not yet match observations (Deal et al., 2016).

A pragmatic solution adopted in several investigations is to add *by hand* an additional ad-hoc term in the chemical evolution equations (denoted with the generic term 'turbulence') that acts towards suppressing the development of chemical abundance gradients, not explicitly connected to a specific physical mechanism. There are various choices in the literature for this $D_{\rm turb}$ diffusion coefficient that contain free parameters calibrated to reproduce some observational constraint (e.g., Richard et al., 2002; VandenBerg et al., 2012).

5.3.3 Rotational Mixing

Detailed stellar evolution calculations are still (for the foreseeable future) possible only with standard 1D modelling. Within the approximation of the so-called *shellular* rotation (see Zahn, 1992), the set of 1D equations of stellar structure for a rotating star are formally identical to the standard nonrotating case, apart from the appearance of two *form factors* f_P and f_T. The only difference is the interpretation of the variables. In the case of rotating models, r is the mean radius of an isobar defined by the value P of the pressure, ρ and T are the volume-averaged density and temperature between two contiguous isobars and L and m are the mass and luminosity inside a given isobar.

A complex issue when modelling rotating stars in 1D is how to describe the transport of angular momentum – which determines the evolution of the angular velocity Ω – and the chemical mixing associated with rotation. These are two sides of the same problem, for rotation triggers hydrodynamical instabilities and large-scale motions of the gas in radiative regions, that result in transport of both angular momentum and chemical elements. An in-depth analysis of rotation-driven instabilities can be found in Maeder (2009).

A first major instability in rotating stars is the so-called *meridional circulation* that moves material inwards from the equator and upwards along the rotational axis towards the poles. An important effect of meridional circulation is that it also advects angular momentum. Local variations of angular velocity with time due to this angular momentum transport, plus the effect of contraction and expansion of the stellar layers (and eventually angular momentum loss from mass loss), will generate – starting, for example, from solid body rotation, usually assumed for pre-MS fully convective stars – a variation of angular rotation velocity with depth in radiative regions. Hence a shear develops between neighbouring layers that leads to an instability. This shear instability stems from the fact that the minimum energy state of a rotating fluid is solid body rotation, and if the star develops differential rotation, it is possible to extract energy by homogenising the velocities through transport of material.

The Solberg–Høiland instability, plus additional instabilities (i.e., Goldreich–Schubert–Fricke, ABCD instabilities) that can develop when equipotentials and isobars do not coincide – so-called baroclinic instabilities – do potentially contribute to the transport of angular momentum and chemicals. Some of these instabilities are implemented in calculations of rotating stellar models, even though their treatment is considered to be much more uncertain than the case of meridional circulation and shear instability (Palacios, 2013).

Angular momentum and element transport in rotating stellar models are implemented in different ways in different stellar evolution codes, with several alternative choices of the functional form of the various instabilities considered, and the free parameters involved (see in-depth discussion in Salaris and Cassisi, 2017). In general, rotational mixing works towards erasing chemical gradients within the star, hence during the MS this causes a continuous slow ingestion of fresh hydrogen into the H-depleted convective core, as well as a slow transport towards the surface of elements whose abundances are increased by H-burning (e.g., He and ^{14}N). The effect of the ingestion of H in the convective core is to increase the MS lifetime compared to the nonrotating counterpart, and to produce slightly larger He-cores at the end of core H-burning.

Figure 5.5 shows an example of the uncertainty in the predictions of rotating stellar models, considering $9M_{\odot}$ models (at the upper end of the intermediate-mass model range) with (for the same initial rotation rate) and without rotation from two different groups (Ekström et al., 2012; Choi et al., 2016). In case of the Choi et al. (2016) models, the effect of rotation on the evolutionary tracks is negligible, contrary to the case of the Ekström

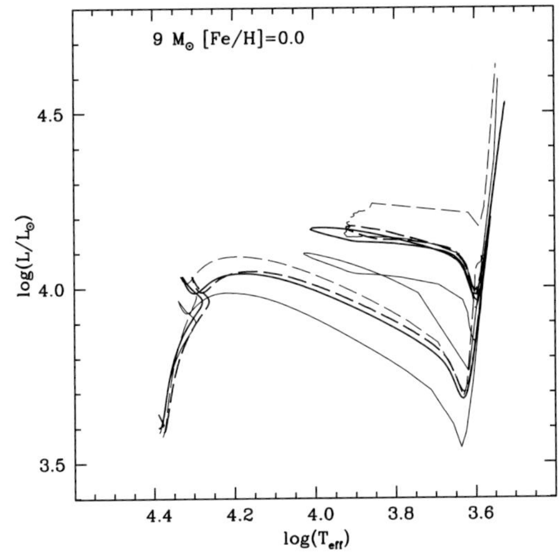

Figure 5.5 Comparison of evolutionary tracks for M = 9 $M_\odot$, initial solar chemical composition, rotating (dashed lines) and nonrotating (solid lines) models, from the Ekström et al. (2012) and Choi et al. (2016) databases (thin and thick lines, respectively). Both rotating models are calculated with an initial rotational velocity $v/v_{crit} = 0.4$, where v_{crit} is the critical velocity.

et al. (2012) models. Variations of surface abundances due to rotational mixing are also much smaller in the Choi et al. (2016) calculations.

5.4 Outlook

It is clear that the description of complex physical processes such as turbulent convection, thermohaline mixing and rotation implemented in stellar evolution calculations is necessarily simplified, with a predictive power in some cases hampered by the use of several free parameters of uncertain calibration. The booming field of asteroseismology is starting to provide new information on the efficiency of element transport processes that can at the very least be used to add further constraints to our current recipes. As an example, Moravveji et al. (2016) studied a $\sim 3.25 M_\odot$ MS rotating star and determined that exponentially decaying CBM (Equation 5.1) has to be preferred to instantanous mixing one, and that mixing in radiative regions is much less efficient than predicted by rotating models. As another example, recent analyses of low-mass core He-burning stars by Bossini et al. (2015), Constantino et al. (2015, 2016) and Bossini et al. (2017) constrained empirically the size of mixed regions beyond the formal Schwarzschild boundary, even though a definitive conclusion on this issue has not been reached yet. The hope is that a synergy between clues from hydrodynamical simulations (Pratt et al., 2016) and the powerful constraints coming from asteroseismic analyses will help to improve our description of element transport in stellar model computations and their predicting power.

References

Angelou, G. C., D'Orazi, V., Constantino, T. N., et al. 2015. Diagnostics of stellar modelling from spectroscopy and photometry of globular clusters. *MNRAS*, **450**, 2423–2440.

Bahcall, J. N., Pinsonneault, M. H., and Wasserburg, G. J. 1995. Solar models with helium and heavy-element diffusion. *Reviews of Modern Physics*, **67**, 781–808.

Böhm-Vitense, E. 1958. Über die Wasserstoffkonvektionszone in Sternen verschiedener Effektivtemperaturen und Leuchtkräfte. Mit 5 Textabbildungen. *ZAP*, **46**, 108.

Bossini, D., Miglio, A., Salaris, M., et al. 2015. Uncertainties on near-core mixing in red-clump stars: effects on the period spacing and on the luminosity of the AGB bump. *MNRAS*, **453**, 2290–2301.

Bossini, D., Miglio, A., Salaris, M., et al. 2017. Kepler red-clump stars in the field and in open clusters: constraints on core mixing. *MNRAS*, **469**, 4718–4725.

Brown, J. M., Garaud, P., and Stellmach, S. 2013. Chemical transport and spontaneous layer formation in fingering convection in astrophysics. *ApJ*, **768**, 34.

Burgers, J. M. 1969. *Flow Equations for Composite Gases*, New York: Academic Press.

Castellani, V., Giannone, P., and Renzini, A. 1971. Overshooting of convective cores in helium-burning horizontal-branch stars. *Astrophys.&Sp.Sci.*, **10**, 340–349.

Castellani, V., Chieffi, A., Tornambe, A., and Pulone, L. 1985. Helium-burning evolutionary phases in population II stars. I Breathing pulses in horizontal branch stars. *ApJ*, **296**, 204–212.

Charbonnel, C., and Lagarde, N. 2010. Thermohaline instability and rotation-induced mixing. I. Low- and intermediate-mass solar metallicity stars up to the end of the AGB. *A&A*, **522**, A10.

Charbonnel, C., and Zahn, J.-P. 2007. Thermohaline mixing: a physical mechanism governing the photospheric composition of low-mass giants. *A&A*, **467**, L15–L18.

Choi, J., Dotter, A., Conroy, C., et al. 2016. Mesa isochrones and stellar tracks (MIST). I. Solar-scaled models. *ApJ*, **823**, 102.

Constantino, T., Campbell, S. W., Christensen-Dalsgaard, J., Lattanzio, J. C., and Stello, D. 2015. The treatment of mixing in core helium burning models. I. Implications for asteroseismology. *MNRAS*, **452**, 123–145.

Constantino, T., Campbell, S. W., Lattanzio, J. C., and van Duijneveldt, A. 2016. The treatment of mixing in core helium burning models. II. Constraints from cluster star counts. *MNRAS*, **456**, 3866–3885.

Deal, M., Richard, O., and Vauclair, S. 2016. Hydrodynamical instabilities induced by atomic diffusion in A stars and their consequences. *A&A*, **589**, A140.

Demarque, P., Sarajedini, A., and Guo, X.-J. 1994. The gap in the color-magnitude diagram of NGC 2420: a test of convective overshoot and cluster age. *ApJ*, **426**, 165–169.

Dorman, B., and Rood, R. T. 1993. On partial mixing zones in horizontal-branch stellar cores. *ApJ*, **409**, 387–403.

Eggenberger, P., Meynet, G., Maeder, A., et al. 2008. The Geneva stellar evolution code. *Astrophys.&Sp.Sci.*, **316**, 43–54.

Eggleton, P. P., Dearborn, D. S. P., and Lattanzio, J. C. 2006. Deep mixing of ^{3}He: reconciling big bang and stellar nucleosynthesis. *Science*, **314**, 1580.

Ekström, S., Georgy, C., Eggenberger, P., et al. 2012. Grids of stellar models with rotation. I. Models from 0.8 to 120 solar masses at solar metallicity (Z = 0.014). *A&A*, **537**, A146.

Freytag, B., Ludwig, H.-G., and Steffen, M. 1996. Hydrodynamical models of stellar convection: the role of overshoot in DA white dwarfs, A-type stars, and the Sun. *A&A*, **313**, 497–516.

Gabriel, M., Noels, A., Montalbán, J., and Miglio, A. 2014. Proper use of Schwarzschild Ledoux criteria in stellar evolution computations. *A&A*, **569**, A63.

Gebran, M., Vick, M., Monier, R., and Fossati, L. 2010. Chemical composition of A and F dwarfs members of the Hyades open cluster. *A&A*, **523**, A71.

Gratton, R. G., Sneden, C., Carretta, E., and Bragaglia, A. 2000. Mixing along the red giant branch in metal-poor field stars. *A&A*, **354**, 169–187.

Gruyters, P., Lind, K., Richard, O., et al. 2016. Atomic diffusion and mixing in old stars. VI. The lithium content of M30. *A&A*, **589**, A61.

Koester, D. 2009. Accretion and diffusion in white dwarfs: new diffusion timescales and applications to GD 362 and G 29-38. *A&A*, **498**, 517–525.

Korn, A., Grundahl, F., Richard, O., et al. 2006. New abundances for old stars: atomic diffusion at work in NGC 6397. *The Messenger*, **125**.

Korn, A. J., Grundahl, F., Richard, O., et al. 2007. Atomic diffusion and mixing in old stars. I. Very large telescope FLAMES-UVES observations of stars in NGC 6397. *ApJ*, **671**, 402–419.

Lattanzio, J. C., Siess, L., Church, R. P., et al. 2015. On the numerical treatment and dependence of thermohaline mixing in red giants. *MNRAS*, **446**, 2673–2688.

Maeder, A. 2009. *Physics, Formation and Evolution of Rotating Stars*, New York: Springer.

Michaud, G., Richer, J., and Richard, O. 2011. Horizontal branch evolution, metallicity, and sdB stars. *A&A*, **529**, A60.

Moravveji, E., Townsend, R. H. D., Aerts, C., and Mathis, S. 2016. Sub-inertial gravity modes in the B8V star KIC 7760680 reveal moderate core overshooting and low vertical diffusive mixing. *ApJ*, **823**, 130.

Mucciarelli, A., Salaris, M., Lovisi, L., et al. 2011. Lithium abundance in the globular cluster M4: from the turn-off to the red giant branch bump. *MNRAS*, **412**, 81–94.

Palacios, A. 2013. Influence of rotation on stellar evolution. Pages 227–287 of: Hennebelle, P., and Charbonnel, C. (eds), *Role and Mechanisms of Angular Momentum Transport during the Formation and Early Evolution of Stars, Evry Schatzman School 2012.* EAS Publications Series, vol. 62. London: EDP Sciences.

Pratt, J., Baraffe, I., Goffrey, T., et al. 2016. Spherical-shell boundaries for two-dimensional compressible convection in a star. *A&A*, **593**, A121.

Quievy, D., Charbonneau, P., Michaud, G., and Richer, J. 2009. Abundances anomalies and meridional circulation in horizontal branch stars. *A&A*, **500**, 1163–1171.

Richard, O., Michaud, G., Richer, J., et al. 2002. Models of metal-poor stars with gravitational settling and radiative accelerations. I. Evolution and abundance anomalies. *ApJ*, **568**, 979–997.

Salaris, M., and Cassisi, S. 2017. Chemical element transport in stellar evolution models. *R. Soc. open sci.*, **4**, 170192.

Salaris, M., Pietrinferni, A., Piersimoni, A. M., and Cassisi, S. 2015. Post first dredge-up [C/N] ratio as age indicator: theoretical calibration. *A&A*, **583**, A87.

Stancliffe, R. J., Fossati, L., Passy, J.-C., and Schneider, F. R. N. 2015. Confronting uncertainties in stellar physics: calibrating convective overshooting with eclipsing binaries. *A&A*, **575**, A117.

Sweigart, A. V. 1990. Horizontal-branch stellar evolution. Pages 1–10 of: Cacciari, C., and Clementini, G. (eds), *Confrontation between Stellar Pulsation and Evolution*. Astronomical Society of the Pacific Conference Series, vol. 11. San Francisco: ADS.

Traxler, A., Garaud, P., and Stellmach, S. 2011. Numerically determined transport laws for fingering ('thermohaline') convection in Astrophysics. *ApJL*, **728**, L29.

Turcotte, S., Richer, J., Michaud, G., Iglesias, C. A., and Rogers, F. J. 1998. Consistent solar evolution model including diffusion and radiative acceleration effects. *ApJ*, **504**, 539–558.

Valle, G., Dell'Omodarme, M., Prada Moroni, P. G., and Degl'Innocenti, S. 2016. Calibrating convective-core overshooting with eclipsing binary systems: the case of low-mass main-sequence stars. *A&A*, **587**, A16.

VandenBerg, D. A., Bergbusch, P. A., Dotter, A., et al. 2012. Models for metal-poor stars with enhanced abundances of C, N, O, Ne, Na, Mg, Si, S, Ca, and Ti, in turn, at constant helium and iron abundances. *ApJ*, **755**, 15.

Ventura, P., and Castellani, M. 2005. Time dependent mixing in He-burning cores: the case of NGC 1866. *A&A*, **430**, 1035–1047.

Viallet, M., Meakin, C., Prat, V., and Arnett, D. 2015. Toward a consistent use of overshooting parametrizations in 1D stellar evolution codes. *A&A*, **580**, A61.

Vick, M., Michaud, G., Richer, J., and Richard, O. 2010. AmFm and lithium gap stars: stellar evolution models with mass loss. *A&A*, **521**, A62.

Vick, M., Michaud, G., Richer, J., and Richard, O. 2013. Population II stars and the Spite plateau: stellar evolution models with mass loss. *A&A*, **552**, A131.

Villante, F. L., Serenelli, A. M., Delahaye, F., and Pinsonneault, M. H. 2014. The chemical composition of the Sun from helioseismic and solar neutrino data. *ApJ*, **787**, 13.

Zahn, J.-P. 1992. Circulation and turbulence in rotating stars. *A&A*, **265**, 115–132.

6

The Symbiotic Stars

ULISSE MUNARI

6.1 Symbiotic Stars: Binaries Accreting from a Red Giant

When Merrill and Humason (1932) discovered CI Cyg and AX Per, the first known *symbiotic stars* (hereafter SySts), they were puzzled (in line with the wisdom of the time, not easily contemplating stellar binarity) by the coexistence in the 'same' star of features belonging to distant corners of the HR diagram: the TiO bands typical of the coolest M giants, the HeII 4686 Å seen only in the hottest O-type stars, and an emission line spectrum matching that of planetary nebulae (hereafter PN). All these features stand out prominently in the spectrum of CI Cyg shown in Figure 6.1 together with its light-curve displaying a large assortment of different types of variability, with the spectral appearance changing in pace (a brighter state usually comes with bluer colors and a lower ionisation).

A great incentive to the study of SySts was provided in the 1980s by the first conference (Friedjung and Viotti, 1982) and monograph (Kenyon, 1986) devoted entirely to them, the first catalog and spectral atlas of known SySts by Allen (1984) and the first simple geometrical modelling of their ionisation front (Seaquist et al., 1984). Allen offered a clean classification criterium for SySts: *a binary star, combining a red giant (RG) and a companion hot enough to sustain HeII (or higher ionisation) emission lines.* The spectral atlas by Munari and Zwitter (2002) shows how the majority of SySts meeting this criterium display in their spectra emission lines of at least the NeV, OVI or FeVII ionisation stages, requiring a minimum photoionisation temperature of 130,000 K (Murset and Nussbaumer, 1994). For some time, a main sequence star accreting at a furious rate ($10^{-5}-10^{-4}$ $M_{\odot}yr^{-1}$) was advocated as the hot component of several symbiotic stars (Kenyon and Webbink, 1984; Mikołajewska and Kenyon, 1992), but this scenario was later abandoned in favor of a WD. Accretion onto a main-sequence star must apply instead to *pre*-SySts systems such as 17 Lep, which are in the first phase of mass transfer (Blind et al., 2011), when the AGB progenitor of the future WD transfers mass to a main sequence companion. As indicated by satellite observations (eg., Munari and Buson, 1994), accretion alone cannot sustain the extreme luminosities (10^3–10^4 $L_{\odot}$) encountered in most SySts (Murset et al., 1991), and – given its high efficiency – nuclear burning at the WD surface must be invoked: burning 1 gr of hydrogen provides in fact 6×10^{18} erg, while accreting the same 1 gr on a 1.3 $M_{\odot}$ WD liberates 6×10^{17} erg, 6×10^{16} erg on a 0.5 $M_{\odot}$ WD and only 2×10^{15} erg on a main-sequence star such as the Sun.

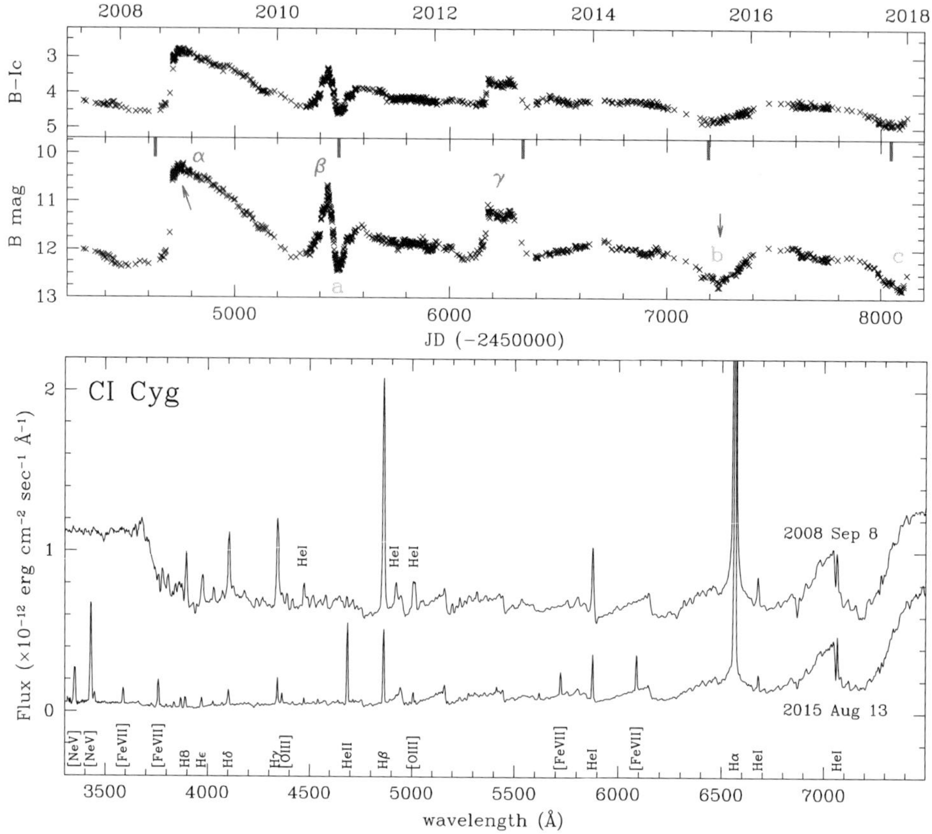

Figure 6.1 Photometric and spectroscopic behaviour of the prototype symbiotic star CI Cyg. *Upper panels:* the arrows mark the time for the two spectra shown, and the thick bars the passage at superior conjunction of the WD in the 855 days' orbit around the M giant companion. The letter 'a' indicates an eclipse of the outbursting WD, while 'b' and 'c' minima in the irradiation-modulated light-curve. α, β and γ mark three separate outbursts of different amplitude, shape and duration. The fact that the $B-I$ color evolution follows the B light curve indicates how the M giant keeps stable while the activity resides entirely with the WD and the circumstellar gas ionised by it. *Bottom panel:* spectrum at outburst peak (8 September 2008) and at irradiation minimum (13 August 2015). Note the huge diversity in ionisation degree (HeII, [NeV], [FeVII]), integrated flux of emission lines and emission in the Balmer continuum (for this and all the following figures in this chapter, spectroscopy is provided by Asiago and Varese telescopes, and photometry by ANS Collaboration).

Homogeneity did not last for long, however, and by the time of the next catalog of SySts (Belczyński et al., 2000), systems with an excitation lower than HeII, and evidence from X-rays for a neutron star (NS) rather than a WD, began percolating through the broadening classification criteria. Currently (Mukai et al., 2016), *any binary where a WD or NS accretes enough material from an RG companion such that this interaction can be detected at some wavelength* is called a SySt. The number of known SySts is rapidly expanding

(~400, Akras et al., 2018), in particular as a consequence of surveys of the Galactic Bulge and Plane (Corradi et al., 2010; Miszalski et al., 2013; Rodríguez-Flores et al., 2014) and of Galaxies in the Local Group (Angeloni et al., 2014; Mikołajewska et al., 2014, 2017).

6.2 Burning Symbiotic Stars

The amount of material burnt at the surface of the WD as been traditionally considered equal to that continously accreted from the RG (eg., Kenyon, 1986), so that neither the nuclear burning switches off (accretion too low), nor the envelope expands to red giant dimension (accretion too high). The WD in burning SySts is, however, radiating close to its Eddington limit, and accretion could fall quite shorter than required to replenish what the WD burnt. Either nuclear burning conditions are therefore met only temporarily (followed by an accreting-only interval to refuel the envelope of the WD), or the Eddington limit is circumvented by discrete accretion episodes such as that of a massive disk dumping mass onto the WD during low-amplitude outbursts. The burning WD emits profusely supersoft X-rays, and indeed several SySts have been detected as supersoft X-ray sources (SSS), including C-1 in the Draco dwarf galaxy (Luna et al., 2013). Most, however, are not detected as SSS sources, the supersoft X-rays being absorbed locally by the abundant circumstellar gas (a situation similar to the early evolution of novae, before the ejecta dilute and turn optically thin to supersoft X-rays from the burning WD at their centre). The low impact made by accretion in burning SySts is confirmed by the widespread absence of both flickering and signature of magnetic-driven accretion (Sokoloski, 2003; Zamanov et al., 2017).

The SySts (mostly those of the burning type) are the only known class of astronomical objects to show OVI 1032, 1036 Å Raman-scattered by neutral hydrogen into a pair of broad emission features at 6825, 7088 Å (Schmid, 1989). A so far unique exception seems to be the Sanduleak's Star in LMC, which partnership with SySts has been questioned (Angeloni et al., 2014). Other Raman-scattered lines from HeII 940, 972 and 1025 Å and NeVII 973 Å have been proposed for SySts (Lee et al., 2014). The simultaneous coexistence of OVI and neutral hydrogen can be offered only by the very extended wind (10^2 au) of an RG orbited within a few photospheric radii by a burning WD.

To accomodate an RG, the orbital separation in a SySt is measured in astronomical units rather than in solar radii as in cataclysmic variables. The orbital periods range mostly from one to four years, with a maximum at 2-3 years and an M5III spectral type for the RG. Most burning SySts in our Galaxy seems to belong to the metal-rich Bulge population (Whitelock and Munari 1992; Gaia will soon tell) and are O-rich (M spectral types) as opposed to C-rich (carbon spectral types) in the Magellanic clouds, a fact related to the lower metallicity of small Megellanic clouds (SMC)/large Megellanic clouds (LMC) and its impact on the amount of carbon brought up by the third dredge-up on the AGB. Chemical abundances has been recently derived for several of the brighest SySts from near-infrared (IR) spectra. They should be treated with caution given the adopted oversimplifications (thin, plane-parallel, static, local thermodynamic equilibrium [LTE] atmospheres) contrasting with the huge complication of real RGB/AGB atmospheres (non-LTE, shocked, wind-supported, macroturbolent and hugely 3D extended). In about 15% of known SySts, the RG is a Mira, of an average M7III spectral type, and with a pulsation period usually quite longer than for field Miras (Whitelock, 2003). To accomodate the Mira well within its Roche lobe so to allow a regular pulsation, the orbital period must be P≥20 yrs. At such a wide orbital separation, no appreciable interaction would have occurred prior to the Mira evolutionary stage, with the binary system classified as an isolated, field RG. The Miras in SySts frequently come with warm dust (D-type SySts, as opposed to S-type with no detectable dust, and D'-type with cooler dust), which is believed to be preferentially located in the shadow cone created by the Mira itself, and possibly causing periodic obscurations along the orbital motion (Whitelock, 2003), even if an alternative location in the collision zone between the winds from the Mira and the WD has been proposed (Hinkle et al., 2013), at least for the Miras in symbiotic novae.

The mass transfer from the RG to the WD of SySts can occur via either capture from wind or Roche-lobe overflow. SySts with Roche-lobe filling RG present ellipsoidal distorted light curves at *I*-band or longer wavelengths, as illustrated in Figure 6.2, with an amplitude depending on orbital inclination and spectral type. For SySts with accreting-only WD, the ellipsoidal modulation dominates all the way down to bluest wavelengths (T CrB in Figure 6.2), while for burning SySts the irradiation by the WD of the facing side of the RG is responsible for the sinusoidal modulation dominating the bluest photometric bands. The amplitude of this modulation is proportional to the temperature and luminosity of the burning WD (cf. IV Vir and higher ionisation LT Del in Figure 6.2). The strong orbital dependence of the emission in the Balmer continuum (primarily responsible for the huge

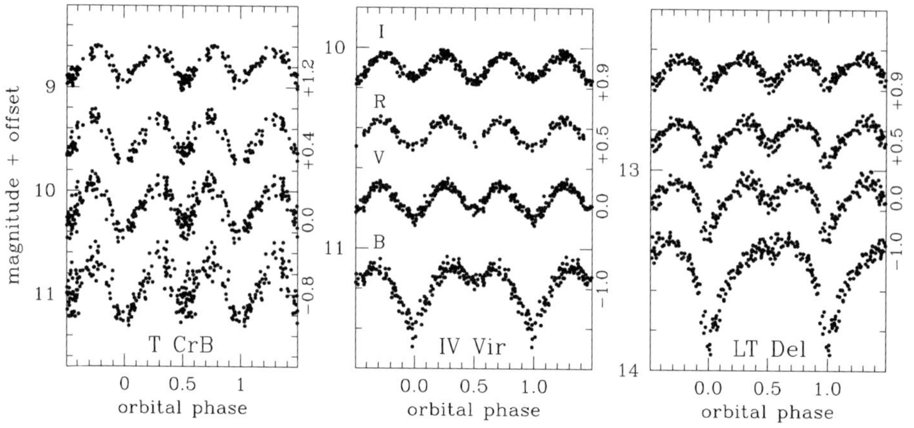

Figure 6.2 *BVRI* light curves of three high-inclination symbiotic stars to illustrate the variable proportion between elipsoidal modulation from red giant filling its Roche lobe (dominating the *I*-band flux; two maxima and two minima per orbital cycle) and irradiation of the same Red Giant by the hot WD companion (dominating the *B* band flux; one maximum and one minimum per orbital cycle).

amplitude seen in the U band) suggests that an important fraction of the irradiation effect resides in the ionised gas between the WD and the RG (Proga et al., 1998), but at least in some cases a direct increase in the surface temperature of the irradiated side of the RG has been documented (Chakrabarty and Roche, 1997; Munari et al., 2016). There are SySts showing no ellipsoidal distortion of the *I* or *JHK* light curves while presenting deep eclipses of the WD during outbursts (eg., FG Ser). Their RG must reside well within the Roche lobe, and the WDs therefore have to accrete from the wind, as it is the case for SySts containing a pulsating Mira variable. Hydrodynamic simulations (Mohamed and Podsiadlowski, 2012) show that the wind can be confined within the RG's Roche lobe and strongly focused towards the binary orbital plane. Such a wind Roche-lobe overflow (WRLOF) can be so efficient to allow the WD to accrete ~50% of the RG's mass loss and not just the few percentages typical of a Bondi–Hoyle–Littleton dynamical cross-section. The WRLOF, as other means of boosting the efficiency of mass accretion from wind (Bisikalo et al., 2006; Pan et al., 2015; Skopal and Cariková, 2015), would also offer a way out to the evolutionary paradox posed by the *yellow* SySts. They are a small group, including both burning and accreting-only cases. Their RG are G/K-type giants, with Halo kinematics and low metallicity ([Fe/H]≤ -1), enriched in *s*-type elements (most notably barium), which are normally brought to surface during third dredge-up at the tip of the AGB. They lack, however, the presence of unstable Tc isotopes and are less luminous than the tip of the AGB, indicating an extrinsic origin of the *s*-type elements, i.e., pollution from the progenitor of the current WD companion (Jorissen et al., 2005; Pereira et al., 2017). Some of these SySts are rotating at a significant fraction of their rotational breakup velocities ($V_{\rm rot} \sin i \geq 100$ km/s), suggesting a massive transfer of both mass and angular momentum from the progenitor of the current WD. The distribution of orbital periods and eccentricities of barium SySts require that dynamically unstable mass transfer by Roche-lobe overflow (and the resulting common-envelope phase with its orbit shrinking and circularisation) is avoided, and massive transfer of mass and angular momentum is achieved via wind.

6.3 Accreting-Only Symbiotic Stars

All-sky surveys as well as pointed X-ray observations (with the *Swift* satellite in particular) are discovering a population of optically unconspicous RG that emits in hard X-rays, a fact requiring them to pair in a binary system with a WD or an NS. Their relatively low X-ray ($\approx$0.1–10 $L_\odot$) and UV ($\approx$1–10 $L_\odot$) luminosities currently limit the serendipitous discovery to systems within ~1 kpc. A WD companion is usually associated with a luminosity larger in UV than in X-rays, while the reverse is true for an NS (given its deepest potential well). SySts are studied as potential progenitors of type Ia supernovae, since the original proposal by Munari and Renzini (1992), whose population synthesis was based on the estimated total number of burning SySts in the Galaxy, i.e., those easier to discover over vast distances thanks to their spectacular emission line spectrum. If burning SyST are just the tip of an iceberg of momentarily quiet, accreting-only SySts, the appeal of the symbiotic channel to SN Ia will be further boosted.

The subtle way these low-key, optically quiet and accreting-only SySts are discovered is well epitomised by SU Lyn, a $V{\sim}8$ mag, M6III giant at about 600 pc distance, completely unnoticed except for an old report about a possible semiregular variability of type SRB (AGB stars with a poorly defined periodicity and low amplitude). Looking for optical

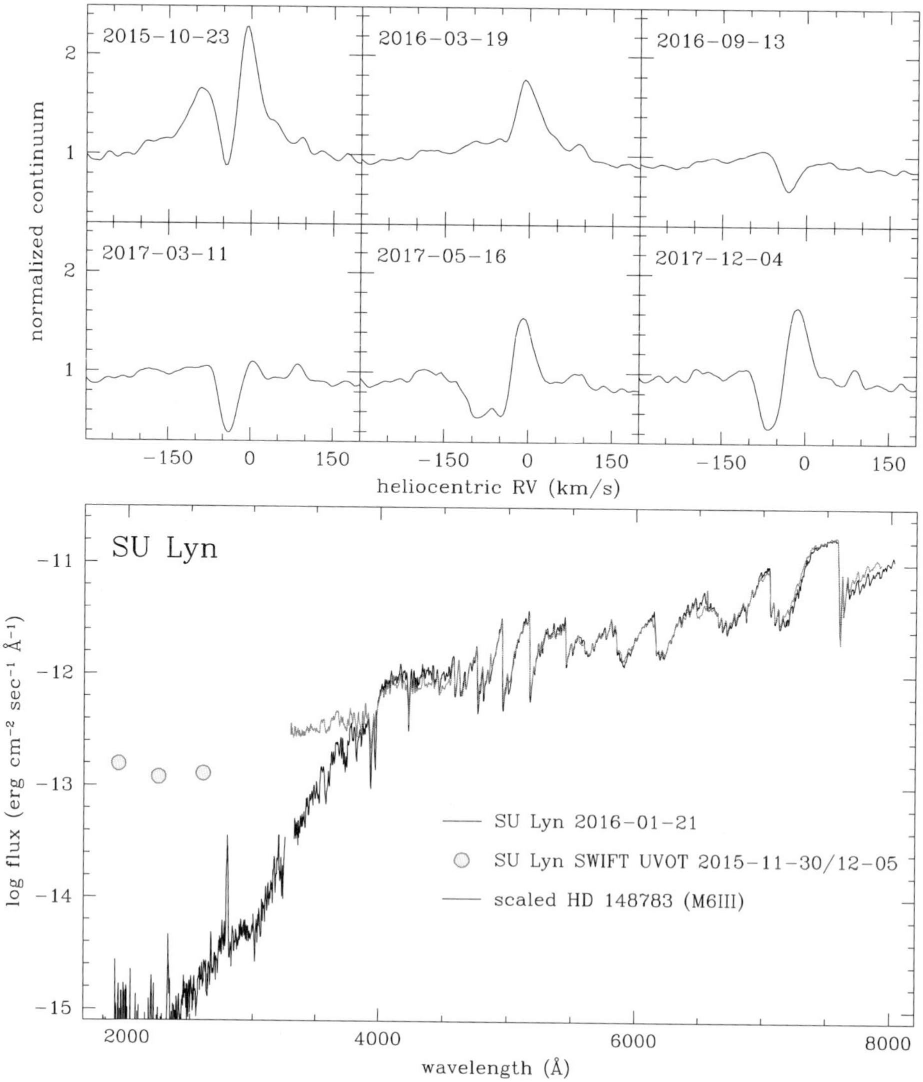

Figure 6.3 *Bottom panel:* low-resolution optical spectra of SU Lyn (grey) reveals nothing different from those of field normal giant of the same spectral type (black). The presence of an accreting WD is betrayed by the extra flux at far-blue ($\lambda \leq 3900$ Å) and UV wavelengths (Swift UVOT dots). *Upper panel:* high-resolution observations show feeble, structured and highly variable emission in Hα (and other lines as well), again disclosing the accreting nature of SU Lyn.

counterparts of *Swift*-BAT sources, Mukai et al. (2016) noted that SU Lyn lied within the error box of one of them. Follow-up observations were organised with *Swift* (to refine the astrometric position of the BAT source and better characterise its UV and X-ray emission) and with the Asiago spectrographs (to investigate if SU Lyn optical spectra could betray peculiarities supporting a physical association with the *Swift*-BAT source). Some results are summarised in Figure 6.3. While the optical spectrum of SU Lyn is identical to that of a normal M6III giant, its bluest part ($\lambda \leq 3900$ Å) shows a flux excess that extends to match the UV excess seen by *Swift* UVOT telescope and the soft and hard X-ray emission observed by *Swift* XRT and BAT instruments. Only high-resolution Echelle spectra can reveal a feeble, structured and quite variable emission in Hα.

Similar tortuous paths affect the discovery of SySts hosting an NS, usually named symbiotic X-ray binaries (or SyXBs; Masetti et al., 2007). SyXBs are quite rare: among the ~200 low-mass X-ray binaries (LMXBs) known in the Galaxy, only ~10 SyXBs cases are currently known. Observationally, these systems are characterised by appreciable X-ray emission ($\sim 10^{32}$–10^{34} erg s^{-1}) positionally associated with an RG star that spectroscopically does not show any abnormal features, with the possible exception of a continuum excess in the blue and ultraviolet ranges (similar to what is illustrated in Figure 6.3 for SU Lyn). The X-ray emission is pulsed (periods from 10^2 to 10^4 seconds, 4U 1954+319 being the slowest at $P_{spin} \sim 18{,}400$ s), indicating that the NS is rotating slowly. The rotation period changes in response to accretion: for 4U 1954+319, Marcu et al. (2011) measured a strong spin-up of -1.8×10^{-4} hr hr^{-1} during outbursts and a spin-down of 2.1×10^{-5} hr hr^{-1} in quiescence. The level of X-ray emission can vary up to four orders of magnitude, suggesting accretion from an inhomogeneous stellar wind possibly coupled with an highly elliptical orbit of the accretor. A notable outlier is GX 1+4, which emits in X-rays up 10^{37} erg s^{-1}, with $P_{spin} \sim 120$ s (Chakrabarty and Roche, 1997), and an optical spectrum (Munari and Zwitter 2002) quite similar to those of burning SySts. A geometrically thin and optically thick accretion disk heavily irradiated by the hard X-rays from the central NS would provide the UV source needed to photoionise the RG wind: plasma diagnostic shows in fact that the gas in GX 1+4 is ionised by thermal UV radiation ($T_{ph} \sim 85{,}000$ K) rather than the nonthermal X-ray power law expected from the central accreting NS.

An handy classification of the main types of X-ray emission seen in SySts has been introduced by Murset et al. (1997) and expanded by Luna et al. (2013) and Nuñez et al. (2016). It is summarised in Table 6.1 with the name of a few well-known SySts in each class, and a compact description of the likely origin for the X-rays.

6.4 Different Types of Outburst in Symbiotic Stars

6.4.1 Z-And or Classical Type

Normal SySts frequently enjoy outburts that usual come in trains of a few individual episodes separated by longer periods spent at quiescence. The example of CI Cyg in Figure 6.1 is indicative: three different maxima (α, β, γ), of declining strength and duration, separated in time by ~1 orbital period. The previous train of multiple maxima ended

Table 6.1 *Different types of X-ray emission observed in symbiotic stars, their likely origin and some of the best-known examples in each class.*

Type	Description	Examples
α	Supersoft, photon energies $\leq$0.4 keV, hydrogen burning on WD surface	AG Dra, Draco C-1, SMC-3
β	Soft, photon energies $\leq$2.4 keV, colliding winds from WD and red giant	Z And, Mira AB, AG Peg
δ	Hard, absorbed, with thermal emission detectable at $\geq$2.4 keV, boundary layer between accretion disk and WD	SU Lyn, 4 Dra, T CrB
β/δ	CHARACTERISTICS of both β and δ type simultaneously present, from colliding winds and disk/WD boundary layer	NQ Gem, CH Cyg MWC 560
γ	Absorbed NS accretor, pulsed by NS spin, optically thick Comptonised plasma	GX 1+4, V934 Her 4U 1954+31

in 1979. Also the spectral evolution depicted in Figure 6.1 is quite typical: compared to quiescence, the ouburst spectrum is characterised by a lower ionisation, a stronger hot continuum veiling the RG molecular spectrum and a large increase in the integrated flux of the emission lines. The amplitude of the outburst is 2 or 3 mag in the *B* band and declines toward the red. The colour evolution of CI Cyg in Figure 6.1 well illustrates how the outburst status is barely detectable in the *I*-band, where the flux is dominated by the RG at all phases. This type of frequent and multiple outbursts is called Z-type or Z-And type from Z Andromedae, a prototype SySt. Jets (frequently seen bipolar in high-resolution spectra) have been observed in about 12 symbiotic stars (for 1/3 of them also spatially resolved; cf. the spectacular images for R Aqr by Schmid et al. (2017)), and in most cases they are associated with Z-And outbursts. The projected jet velocities are of the order of 1000−1500 km/s, equivalent to the escape velocity from the region closest to the central WD, with the noteworthy exception of MWC 560, where velocities $V_{\rm ej} \geq 6{,}000$ km s^{-1} were observed (Tomov et al., 1990). In response to the great differences seen in the jets from one object to another, a variety of launching mechanisms have been proposed and modelled (e.g., Stute and Sahai, 2007; Skopal et al., 2009; Tomov et al., 2011). Whereas RGs in SySts rotate faster than field RGs and appear synchronised with orbital period ($P_{\rm rot} \simeq P_{\rm orb}$), those in systems emitting jets seem to rotate faster at $P_{\rm rot} < P_{\rm orb}$ (Zamanov and Stoyanov, 2012).

Considering the possible causes of Z-And type outburst, a great variety of different mechanisms have been invoked (e.g., Bisikalo et al., 2006; Skopal et al., 2011; Tomov et al., 2011; Ramsay et al., 2016; de Val-Borro et al., 2017) such as a sudden increase in the mass-transfer rate from the RG, either triggered by intrinsic variability of the RG or its passage at periastron; the formation of an optically thick, cool, disk-shaped zone around the WD equator as a consequence of enhanced wind from the WD; an enhanced wind from the WD that leads instead to the disruption of the inner part of the accretion disk with the formation of hollow cones around the WD axis of rotation and thus to the appearance of collimated outflows; changes in the kinematical regime of colliding winds from the WD and the RG, etc. A coordinated effort between X-ray, ultraviolet, optical and radio observations to follow in detail and over all the relevant phases the Z-And outbursts of at least a few SySts seems required to rise firm constraints useful in guiding future modelling efforts.

Basically, the explanations for Z-And type outbursts tend to cluster into two broad categories: (a) release of potential energy from extra-accreted matter, or (b) shift to longer wavelengths of the emission from WD burning shell of the WD, expanding in radius as a response to an increase in the mass accretion rate. Both modes could occur in succession in the same object, as illustrated in Figure 6.4 by the 2016/2017 outburst of StHα 169.

6.4.2 Symbiotic Novae

A very few ($\sim$10) symbiotic novae (SyN) have been seen to erupt in historical times in our Galaxy, while others were already in outburst when discovered as SySts. They should not be confused with the novae erupting in symbiotic binaries described later in this chapter. The outbursts of SyN last about a century and are of large amplitude, and only one eruption has been recorded, with the possible exception of BF Cyg that shortly after returning to quiescence from the outburst initiated in 1894 (Leibowitz and Formiggini, 2006) started a new SyN cycle in 2006 and currently is still at maximum brightness.

The typical photometric and spectroscopic evolution of SyN is illustrated in Figure 6.5. A large rise in brightness takes the star, in about $\sim$1 year, from faint anonymity to bright spotlight, from where an extremely slow decline needs about a century to return the system to quiescence level. The spectral evolution is equally slow. During the rise towards maximum brightness, the spectrum cools up to an F-type supergiant with feeble emission lines limited to Balmer and FeII. During the decline, the F-type continuum weakens, a nebular continuum takes over and the emission lines grow in intensity and ionisation degree. In a few cases (e.g., HM Sge and V1016 Cyg), the phase dominated by the F-type supergiant continuum is probably too short and, when transiting at maximum brightness, the spectrum of the SyN is already dominated by a nebular continuum and strong emission lines.

An SyN outburst could even be the event that initiates a symbiotic cycle in the life of a RG+WD binary, coming after a long and quiet period of accretion only. Under the high mass-transfer rates allowed by WRLOF or by plain Roche-lobe filling, the material should pile up in nondegenerate form on the surface of a nonmassive and hot WD. Upon reaching conditions for hydrogen burning, this will proceed under thermal equilibrium, avoiding the thermonuclear runaway (TNR) of a classical nova and the consequent violent mass ejection. To adjust to the large nuclear luminosity produced at its base, the nondegenerate envelope expands to supergiant dimensions. The absence of massive ejection retains most of the mass in the WD envelope, which keeps burning under stable conditions for a long time, in excess of the $\sim$century an SyN takes to return to quiescence. AG Peg has only recently

Hinkle, K. H., Fekel, F. C., Joyce, R. R., and Wood, P. 2013. Infrared spectroscopy of symbiotic stars. IX. D-type symbiotic novae. *ApJ*, **770**, 28.

Jorissen, A., Začs, L., Udry, S., Lindgren, H., and Musaev, F. A. 2005. On metal-deficient barium stars and their link with yellow symbiotic stars. *A&A*, **441**, 1135–1148.

Kenyon, S. J. 1986. The symbiotic stars. Cambridge: Cambridge University Press.

Kenyon, S. J., and Webbink, R. F. 1984. The nature of symbiotic stars. *ApJ*, **279**, 252–283.

Lee, H.-W., Heo, J.-E., and Lee, B.-C. 2014. Raman-scattered Ne VII 973 at 4881 Å in the symbiotic star V1016 Cygni. *MNRAS*, **442**, 1956–1962.

Leibowitz, E. M., and Formiggini, L. 2006. Multiperiodic variations in the last 104-yr light curve of the symbiotic star BF Cyg. *MNRAS*, **366**, 675–681

Luna, G. J. M., Sokoloski, J. L., Mukai, K., and Nelson, T. 2013. Symbiotic stars in X-rays. *A&A*, **559**, A6.

Marcu, D. M., et al. 2011. The 5 Hr Pulse Period and Broadband Spectrum of the Symbiotic X-Ray Binary 3A 1954+319. *ApJL*, **742**, L11.

Masetti, N., et al. 2007. IGR J16194-2810: a new symbiotic X-ray binary. *A&A*, **470**, 331–337

Merrill, P. W., and Humason, M. L. 1932. A bright line of ionized helium, λ 4686, in three stellar spectra with titanium bands. *PASP*, **44**, 56–57.

Mikołajewska, J., and Kenyon, S. J. 1992. On the nature of the symbiotic binary AX Persei. *AJ*, **103**, 579-592.

Mikołajewska, J., Caldwell, N., and Shara, M. M. 2014. First detection and characterization of symbiotic stars in M31. *MNRAS*, **444**, 586–599.

Mikołajewska, J., Shara, M. M., Caldwell, N., Iłkiewicz, K., and Zurek, D. 2017. A survey of the Local Group of galaxies for symbiotic binary stars. I. *MNRAS*, **465**, 1699–1710.

Miszalski, B., Mikołajewska, J., and Udalski, A. 2013. Symbiotic stars and other Hα emission-line stars towards the Galactic bulge. *MNRAS*, **432**, 3186–3217.

Mohamed, S., and Podsiadlowski, P. 2012. Mass transfer in Mira binaries. *BalA*, **21**, 88–96.

Mukai, K., et al.. 2016. SU Lyncis, a hard X-ray bright M giant: clues point to a large hidden population of symbiotic stars. *MNRAS*, **461**, L1–L5.

Munari, U., and Buson, L. M. 1994. The ultraviolet spectra of the symbiotic stars MWC 960, FN Sgr, SS 29 and Draco C-1. *A&A*, **287**, 87–94.

Munari, U., and Renzini, A. 1992. Are symbiotic stars the precursors of type Ia supernovae? *ApJLett*, **397**, L87–L90.

Munari, U., and Zwitter, T. 2002. A multi-epoch spectrophotometric atlas of symbiotic stars. *A&A*, **383**, 188–196.

Munari, U., Dallaporta, S., and Cherini, G. 2016. The 2015 super-active state of recurrent nova T CrB and the long term evolution after the 1946 outburst. *New Astronomy*, **47**, 7–15.

Munari, U., et al. 2011. The 2010 nova outburst of the symbiotic Mira V407 Cyg. *MNRAS*, **410**, L52–L56.

Murset, U., and Nussbaumer, H. 1994. Temperatures and luminosities of symbiotic novae. *A&A*, **282**, 586–604.

Murset, U., Nussbaumer, H., Schmid, H. M., and Vogel, M. 1991. Temperature and luminosity of hot components in symbiotic stars. *A&A*, **248**, 458–474.

Murset, U., Wolff, B., and Jordan, S. 1997. X-ray properties of symbiotic stars. *A&A*, **319**, 201–210.

Nuñez, N. E., Nelson, T., Mukai, K., Sokoloski, J. L., and Luna, G. J. M. 2016. Symbiotic stars in X-rays. *ApJ*, **824**, 23.

Pan, K.-C., Ricker, P. M., and Taam, R. E. 2015. Simulations of the symbiotic recurrent nova V407 CYG. I. Accretion and shock evolutions. *ApJ*, **806**, 27.

Pereira, C. B., Baella, N. O., Drake, N. A., Miranda, L. F., and Roig, F. 2017. High-resolution optical spectroscopic observations of four symbiotic stars: AS 255, MWC 960, RW Hya, and StHα 32. *ApJ*, **841**, 50.

Proga, D., Kenyon, S. J., and Raymond, J. C. 1998. Illumination in symbiotic binary stars: non-LTE photoionization models. II. WIND case. *ApJ*, **501**, 339–356.

Ramsay, G., Sokoloski, J. L., Luna, G. J. M., and Nuñez, N. E. 2016. Swift observations of the 2015 outburst of AG Peg. *MNRAS*, **461**, 3599–3606.

Rodríguez-Flores, E. R., et al. 2014. IPHAS and the symbiotic stars. III. *A&A*, **567**, A49.

Schmid, H. M. 1989. Identification of the emission bands at 6830, 7088 A. *A&A*, **211**, L31–L34.

Schmid, H. M., et al. 2017. SPHERE/ZIMPOL observations of R Aqr. I. Imaging of the stellar binary and the innermost jet clouds. *A&A*, **602**, A53.

Seaquist, E. R., Taylor, A. R., and Button, S. 1984. A radio survey of symbiotic stars. *ApJ* **284**, 202–210.

Skopal, A., and Cariková, Z. 2015. Wind mass transfer in S-type symbiotic binaries. I. Focusing by the wind compression model. *A&A*, **573**, A8.

Skopal, A., et al. 2011. Formation of a disk structure in the symbiotic binary AX Persei during its 2007-10 precursor-type activity. *A&A*, **536**, A27.

Skopal, A., Pribulla, T., Budaj, J., et al. 2009. Transient jets in symbiotic prototype Z Andromedae. *ApJ*, **690**, 1222–1235.

Sokoloski, J. L. 2003. Rapid variability as a diagnostic of accretion and nuclear burning in symbiotic stars and supersoft X-ray sources. *ASPC*, **303**, 202–217

Starrfield, S., Iliadis, C., and Hix, W. R. 2016. The thermonuclear runaway and the classical nova outburst. *PASP*, **128**, 051001.

Stute, M., and Sahai, R. 2007. Hydrodynamical simulations of the jet in the symbiotic star MWC 560. III. Application to X-ray jets in symbiotic stars. *ApJ*, **665**, 698–706.

Tomov, N. A., Bisikalo, D. V., Tomova, M. T., and Kil'Pio, E. Y. 2011. Interpretation of the line spectrum of classical symbiotic stars. *AIPCS*, **1356**, 35–44.

Tomov, T., Kolev, D., Zamanov, R., Georgiev, L., and Antov, A. 1990. MWC560: a unique astrophysical object. *Nature*, **346**, 637.

Tomov, T. V., Stoyanov, K. A., and Zamanov, R. K. 2016. AG Pegasi: now a classical symbiotic star in outburst? *MNRAS*, **462**, 4435–4441.

Whitelock, P. A. 2003. A comparison of symbiotic and normal Miras (invited review talks). *ASPC*, **303**, 41–56

Whitelock, P. A., and Munari, U. 1992. Photometric properties of symbiotic stars and the nature of the cool component. *A&A*, **255**, 171–180.

Zamanov, R. K., and Stoyanov, K. A. 2012. Rotation of the red giants and white dwarfs in symbiotic binary stars. *Bulg.A.J*, **18**, 41–52

Zamanov, R. K., et al. 2017. Discovery of optical flickering from the symbiotic star EF Aquilae. *AN*, **338**, 680–685.

7

Binary Post-AGB Stars as Tracers of Stellar Evolution

HANS VAN WINCKEL

7.1 Introduction

The final evolution of low- and intermediate-mass stars (0.8–8 $M_\odot$) is a rapid transition from the Asymptotic Giant Branch (AGB) over the post-AGB phase towards the planetary nebula (PN) stage. When most of the stellar envelope is removed by an intense mass-loss rate (up to 10^{-4} $M_\odot$ /yr) at the end of the AGB, the star enters the post-AGB phase of stellar evolution. During this phase, the remaining stellar envelope begins to shrink and the effective temperature starts to increase with almost constant luminosity. If the temperature increases on a timescale shorter than the dispersion time of the matter previously ejected by the star, a PN will appear, as the result of the ionisation of the circumstellar shell (Balick and Frank, 2002; Herwig, 2005). During the AGB-to-PN transition, some objects show a resolved reflection nebula, and these are referred to as protoplanetary nebula or pre-planetary nebula (Sahai et al., 2007). These protoplanetary nebulae are only a subset of post-AGB stars, as the latter also includes objects that will not ionise their ejected matter as it is too dispersed. It is fair to say that the samples of protoplanetary nebulae and post-AGB stars, in general, do not connect very well on evolutionary tracks with the AGB stars, nor with the planetary nebulae (PNe). The diversity seen in individual examples is not well understood in the framework of stellar evolution theory (e.g., Van Winckel, 2003).

One of the most important research questions regarding the final evolution of low- and intermediate-mass stars is the impact of binarity (see the recent review by De Marco and Izzard, 2017, and references therein). Also, this book provides ample illustration of research results focussing on this question. Binary interaction alters the intrinsic properties of the evolved star (such as pulsations, mass loss, dust formation, circumstellar envelope morphology, etc.) and plays a dominant role in determining its ultimate fate. A plethora of peculiar objects ranging from the spectacular thermonuclear novae, Type Ia supernovae, subluminous supernovae, gravitational wave sources, etc., to less energetic systems such as subdwarf B stars, barium stars, cataclysmic variables, bipolar PNe, etc., result from mass transfer in binary stars.

During evolution of the low- and intermediate-mass star, binary interactions will increase when one of the binary components evolves to giant dimensions. The dynamical behaviour of this interaction is determined by the balance between the Roche lobe and the stellar radius of the red giant. Binaries with a main sequence separation of less than about two astronomical units (au) will be tidally captured somewhere on the red giant branch (RGB) or AGB. When unstable mass transfer ensues, a common-envelope (CE) event occurs, resulting in a dramatic shortening of the orbital separation. This might lead to a complete stellar merger but, if this is avoided, the final period of these binaries after the giant phase is typically in the order of days, depending on the envelope's binding energy. When the initial orbit is wider, the interaction is thought to be via wind accretion or wind shaping, and the primary component will evolve off the giant branch with a very wide orbit. When the originally lower-mass component evolves also onto the giant branch, interaction may take place again, but this time with a compact companion. This theoretical scenario is borne out by population-synthesis models (e.g., Han et al., 2002, 2003). The population-synthesis model normalised to the ellipsoidal binaries on the red giant phase, predicts the final period distribution of evolved binaries to be bimodal in which the CE channel results in short-period binaries (P $\sim$ 1 day) and the wind accretion channel results in wider systems (P $>$1,000 days) (Nie et al., 2012). The orbital periods of around 1,000 days are least predicted and lie in the middle of this bimodal distribution.

This is, however, in stark contrast to what is observed in post-AGB binaries, and here we focus on the observed properties of these optically bright objects. The reasons for this strong discrepancy between the observed and predicted period distributions are twofold. Firstly, a wide range of binary interactions are not understood from first principles or are even missing in the binary evolution models. The theoretical models are therefore subject to many uncertainties, such as the efficiency of envelope ejection (e.g., Toonen and Nelemans, 2013); the unknown physical description of the common-envelope phase (e.g., Izzard et al., 2012; Ivanova et al., 2013); the postulated increase of the mass-loss prior to contact as to lower the envelope binding energy (e.g., Chen et al., 2011; Abate et al., 2013); the badly understood impact of radiation pressure on the shape of the classical Roche potential (Dermine et al., 2009); and the assumed mass transfer efficiency and its orbital phase dependency. Secondly, there is also an observational challenge as many parameters involved in predicting the outcome of the binary evolution channels are not well constrained due to the lack of observational data, where detailed studies of individual objects prevail over systematic and time-consuming surveys that are more complete.

In this chapter, we focus on binary post-AGB stars and their observational properties. In Section 7.2, we focus on their spectral energy distribution (SED), as this turned out to be very characteristic. We report on the orbital properties in Section 7.3 and review all observational evidence for the presence of stable circumbinary discs as a common feature of all the binaries in Section 7.4. Jets and their origin form the core of Section 7.5. The impact of the feedback from the disc on the primary star is given in Section 7.6. We end in Section 7.7 by naming a few challenges ahead to come to a better understanding of binary stellar evolution of low-and intermediate mass stars.

7.2 Spectral Energy Distribution

While the first binary post-AGB stars were serendipitously detected (e.g., Waelkens et al., 1996), it turned out that their SEDs had some common but distinct properties. This is

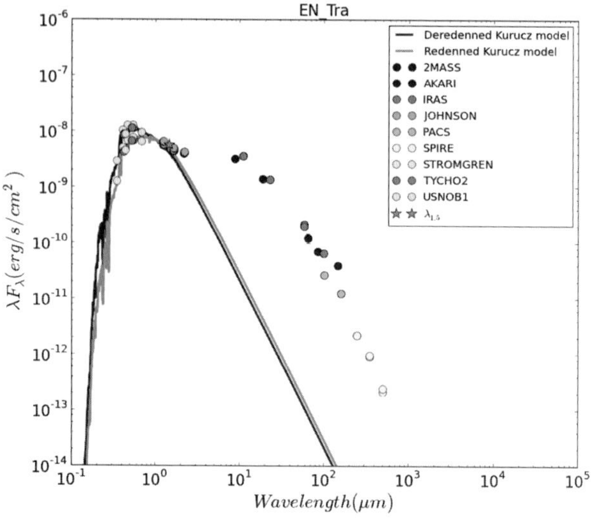

Figure 7.1 The SED of 89 Her (Hillen et al., 2013) as an example of an optically bright post-AGB binary.

illustrated in Figure 7.1, where a typical SED of a post-AGB binary is displayed. The energetics display a clear near-infrared excess by thermal emission of dust, indicating that this circumstellar dust must be close to the central star, near sublimation temperature. The peak of the dust excess is around 10 μm and at the long wavelength tail, the spectral index follows the Rayleigh jeans slope up to submillimetre wavelengths. It is now well established that these features in the SED indicate the presence of a stable compact disc in the system (see Section 7.4). We call this type of SEDs *disc-type SEDs*.

These specific SED characteristics, in combination with the availability of all-sky infrared (IR) surveys of Infrared Astronomical Satellite (IRAS), Wide-Field Infrared Survey Explorer (WISE) and Akari allowed for efficient searches for similar Galactic systems (e.g., De Ruyter et al., 2006; Gezer et al., 2015) and in total about 85 such Galactic sources have been identified so far. These SEDs are in clear contrast to the ones of optically bright post-AGB stars with detached shells: the shell of gas and dust, which was expelled when the object was still on the AGB, is cooling and expanding, resulting in an SED without a near-IR excess and a peak in the SED typically around 30 to 60 μm (Van Winckel, 2003).

Moreover, the Spitzer surveys in the Large and Small Magellanic Clouds (Meixner et al., 2006; Bolatto et al., 2007; Gordon et al., 2011) allowed for a systematic approach in the search for post-AGB stars also in these galaxies. After a photometric selection (van Aarle et al., 2011; Kamath et al., 2014), a spectral survey was performed to locate similar evolved sources and to differentiate them from interlopers (Kamath et al., 2014, 2015). The objects with distinct near-IR excess represent more than half of the population of optically bright post-AGB stars. Objects with similar SEDs also appear at lower luminosities, indicating that the central evolved star is a post–Red Giant Branch (post-RGB) stars, rather than post-AGB stars (Kamath et al., 2016).

We conclude that within the known samples of the rare post-AGB stars, disc SEDs are very frequently detected both in the Galaxy and the Magellanic Clouds.

7.3 Orbital Properties

The binary nature of individual objects (Van Winckel et al., 1995; Waelkens et al., 1996; Gonzalez and Wallerstein, 1996; Maas et al., 2002) inspired a systematic monitoring programme of Galactic evolved sources using our own HERMES spectrograph (Raskin et al., 2011) mounted on the Flemish 1.2 m Mercator telescope. This has resulted in the discovery of many evolved binaries with unexpected periods between 100 and 2,000 days within the sample of sources with disc SEDs (e.g., Van Winckel et al., 2009; Gorlova et al., 2012; Gezer et al., 2015; Manick et al., 2017). The $a \sin i$ is typically $\sim$1 au. The luminous evolved component has a likely unevolved companion with a very minor contribution to the energy budget (e.g., Van Winckel et al., 2009). Figure 7.2 is illustrative of this. One of the most striking results is a clear lack of spiralled-in systems at short periods in the order of days. This is not a observational bias: these would be the most easy to detect in radial velocity monitoring programmes. Another striking observation is the rare occurrence of zero-eccentricity systems as many objects have significant eccentricities.

The global picture that emerges from the orbital elements is that the star evolved in a system that is too small to accommodate a full-grown RGB or AGB star. During a poorly understood phase of strong interaction, the system did not suffer a dramatic spiral in. The observed orbits (see Figure 7.2) show that the common envelope was either very rapidly

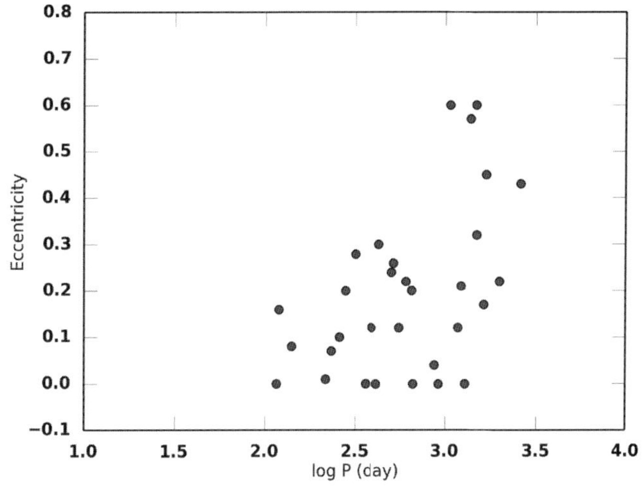

Figure 7.2 The e-log(P) diagram of post-AGB binaries.

expelled or somehow avoided. We postulate that this binary interaction phase truncated the AGB or RGB evolution. All the ∼30 orbits found until now and their nonzero eccentricities fall in a period regime *not* predicted by standard binary channels. Moreover, the tidal circularisation was not effective and most objects have distinct eccentric orbits. Eccentricity pumping mechanisms need to be invoked to explain the eccentricities. As the dust sublimation radii for these sources are well beyond the orbit, the dusty discs are circumbinary.

Finding binaries among post-AGB stars with an expanding shell (shell-sources), on the contrary, turned out to be a cumbersome challenge: despite radial-velocity monitoring efforts for decades, confirmed spectroscopic binaries were not detected yet in post-AGB stars with such detached shells, which puts strong constraints on the companion mass and/or possible orbits (Hrivnak et al., 2011, 2017). Interestingly, these detached shells are often resolved and show bipolar and even multipolar geometries (Hrivnak et al., 2017), which are often associated with binary interaction physics. If at all, these shell objects are likely very wide systems.

Spectroscopic binaries in post-AGB stars are uniquely connected to their SED properties: almost all binaries known so far have disc-SEDs. Very few exceptions exist and these have a very small dust excess, which seems to be the relic of the circumbinary disc.

7.4 Discs: Resolved from the Inner Edge to the Outer Radius

7.4.1 Near- and Mid-IR Interferometry

It is now well established that this type of disc-SED is a good observational proxy for the presence of a stable compact circumbinary disc. As the infrared emission is compact and the objects are typically at a distance of several kiloparsecs, interferometric techniques are needed to resolve them. Applications of high spatial-resolution techniques with limited uv-coverage (Deroo et al., 2006, 2007; Hillen et al., 2013, 2014, 2015, 2017) are unveiling the very compact infrared emission regions. There is by now ample evidence that these circumbinary discs show many similarities with the passive protoplanetary discs (PPDs) around young stars (e.g., Dullemond and Monnier, 2010; Menu et al., 2015).

The spatial information obtained by interferometric experiments is combined with detailed physical modelling via radiative transfer calculations, in which the transport of radiation in three dimensions is treated self-consistently. The vertical scale of the disc is computed under the assumption of hydrostatic equilibrium. For the opacity calculations, the opacities of silicates are typically used (Min et al., 2007) with a variable size distribution of the grains. Turbulent mixing and vertical settling of grains can be included self-consistently (Mulders and Dominik, 2012), as it strongly affects the structure of the disc, and therefore the observables. The good agreement between the best models and the interferometric observations in the preceding references gives strong support to the physical interpretation of the circumstellar material: the structure is that of a dusty settled disc. The best-fit grain size distribution shows that grain growth is significant. Submicron-sized particles hardly contribute to the total opacity in this disc, while millimetre-sized grains need to be included to fit the submillimetre fluxes.

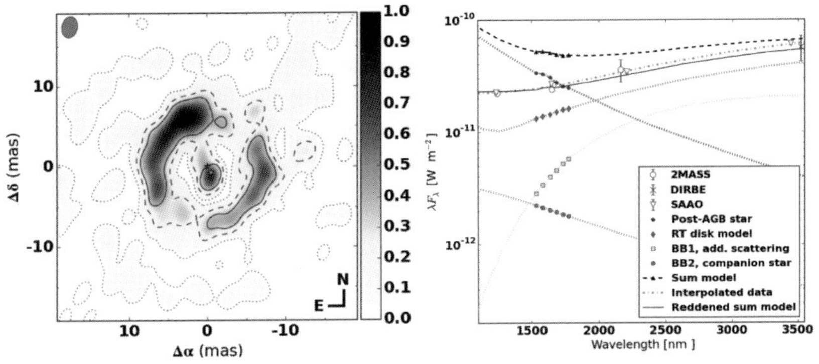

Figure 7.3 The interferometric image in the H-band of IRAS08544-4431 on the left. The central luminous post-AGB star is removed for clarity (star). The inner rim of the circumbinary disc is well resolved, and the off-centre flux contribution is interpreted as coming from the accretion disc around the companion (Hillen et al., 2016). On the right, the dissection of the SED is shown with all spatially resolved contributing components. Reproduced with permission from *Astronomy & Astrophysics*, ©ESO.

A first interferometric imaging experiment, which is possible provided the uv-plane is covered extensively, was presented in Hillen et al. (2016). Thanks to the unprecedented spatial resolution, all important building blocks (circumbinary disc, central binary, accretion disc around the companion) could be spatially resolved and their flux contribution could be quantified, which is illustrated in Figure 7.3. The most apparent is the inner rim of the circumbinary disc, which is indeed at the sublimation radius of the star. For clarity, the central bright star, which is unresolved, is removed from the image. The companion itself is too faint to contribute significantly to the total H-band flux, but it is the accretion disc around the companion that is resolved as an additional contributor to the centre of the image, slightly displaced from the luminous post-AGB star. The physical model of the circumbinary disc assumes a settled disc in hydrostatic equilibrium, with an axisymmetric surface density. While the model fits many (interferometric) observables, the real data show more structure.

Just like for pre–main-sequence stars across a wide range of mass and luminosity (Dullemond and Monnier, 2010; Kraus et al., 2010), spectrointerferometry proves very powerful to study the physics of settled discs.

7.4.2 CO Interferometry

The most conclusive evidence for stability comes from the detection of resolved Keplerian velocity fields in three systems so far: the Red Rectangle, AC Her and IW Car. These were spatially resolved in CO using the ALMA and the Plateau De Bure interferometers (Bujarrabal et al., 2013, 2015, 2017). Moreover, for 89 Her the inner CO disc is found to be not resolved but likely in Keplerian rotation as well (Bujarrabal et al., 2007).

In addition to the Keplerian velocity, a slowly expanding component was resolved in the maps of the Red Rectangle, 89 Her and IW Car, showing a bipolar low-velocity outflow. Given its structure and kinematics, these outflows likely originate from the disc. The total angular momentum of the disc of the Red Rectangle is significant and suggests that, if the orbit is the source of angular momentum, the binary should have shrunk considerably (Bujarrabal et al., 2016). By comparing the total mass of the CO disc with the mass loss as detected in CO, the estimated lifetimes are between ~8,000 and 10,000 year (Bujarrabal et al., 2016, 2017).

Single-dish observations confirm that rotation must be widespread among other disc sources as well: the CO rotational lines are very narrow, indicating that Keplerian rotation rather than expansion is detected in all the sources of the survey (Bujarrabal et al., 2013). Clearly more systems need to be investigated in detail.

7.4.3 Longevity

The discs have other distinct observational features that are interpreted as indicators of longevity. Infrared spectroscopic observations of dust features show that strong dust-grain processing is needed to explain the profiles. The processing leads to a high degree of crystallinity in the silicate bands (e.g., Molster et al., 2002; Gielen et al., 2007, 2011).

It is remarkable that in all systems, Galactic and from the Magellanic Clouds alike, the dust is oxygen rich, and magnesium-rich end members of the silicates prevail. Additionally, some objects show a carbon-rich component such as the Extended Red Emission (ERE) in the Red Rectangle (e.g., Cohen et al., 2004) and polycyclic aromatic hydrocarbons (PAH) emission in a minority of objects.

Another indicator for longevity is the presence of large grains. In order to explain the contrast and width of the silicate bands (e.g., Gielen et al., 2007, 2011; Arneson et al., 2017), micron-sized grains are needed. The presence of even larger grains, to the centimetre regime, are needed to explain the spectral index at submillimetre wavelengths (e.g., Jura et al., 1997; De Ruyter et al., 2005; Gielen et al., 2007; Hillen et al., 2014, 2015).

7.5 Outflows and Jets

Another important structural component is detected only indirectly. This is best illustrated by showing orbital phase-resolved high-resolution spectra (Figure 7.4). The H_α profile turns into a P Cygni profile only when the unseen companion is in front of the luminous primary, which is the post-AGB star. The interpretation of this behaviour is that a high-velocity outflow or jet originates around the companion and is only detected when continuum photons of the primary are scattered out of the line of sight (Thomas et al., 2013; Gorlova et al., 2012, 2015). When a good phase coverage is reached, the wide opening angle as well as the latitudinally dependent velocity law can be deduced from the spectra (Thomas et al., 2013; Bollen et al., 2017). For the objects that are well studied, the outflow velocity is larger in the polar direction, with a high velocity and low density on the jet axes, and a

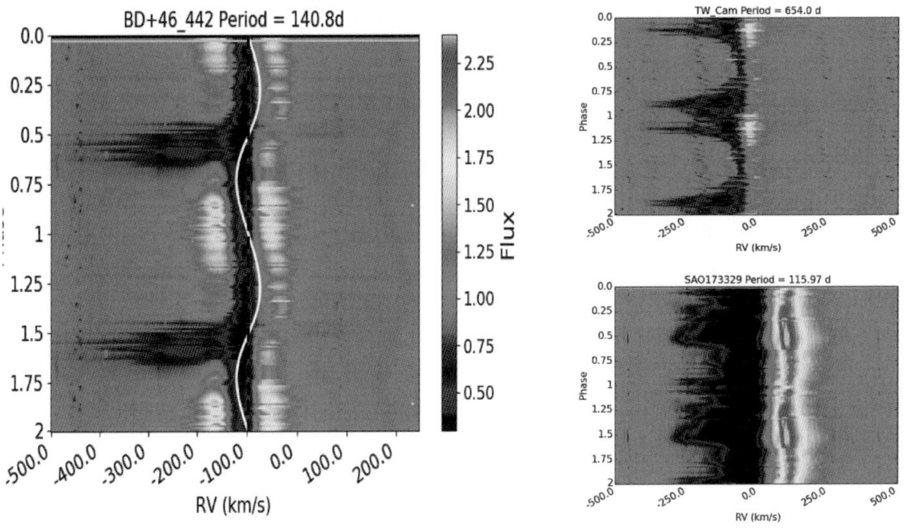

Figure 7.4 Dynamic H-alpha profiles folded on the orbital period. On the left, the data of BD+46°442 (Bollen et al., 2017) folded on its 140.7 days orbital period. The full line represents the movement of the primary. The absorption at high velocity appears when the companion is in front of the primary (phase 0.5). On the right, two other illustrative examples. At the top, the H_α profiles of the RV Tauri pulsator TW Cam are displayed, folded on the orbital period of 654 days. The pulsations are seen in the photospheric lines (Manick et al., 2017). The absorption component appears when the companion is in front (phase 1.0). The bottom shows the profiles of SAO 173329 with an orbital period of 116 days. Here we look almost pole-on, which means the high-velocity outflow is always in the line of sight to the primary. Reproduced with permission from *Astronomy & Astrophysics*, ©ESO.

lower velocity, denser outflow at the conical edge. The deprojected velocity corresponds to the escape velocity of a mainsequence star, rather than from a compact object.

The jets are not strongly collimated, but are launched likely from a circum-companion accretion disc. This disc is seen outside conjunction via the shell profiles of the H_α emission, which can be used to map in velocity space, via Doppler tomography (Bollen et al., 2017). Whether the circum-companion accretion disc is fed by direct mass transfer from the primary or via accretion from the circumbinary disc is not constrained yet.

While modelled and published only in a limited number of systems until now, similar orbital phase-dependent H_α profiles are now commonly observed in post-AGB binaries (see Figure 7.4). When a system is seen nearly pole-on, the P-Cygni line profile is seen at all orbital phases as the wide jet is always in the line of sight. Post-AGB binaries provide ideal test beds to study jet formation and launching mechanisms over a wide range of orbital periods and mass accretion regimes.

7.6 Feedback from Circumbinary Discs

Many post-AGB stars display a chemical anomaly in their photosphere called 'depletion' (Van Winckel, 2003; Rao and Giridhar, 2014; Venn et al., 2014). The abundance trends in

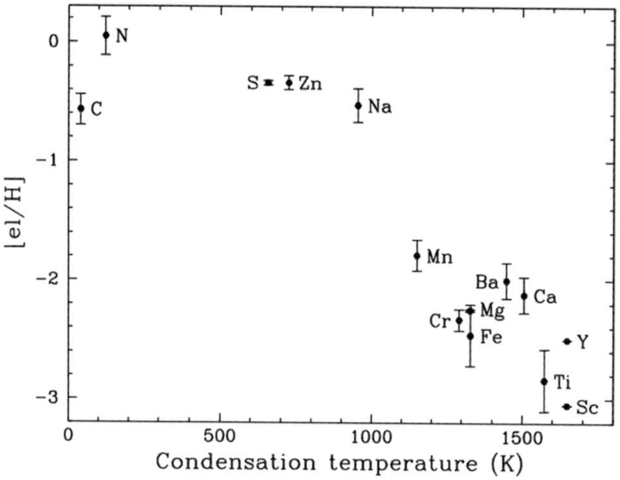

Figure 7.5 A depleted photosphere has a chemical anomaly: the photospheric elemental abundances correlate with the dust condensation temperatures. (The figure gives the abundances of MACHO 82.8405.15 adapted from Gielen et al., 2009. Reproduced with permission from *Astronomy & Astrophysics*, ©ESO).

such atmospheres resemble the gas phase of the interstellar medium: refractory elements are underabundant, while volatiles retain their original abundances (see Figure 7.5). Good tracers of depletion are the [S/Ti] and [Zn/Ti] or [Zn/Fe] abundance ratios. The *s*-process elements are also refractory (Lodders, 2003) so the eventual enrichment of AGB nucleosynthesis products is masked by the depletion process.

The process to acquire this chemical anomaly is not completely understood yet. Dust formation leads to a chemical fractionation in the circumstellar environment. To obtain a depleted photosphere, the radiation pressure on circumstellar dust grains must fractionate the dust from the gas. The cleaned gas is then reaccreted on to the stellar surface.

Such depletion patterns are intimately linked to the presence of a disc around the post-AGB star, and indicate that gas, cleaned from the refractory elements by dust formation, is accreted to the primary (e.g., Giridhar et al., 2005; Gezer et al., 2015, and references therein). The depletion process can be very efficient, and in some cases the original Fe content is reduced to [Fe/H]=-3 to -4, while the nonrefractory elements such as S and Zn retained their original abundances (Van Winckel et al., 1995; Venn et al., 2014). Depleted atmospheres are also found in post-AGB stars in the LMC and SMC (Reyniers and Van Winckel, 2007; Gielen et al., 2009).

The longevity of the disc (Waters et al., 1992) seems to be a needed but not sufficient condition for the process to occur. As shown in a more systematic comparison in Gezer et al. (2015), depleted photospheres are strictly limited to objects with a disc. For some exceptions, there is good observational evidence of a former disc. If the accretion of mass is similar or larger than the envelope reduction due to external mass loss and core growth, the post-AGB lifetime will be lengthened.

7.7 Conclusions

The specific SED characteristics of established binary post-AGB stars allow for a very efficient search for similar disc-SEDs among the evolved optically bright stars. Moreover, detailed and long-term radial velocity monitoring programmes have shown that the central stars with disc-SEDs are indeed binaries with typical orbital periods between 100 and 2,000 days (Figure 7.2), which are orbits least predicted by stellar binary evolution models. No spiralled-in systems are found, despite the orbital sizes that are too small to accommodate an AGB star. While spiralled-in systems are commonly observed in central stars of PNe (e.g., Jones and Boffin, 2017; see also Chapter 8), post-AGB stars with discs evolve on tracks which avoid spiralling-in. Post-AGB binaries are by no means the only evolved binaries with orbits that are not well understood.

The discs around evolved binaries are scaled-up versions of protoplanetary discs, because the ratio of luminosity versus mass is much higher. They have a physically larger dust-free inner region of several astronomical units. It is remarkable that a hydrostatic model dominated by dust sublimation physics can represent the morphology and energetics of the circumbinary disc even better than it does for discs around young stars, for which these models were originally developed (Hillen et al., 2017). Distances are typically poorly constrained, except for the disc sources in the LMC and SMC, but for the latter, orbital properties are as yet not derived (Kamath et al., 2014, 2015). This does hamper the comparison with binary models. Constraining the distances will become possible when Gaia accumulates enough data that enable differentiation between parallax movements and orbital movements.

There is increasing evidence that the presence of a circumbinary disc has a lasting effect on the dynamical evolution of the binary orbit, as resonant coupling between the binary and the disc can pump up the eccentricity (Artymowicz et al., 1991; Lubow, 2010; Dermine et al., 2013; Vos et al., 2015). This provides a way to explain the commonly observed large eccentricities. The integrated effect depends strongly on the mass and longevity of the disc, as well as on the radial mass distribution within the disc. Quantification of this eccentricity-pumping effect is hence only possible when the evolution and timescale of the discs are understood. Moreover, the total angular momentum stored in the discs are, so far, determined in a few systems only (Bujarrabal et al., 2017).

Jets are seen as a major shaping agent in PNe (Balick and Frank, 2002; De Marco and Izzard, 2017) and are thought to shape the multipolar and often complex reflection nebulae in protoplanetary stars (Sahai et al., 2007). Grazing jets (Shiber et al., 2017; Soker, 2017) are potentially important for avoiding common-envelope phase in many binary systems. Despite their importance, jet physics and launching mechanisms are still poorly understood. The optically bright post-AGB binaries offer ideal test beds for jets and their creation processes. The circumcompanion accretion discs where they are launched can be fed either by the primary or by the circumbinary dusty disc or both. The reflection nebulae, which are shell sources, are often very multipolar and structured, but for these, the monitoring data comply only with either single stars, very wide systems or very low-mass companions (Hrivnak et al., 2017). These shell systems are very different from the objects with disc-SEDs.

We conclude that stable circumbinary discs are integral component of many post-AGB binaries and are likely playing a lead role in the evolution of the systems. We can postulate the disc creation, evolution and evaporation play a lead role in all systems with a former AGB star. Detailed studies of the structure and evolution of these discs and their impact on the orbital evolution are therefore badly needed.

Acknowledgements

It is a pleasure to acknowledge the many colleagues, post-docs and PhD students who contributed recently to the research on post-AGB binaries: Christoffel Waelkens, Devika Kamath, Peter Wood, Valentin Bujarrabal, Orsola De Marco, Michel Hillen, Alain Jorissen, Sophie van Eck, Lionel Siess, Onno Pols, Gijs Nelemans, Bruce Hrivnak, Brent Miszalski, Rajeev Manick, Ana Escorza, Shreeya Shetye, Glenn-Micheal Oomen, Dylan Bollen, Ilknur Gezer and Joonas Saario.

References

Abate, C., Pols, O. R., Izzard, R. G., Mohamed, S. S., and de Mink, S. E. 2013. Wind Roche-lobe overflow: application to carbon-enhanced metal-poor stars. *A&A*, **552**, A26.

Arneson, R. A., Gehrz, R. D., Woodward, C. E., et al. 2017. A SOFIA FORCAST Grism study of the mineralogy of dust in the winds of proto-planetary nebulae: RV Tauri stars and SRd variables. *ApJ*, **843**, 51.

Artymowicz, P., Clarke, C. J., Lubow, S. H., and Pringle, J. E. 1991. The effect of an external disk on the orbital elements of a central binary. *ApJ L*, **370**, L35–L38.

Balick, B., and Frank, A. 2002. Shapes and shaping of planetary nebulae. *ARA&A*, **40**, 439–486.

Bolatto, A. D., Simon, J. D., Stanimirović, S., et al. 2007. The Spitzer Survey of the Small Magellanic Cloud: S^3MC imaging and photometry in the mid- and far-infrared wave bands. *ApJ*, **655**, 212–232.

Bollen, D., Van Winckel, H., and Kamath, D. 2017. Jet creation in post-AGB binaries: the circum-companion accretion disc around BD+46°442. *A&A*, **607**, A60.

Bujarrabal, V., Castro-Carrizo, A., Alcolea, J., and Van Winckel, H. 2015. Detection of Keplerian dynamics in a disk around the post-AGB star AC Herculis. *A&A*, **575**, L7.

Bujarrabal, V., Castro-Carrizo, A., Alcolea, J., et al. 2013. ALMA observations of the Red Rectangle, a preliminary analysis. *A&A*, **557**, L11.

Bujarrabal, V., Castro-Carrizo, A., Alcolea, J., et al. 2016. Further ALMA observations and detailed modeling of the Red Rectangle. *A&A*, **593**, A92.

Bujarrabal, V., Van Winckel, H., Neri, R., et al. 2007. The nebula around the post-AGB star 89 Herculis. *A&A*, **468**, L45–L48.

Bujarrabal, V., Castro-Carrizo, A., Alcolea, J., et al. 2017. A second post-AGB nebula that contains gas in rotation and in expansion: ALMA maps of IW Carinae. *A&A*, **597**, L5.

Chen, X., Han, Z., and Tout, C. A. 2011. Tidally enhanced stellar wind: a way to make the symbiotic channel to Type Ia supernova viable. *ApJ L*, **735**, L31.

Cohen, M., Van Winckel, H., Bond, H. E., and Gull, T. R. 2004. Hubble space telescope imaging of HD 44179, the Red Rectangle. *AJ*, **127**, 2362–2377.

De Marco, O., and Izzard, R. G. 2017. Dawes Review 6: The impact of companions on stellar evolution. *PASA*, **34**, 1–40.

De Ruyter, S., Van Winckel, H., Dominik, C., Waters, L. B. F. M., and Dejonghe, H. 2005. Strong dust processing in circumstellar discs around 6 RV Tauri stars: are dusty RV Tauri stars all binaries? *A&A*, **435**, 161–166.

De Ruyter, S., Van Winckel, H., Maas, T., Lloyd Evans, T., Waters, L. B. F. M., and Dejonghe, H. 2006. Keplerian discs around post-AGB stars: a common phenomenon? *A&A*, **448**, 641–653.

Dermine, T., Izzard, R. G., Jorissen, A., and Van Winckel, H. 2013. Eccentricity-pumping in post-AGB stars with circumbinary discs. *A&A*, **551**, A50.

Dermine, T., Jorissen, A., Siess, L., and Frankowski, A. 2009. Radiation pressure and pulsation effects on the Roche lobe. *A&A*, **507**, 891–899.

Deroo, P., Acke, B., Verhoelst, T., et al. 2007. AMBER and MIDI interferometric observations of the post-AGB binary IRAS 08544-4431: the circumbinary disc resolved. *A&A*, **474**, L45–L48.

Deroo, P., van Winckel, H., Min, M., et al. 2006. Resolving the compact dusty discs around binary post-AGB stars using N-band interferometry. *A&A*, **450**, 181–192.

Dullemond, C. P., and Monnier, J. D. 2010. The inner regions of protoplanetary disks. *ARA&A*, **48**, 205–239.

Gezer, I., Van Winckel, H., Bozkurt, Z., et al. 2015. The WISE view of RV Tauri stars. *MNRAS*, **453**, 133–146.

Gielen, C., Bouwman, J., van Winckel, H., et al. 2011. Silicate features in Galactic and extragalactic post-AGB discs. *A&A*, **533**, A99.

Gielen, C., Van Winckel, H., Reyniers, M., et al. 2009. Chemical depletion in the Large Magellanic Cloud: RV Tauri stars and the photospheric feedback from their dusty discs. *A&A*, **508**, 1391–1402.

Gielen, C., Van Winckel, H., Waters, L. B. F. M., Min, M., and Dominik, C. 2007. Dust-grain processing in circumbinary discs around evolved binaries: the RV Tauri spectral twins RU Centauri and AC Herculis. *A&A*, **475**, 629–637.

Giridhar, S., Lambert, D. L., Reddy, B. E., Gonzalez, G., and Yong, D. 2005. Abundance analyses of field RV Tauri stars. VI. An extended sample. *ApJ*, **627**, 432–445.

Gonzalez, G., and Wallerstein, G. 1996. ST Pup: a binary Type II Cepheid with a peculiar chemical composition. *MNRAS*, **280**, 515–536.

Gordon, K. D., Meixner, M., Meade, M. R., et al. 2011. Surveying the agents of Galaxy evolution in the tidally stripped, low metallicity Small Magellanic Cloud (SAGE-SMC). I. Overview. *AJ*, **142**, 102.

Gorlova, N., Van Winckel, H., Gielen, C., et al. 2012. Time-resolved spectroscopy of BD+46^{deg}442: gas streams and jet creation in a newly discovered evolved binary with a disk. *A&A*, **542**, A27.

Gorlova, N., Van Winckel, H., Ikonnikova, N. P., et al. 2015. IRAS 19135+3937: an SRd variable as interacting binary surrounded by a circumbinary disc. *MNRAS*, **451**, 2462–2478.

Han, Z., Podsiadlowski, P., Maxted, P. F. L., and Marsh, T. R. 2003. The origin of subdwarf B stars. II. *MNRAS*, **341**, 669–691.

Han, Z., Podsiadlowski, P., Maxted, P. F. L., Marsh, T. R., and Ivanova, N. 2002. The origin of subdwarf B stars. I. The formation channels. *MNRAS*, **336**, 449–466.
Herwig, F. 2005. Evolution of asymptotic giant branch stars. *ARA&A*, **43**, 435–479.
Hillen, M., de Vries, B. L., Menu, J., et al. 2015. The evolved circumbinary disk of AC Herculis: a radiative transfer, interferometric, and mineralogical study. *A&A*, **578**, A40.
Hillen, M., Kluska, J., Le Bouquin, J.-B., et al. 2016. Imaging the dust sublimation front of a circumbinary disk. *A&A*, **588**, L1.
Hillen, M., Menu, J., Van Winckel, H., et al. 2014. An interferometric study of the post-AGB binary 89 Herculis. II. Radiative transfer models of the circumbinary disk. *A&A*, **568**, A12.
Hillen, M., Van Winckel, H., Menu, J., et al. 2017. A mid-IR interferometric survey with MIDI/VLTI: resolving the second-generation protoplanetary disks around post-AGB binaries. *A&A*, **599**, A41.
Hillen, M., Verhoelst, T., Van Winckel, H., et al. 2013. An interferometric study of the post-AGB binary 89 Herculis. I. Spatially resolving the continuum circumstellar environment at optical and near-IR wavelengths with the VLTI, NPOI, IOTA, PTI, and the CHARA Array. *A&A*, **559**, A111.
Hrivnak, B. J., Lu, W., Bohlender, D., et al. 2011. Are proto-planetary nebulae shaped by a binary? Results of a long-term radial velocity study. *ApJ*, **734**, 25.
Hrivnak, B. J., Van de Steene, G., Van Winckel, H., et al. 2017. Where are the binaries? Results of a long-term search for radial velocity binaries in proto-planetary nebulae. *ApJ*, **846**, 96.
Ivanova, N., Justham, S., Chen, X., et al. 2013. Common envelope evolution: where we stand and how we can move forward. *A&A Rev.*, **21**, 59.
Izzard, R. G., Hall, P. D., Tauris, T. M., and Tout, C. A. 2012 (Aug.). Common envelope evolution. Pages 95–102 of: *IAU Symposium. IAU Symp.*, **283**. Cambridge: Cambridge University Press.
Jones, D., and Boffin, H. M. J. 2017. Binary stars as the key to understanding planetary nebulae. *Nature Astronomy*, **1**, 0117.
Jura, M., Turner, J., and Balm, S. P. 1997. Big grains in the Red Rectangle? *ApJ*, **474**, 741.
Kamath, D., Wood, P. R., and Van Winckel, H. 2014. Optically visible post-AGB/RGB stars and young stellar objects in the Small Magellanic Cloud: candidate selection, spectral energy distributions and spectroscopic examination. *MNRAS*, **439**, 2211–2270.
Kamath, D., Wood, P. R., and Van Winckel, H. 2015. Optically visible post-AGB stars, post-RGB stars and young stellar objects in the Large Magellanic Cloud. *MNRAS*, **454**, 1468–1502.
Kamath, D., Wood, P. R., Van Winckel, H., and Nie, J. D. 2016. A newly discovered stellar type: dusty post-red giant branch stars in the Magellanic Clouds. *A&A*, **586**, L5.
Kraus, S., Hofmann, K.-H., Menten, K. M., et al. 2010. A hot compact dust disk around a massive young stellar object. *Nature*, **466**, 339–342.
Lodders, K. 2003. Solar system abundances and condensation temperatures of the elements. *ApJ*, **591**, 1220–1247.
Lubow, S. H. 2010. Eccentricity growth rates of tidally distorted discs. *MNRAS*, **406**, 2777–2786.
Maas, T., Van Winckel, H., and Waelkens, C. 2002. RU Cen and SX Cen: two strongly depleted RV Tauri stars in binary systems. The RV Tauri photometric b phenomenon and binarity. *A&A*, **386**, 504–516.
Manick, R., Van Winckel, H., Kamath, D., Hillen, M., and Escorza, A. 2017. Establishing binarity amongst Galactic RV Tauri stars with a disc. *A&A*, **597**, A129.

Meixner, M., Gordon, K. D., Indebetouw, R., et al. 2006. Spitzer Survey of the Large Magellanic Cloud: surveying the agents of a Galaxy's evolution (SAGE). I. Overview and initial results. *AJ*, **132**, 2268–2288.

Menu, J., van Boekel, R., Henning, T., et al. 2015. The structure of disks around intermediate-mass young stars from mid-infrared interferometry. Evidence for a population of group II disks with gaps. *A&A*, **581**, A107.

Min, M., Waters, L. B. F. M., de Koter, A., et al. 2007. The shape and composition of interstellar silicate grains. *A&A*, **462**, 667–676.

Molster, F. J., Waters, L. B. F. M., Tielens, A. G. G. M., Koike, C., and Chihara, H. 2002. Crystalline silicate dust around evolved stars. III. A correlations study of crystalline silicate features. *A&A*, **382**, 241–255.

Mulders, G. D., and Dominik, C. 2012. Probing the turbulent mixing strength in protoplanetary disks across the stellar mass range: no significant variations. *A&A*, **539**, A9.

Nie, J. D., Wood, P. R., and Nicholls, C. P. 2012. Predicting the fate of binary red giants using the observed sequence E star population: binary planetary nebula nuclei and post-RGB stars. *MNRAS*, **423**, 2764–2780.

Rao, S. S., and Giridhar, S. 2014. Chemical compositions of RV Tauri stars and related objects. *Rev. Mex. Astron. Astrofis.*, **50**, 49–65.

Raskin, G., van Winckel, H., Hensberge, H., et al. 2011. HERMES: a high-resolution fibre-fed spectrograph for the Mercator telescope. *A&A*, **526**, A69.

Reyniers, M., and Van Winckel, H. 2007. First detection of photospheric depletion in the Large Magellanic Cloud. *A&A*, **463**, L1–L4.

Sahai, R., Morris, M., Sánchez Contreras, C., and Claussen, M. 2007. Preplanetary nebulae: a Hubble space telescope imaging survey and a new morphological classification system. *AJ*, **134**, 2200–2225.

Shiber, S., Kashi, A., and Soker, N. 2017. Simulating the onset of grazing envelope evolution of binary stars. *MNRAS*, **465**, L54–L58.

Soker, N. 2017. Energizing the last phase of common-envelope removal. *MNRAS*, **471**, 4839–4843.

Thomas, J. D., Witt, A. N., Aufdenberg, J. P., et al. 2013. Geometry and velocity structure of HD 44179's bipolar jet. *MNRAS*, **430**, 1230–1237.

Toonen, S., and Nelemans, G. 2013. The effect of common-envelope evolution on the visible population of post-common-envelope binaries. *A&A*, **557**, A87.

van Aarle, E., van Winckel, H., Lloyd Evans, T., Ueta, T., Wood, P. R., and Ginsburg, A. G. 2011. The optically bright post-AGB population of the LMC. *A&A*, **530**, A90.

Van Winckel, H. 2003. Post-Agb stars. *ARA&A*, **41**, 391–427.

Van Winckel, H., Lloyd Evans, T., Briquet, M., et al. 2009. Post-AGB stars with hot circumstellar dust: binarity of the low-amplitude pulsators. *A&A*, **505**, 1221–1232.

Van Winckel, H., Waelkens, C., and Waters, L. B. F. M. 1995. The extremely iron-deficient 'Post-AGB' stars and binaries. *A&A*, **293**, L25–L28.

Venn, K. A., Puzia, T. H., Divell, M., et al. 2014. Searching for dust around hyper metal poor stars. *ApJ*, **791**, 98.

Vos, J., Østensen, R. H., Marchant, P., and Van Winckel, H. 2015. Testing eccentricity pumping mechanisms to model eccentric long-period sdB binaries with MESA. *A&A*, **579**, A49.

Waelkens, C., Van Winckel, H., Waters, L. B. F. M., and Bakker, E. J. 1996. Variability and nature of the binary in the Red Rectangle nebula. *A&A*, **314**, L17–LL20.

Waters, L. B. F. M., Trams, N. R., and Waelkens, C. 1992. A scenario for the selective depletion of stellar atmospheres. *A&A*, **262**, L37–L40.

8

The Importance of Binarity in the Formation and Evolution of Planetary Nebulae

DAVID JONES

8.1 Introduction

Planetary nebulae (PNe) are glowing shells of gas and dust surrounding (pre-)white dwarfs (pre-WD). The canonical formation scenario for PNe is a single star of low to intermediate initial mass, which has recently evolved off the AGB (and is thus a pre–white dwarf), ionising its own slow-moving AGB ejecta that has been swept up into a shell by a fast, tenuous wind originating from the central star (Kwok et al., 1978). This model, frequently referred to as the Interacting Stellar Winds (ISW) model, successfully reproduces a wide range of PN properties; however, it fails to account for the plethora of elaborate, often highly axisymmetric, morphologies observed (Balick and Frank, 2002).

Many additional ingredients to the ISW model have been considered, including rapid rotation on the AGB and strong magnetic fields in single stars (Nordhaus et al., 2007; García-Segura et al., 2014), however only one remains a viable candidate for producing the most axisymmetric structures found in PNe – central star binarity (e.g., Jones and Boffin, 2017a). The idea that some PNe might host binary central stars is not a new one, with such systems being considered key evidence for the existence of the common-envelope (CE) phase first proposed in 1976 by Paczynski (1976) and the discovery of the first post-CE binary central star shortly after (Bond, 1976). However, it is only recently that the importance of binarity, not only in the shaping of PNe but also in their formation, has begun to be truly appreciated.

In a wider context, many binary phenomena experienced by low- and intermediate-mass stars represent evolutionary phases that follow the production of a PN, meaning that PNe with binary central stars offer an important window to study the origins of such objects, including Barium stars, cataclysmic variables, stellar mergers, novae and, perhaps most importantly, the cosmologically important supernovae type Ia.

8.2 Close-Binary Systems

While many forms of binary interaction will influence the evolution of the system, particularly in terms of the stellar mass-loss rates, none will have such a dramatic effect as the occurrence of a CE phase. Similarly, the implications of a CE phase on the large-scale morphology of any resulting PN are perhaps the most easily understood. Via the conservation of angular momentum, the ejected CE will always be preferentially deposited into the binary orbital plane, resulting in a clearly axisymmetric structure. It then stands to reason that the interaction between the ejected envelope and a subsequent wind, originating from the now exposed pre-WD core, will form a similarly axisymmetric structure, providing a natural explanation for bipolar PNe (Nordhaus and Blackman, 2006).

8.2.1 Binary Fraction

Perhaps the clearest obstacle to accepting the importance of central star binarity as the key ingredient in the shaping of PNe has been the lack of a robust measure of the binary fraction amongst the central stars of PNe (CSPNe). Several estimates of the close-binary fraction had been made by compiling the results of multiple surveys, each of which employing different observing techniques with differing sensitivities making accounting for biases impossible (e.g., Bond, 2000), but it was not until the OGLE survey that a more definitive fraction could be derived. Miszalski et al. (2009a) used photometry from the OGLE microlensing survey to search for photometric variability in the central stars of some 149 PNe, deriving a detectable close-binary fraction of 12–21% and more than doubling the sample of known binary CSPN in the process. This figure can be considered a hard lower limit to the true binary fraction given that it only represents the binary fraction detectable by the survey (Jones and Boffin, 2017a), which is highly sensitive to secondary type (lower mass secondaries will go undetected), orbital separation/period (long period, wider binaries will go undetected) as well as apparent magnitude (fainter CSPNe or those contaminated by bright nebulae were not recoverable). However, this fraction is indeed capable of accounting for the most axisymmetric PNe, with a similar fraction of the total PN population showing such bipolar morphologies (Frew et al., 2016).

Surveys based on, for example, radial velocity variability (more sensitive to longer orbital periods, e.g., De Marco et al., 2004) or infrared excesses due to the contribution of a lower temperature companion to the total spectral energy distribution of the CSPN (more sensitive to lower-mass companions, e.g., Douchin et al., 2015) tend to derive much higher binary fractions (approaching 100%) but with far greater uncertainties. The Kepler mission, which provides space-based photometry with a precision impossible for a ground-based survey, offers a unique opportunity to probe further the photometrically detectable binary fraction. The first Kepler field contained only five CSPNe with useable data; however, four of these were found to present variability consistent with binarity (or, rather, binary evolution – one was determined to be a merger product), indicative that an analysis of data from later fields containing more CSPNe may lead to a similarly large binary fraction (De Marco et al., 2015). Consequently, it is now clear that the binary fraction amongst CSPNe is appreciable, at least ~20% but probably even greater. Furthermore, this fraction does not include systems that were initially binary but merged during the CE, the total number of which is rather uncertain. However, there is some evidence that such systems do go on to form PNe with morphologies similar to those that survived the CE (Miszalski et al., 2012b, and references therein). Ultimately, the total number of known post-CE binary CSPNe with solid period determinations is a little more than 50, with all those systems being listed in Table 8.1 (with a more detailed version continually maintained by the author at http://drdjones.net/bCSPN).

Table 8.1 *List of post-CE PNe with confirmed orbital periods (a more detailed version is continually maintained by the author at http://drdjones.net/bCSPN).*

PN G	Common name	Period (days)	Reference
000.2–01.9	M 2-19	0.67	Miszalski et al. (2009a)
000.5–03.1a	MPA J1759-3007	0.50	Miszalski et al. (2009a)
000.6–01.3	Bl 3-15	0.27	Miszalski et al. (2009a)
000.9–03.3	PHR J1801-2947	0.32	Miszalski et al. (2009a)
001.2–02.6	PHR J1759-2915	1.10	Miszalski et al. (2009a)
001.8–02.0	PHR J1757-2824	0.80	Miszalski et al. (2009a)
001.9–02.5	PPA J1759-2834	0.31	Miszalski et al. (2009a)
005.0+03.0	Pe 1-9	0.14	Miszalski et al. (2009a)
005.1–08.9	Hf 2-2	0.40	Hillwig et al. (2016a)
009.6+10.5	Abell 41	0.23	Jones et al. (2010)
017.3–21.9	Abell 65	1.00	Hillwig et al. (2015b)
034.5–06.7	NGC 6778	0.15	Miszalski et al. (2011b)
049.4+02.4	Hen 2-428	0.18	Santander-García et al. (2015)
053.8–03.0	Abell 63	0.46	Afşar and Ibanoğlu (2008)
054.2–03.4	The Necklace	1.16	Corradi et al. (2011)
055.4+16.0	Abell 46	0.47	Afşar and Ibanoğlu (2008)
058.6–03.6	Nova Vul 2007	0.07	Rodríguez-Gil et al. (2010)
068.1+11.0	ETHOS 1	0.53	Miszalski et al. (2011a)
075.9+11.6	AMU 1	2.93	De Marco et al. (2015)
076.3+14.1	Patchick 5	1.12	De Marco et al. (2015)
086.9–03.4	Ou 5	0.36	Corradi et al. (2014)
135.9+55.9	TS01	0.16	Tovmassian et al. (2010)
136.3+05.5	HFG 1	0.58	Exter et al. (2005)
144.8+65.8	LTNF 1	2.29	Ferguson et al. (1999)
-	GK Per	1.90	Bode et al. (1987)
215.6+03.6	NGC 2346	15.99	Brown et al. (2019)
220.3–53.9	NGC 1360	141.6	Miszalski et al. (2018)
222.8–04.2	PM 1-23	1.26	Manick et al. (2015)
242.6–11.6	M 3-1	0.13	Jones et al. (2019)
253.5+10.7	K 1-2	0.68	Exter et al. (2003)
259.1+00.9	Hen 2-11	0.61	Jones et al. (2014)
283.9+09.7	DS 1	0.36	Hilditch et al. (1996)
290.5+07.9	Fg 1	1.20	Boffin et al. (2012)
307.2–03.4	NGC 5189	4.04	Manick et al. (2015)
307.5–04.9	MyCn 18	18.15	Miszalski et al. (2018)
329.0+01.9	Sp 1	2.91	Hillwig et al. (2016b)

Table 8.1 *(cont.)*

PN G	Common name	Period (days)	Reference
332.5–16.9	HaTr 7	0.32	Hillwig et al. (2016c)
335.2–03.6	HaTr 4	1.74	Hillwig et al. (2016a)
337.0+08.4	ESO 330-9	0.30	Hillwig et al. (2016c)
338.1–08.3	NGC 6326	0.37	Miszalski et al. (2011b)
338.8+05.6	Hen 2-155	0.15	Jones et al. (2015)
341.6+13.7	NGC 6026	0.53	Hillwig et al. (2010)
349.3–01.1	NGC 6337	0.17	Hillwig et al. (2010)
354.5–03.9	Sab 41	0.30	Miszalski et al. (2009a)
355.3–03.2	PPA J1747-3435	0.23	Miszalski et al. (2009a)
355.7–03.0	Hen 2-283	1.13	Miszalski et al. (2009a)
357.0–04.4	PHR J1756-3342	0.27	Miszalski et al. (2009a)
357.1–05.3	BMP J1800-3407	0.15	Miszalski et al. (2009a)
357.6–03.3	H 2-29	0.24	Miszalski et al. (2009a)
358.7–03.0	K 6-34	0.20	Miszalski et al. (2009a)
358.8+04.0	Th 3-15	0.15	Soszyński et al. (2015)
359.1–02.3	M 3-16	0.57	Miszalski et al. (2009a)
359.5–01.2	JaSt 66	0.28	Miszalski et al. (2009a)

8.2.2 The Link between Morphology and Central Star Binarity

With the greatly increased number of known binary CSPNe following the work of Miszalski et al. (2009a), it was possible to assess the sample for prevalent morphological characteristics, which might reasonably be used to set them apart from the rest of the PN population (i.e., those where no evidence of binarity was detected). Just as expected from the models of CE ejection, post-CE PNe show a strong tendency to present with canonical bipolar structures (likely more than half are bipolar once inclination effects are accounted for; Miszalski et al., 2009b). This strong connection between bipolarity and central star binarity has since been confirmed beyond doubt – in *every* case, where the two have been derived, the nebular symmetry axis is found to lie perpendicular to the orbital plane of the binary (Hillwig et al., 2016b), including in the Fine Ring Nebula (Sp 1; Figure 8.1), which was initially believed to be spherical but has since been shown to be a near pole-on bipolar structure (Jones et al., 2012).

Intriguingly, Miszalski et al. (2009b) also found a strong correlation between central star binarity and the presence of equatorial rings/torii, jets and low-ionisation filamentary structures (see Figures 8.2 and 8.3). Equatorial torii (e.g., Abell 41, Jones et al., 2010) and knotty waists (e.g., the Necklace, Corradi et al., 2011) can clearly be explained under the same prescription as the bipolar structures, if the amount of CE material deposited in the orbital plane is sufficiently high (knots can arise from the interaction of a photoionising wind with an inhomogeneous, clumpy torus; Miszalski et al., 2009b). Similarly, jets are a

Figure 8.1 Image of PN Shapley 1 (image credit: ESO). Upon first inspection, the PN appears to present a roughly spherical structure but is, in fact, a near-pole-on bipolar structure, the inclination of which is consistent with that of its central binary star (Jones et al., 2012; Hillwig et al., 2016b).

Figure 8.2 Image of PN Fleming 1 showing its remarkable pair of precessing jets, which are thought to have been produced by an intense episode of mass transfer just prior to the CSPN entering the CE phase (Boffin et al., 2012). Image credit: ESO/H. Boffin.

Figure 8.3 Image of PN NGC 6326 highlighting the filamentary structures found by Miszalski et al. (2009b) to be typical of central star binarity (Miszalski et al., 2011b). Image credit: ESA/Hubble and NASA.

natural result of mass transfer between the binary components, and while most hydrodynamical models predict that accretion onto the secondary is minimal during the CE (Ricker and Taam, 2012; MacLeod and Ramirez-Ruiz, 2015), there is now growing evidence that a significant amount of material might be accreted before entering into the CE (see Section 8.2.4). Low-ionisation filaments not aligned in either the polar or equatorial directions, such as those of NGC 6326 (see Figure 8.3 and Miszalski et al., 2011b), are perhaps more difficult to understand, but are thought to form through a wind–shell interaction whereby modest inhomogeneities in the shell (which would, in these systems, be the ejected CE) are enhanced by the influence of an encroaching fast, low-density wind (Steffen et al., 2013).

In order to yet further increase the known binary CSPN population, whilst minimising the required investment of observing time, these newly identified morphological features have all been used to select PNe for targeted follow-up leading to an accelerated rate of discovery (e.g., Miszalski et al., 2011a,b; Jones et al., 2014, 2015, 2019). As the sample of known binary stars has grown (as well as the sample of those that have been the subject of detailed studies based on multiband photometric and radial velocity observations), several interesting surprises have been revealed, some of which will be discussed in Sections 8.2.3 and 8.2.4.

8.2.3 The Link to Type Ia Supernovae

Perhaps the most intriguing result that can be derived from the current sample of known post-CE CSPNE is the apparent overabundance of binary systems comprised of two white dwarfs, with approximately 20% of all known post-CE CSPNe found to be such double degenerate (DD) binaries (Jones and Boffin, 2017a). This high fraction of post-CE DD CSPNe is at odds with population synthesis models, perhaps indicating that the expected birth rate of such systems could be much higher than previously suspected (De Marco et al., 2008). Such a finding has important consequences for our understanding of the CE phase given that many of these systems would have had to survive two CEs (Tovmassian et al., 2010). Under the classical prescription of the CE, whereby some fraction (α) of the orbital energy is used to unbind and eject the envelope of the giant star (De Marco et al., 2011), it stands to reason that surviving two such events presents more of a challenge. For low-mass companions, this prescription shows that orbital energy alone would not be enough to eject the envelope during even the first CE, indicating that some other energy source, thought by some to be the recombination energy of the envelope itself, is required (Nandez and Ivanova, 2016). For more massive companions, the ejection of the first CE is easily understood, but how the resulting spiralled-in binary could survive a second CE is more challenging to explain, but cannot be excluded given that the energy released for a given change in separation is much greater at shorter periods. It is, however, problematic that, due to the now shortened orbital period, the second CE is significantly more likely to occur when the companion is ascending the RGB (rather than the AGB), when its envelope is more bound. It has been suggested that, under certain circumstances, the first CE can be avoided with the initially more massive component of the binary losing the majority of its envelope via a period of stable, nonconservative mass transfer, and the system then evolving to DD through a single CE (Woods et al., 2012). Similarly, a grazing envelope (GE) evolution, whereby the giant radius and orbital separation shrink simultaneously, could also provide a means to unbind the envelope of either (or both) star(s), allowing the system to become DD before merging (Soker, 2015).

Studying DD CSPNe offers a unique window into the formation of these systems, as the presence of a PN indicates that they have only recently left the final (or perhaps only) CE phase of their evolution, and that the system has not yet had the time to adjust thermally. Furthermore, understanding the formation processes for DD binaries has great implications in the study of supernovae type Ia (SNe Ia) given that the merger of such systems is one of the proposed pathways for their formation (Iben and Tutukov, 1984). Indeed, several recent studies have suggested that a DD merger may be responsible for most, if not all, SNe Ia (Maoz et al., 2014; Maoz and Hallakoun, 2017), an hypothesis that is perhaps supported most strongly by the observed delay-time distribution of observed SNe Ia (Maoz and Badenes, 2010). Understanding the origins of these SNe Ia is of critical importance given their role in the discovery of the accelerated expansion of the Universe (Reiss et al., 1998; Schmidt et al., 1998; Perlmutter et al., 1999), which was ultimately awarded the 2011 Nobel Prize in Physics. However, in spite of their cosmological importance, it is still unclear why SNe Ia present with the same brightness (once corrected for the shape of their post-explosion light curves; Phillips, 1993), and hence how they can be such astonishingly precise standard candles.

The best candidate SN Ia progenitor found to date is the CSPN of Hen 2-428, shown by Santander-García et al. (2015) to comprise a DD binary, the total mass of which is above the Chandrasekhar limit and that will merge in less than a Hubble time. Other DD CSPNe are also found to have short merging times but with a total mass less than the Chandrasekhar mass (e.g., TS 01; Tovmassian et al., 2010). These may not result in classical SNe Ia, but will clearly be observed as transients and perhaps produce stellar classes such as R Coronae Borealis stars (Clayton et al., 2011). Beyond the DD scenario for SN Ia formation, it is important to note that some of the strongest candidates for the single degenerate (SD) pathway, whereby a massive WD reaches the Chandrasekhar mass via accretion from a companion, are also found inside PNe (e.g., Nova Vul 2007; Rodríguez-Gil et al., 2010). Finally, a significant fraction of SNe Ia is found to explode in circumstellar environments consistent with the presence of a previously ejected PN (Tsebrenko and Soker, 2015), indicating that, irrespective of the dominant evolutionary pathway towards SNe Ia, CSPNe will be key to understanding their formation (as well as those of other astrophysical transients and merger products).

8.2.4 Pre–Common-Envelope Mass Transfer

Further to the apparent overabundance of DD systems, studies of the single degenerate systems – those comprised of a WD and a main-sequence (MS) star – have also brought surprising results. All systems with an MS secondary, that have been the subject of detailed modelling based on both light and radial velocity curves, are found to present with companions that are inflated with respect to their field star counterparts (Jones et al., 2015). While the total number of systems modelled is rather low, this is still a striking result given the levels of inflation (in some cases, the measured radii are a factor of two or more greater than those predicted by zero age main-sequence models; see Table 8.2). It is now generally accepted that this inflation is a result of rapid mass transfer either during or just prior to the CE phase. This rapid mass transfer knocks the secondaries out of thermal equilibrium, causing their radii to grow, and given that their thermal timescales (millions of years) are much larger than the timescales for CE ejection and PN visibility (tens of thousands of years), they remain out of equilibrium. The high levels of irradiation in these systems, as highlighted by the tendency of these MS stars to present with higher temperatures than single-star counterparts (see Table 8.2), probably also act to maintain the secondaries out of thermal equilibrium but are unlikely to be the primary cause of the inflation itself (De Marco et al., 2008).

Further support for this hypothesised phase of intense mass transfer comes from the observed jet-like structures in many post-CE PNe. Jets are believed to be a natural consequence of accretion and, as such, the clear connection between these structures and central

Table 8.2 *Parameters of main-sequence secondary stars in post–common-envelope planetary nebula central stars (Jones et al., 2015, and references therein).*

PN	Mass ($M_\odot$)	Radius ($R_\odot$)	Temperature (kK)
Abell 46	0.15 ± 0.02	0.46 ± 0.02	3.9 ± 0.4
Abell 63	0.29 ± 0.03	0.56 ± 0.02	6.1 ± 0.2
Abell 65	0.22 ± 0.04	0.41 ± 0.05	5.0 ± 1.0
DS 1	0.23 ± 0.01	0.40 ± 0.01	3.4 ± 1.0
Hen 2-155	0.13 ± 0.02	0.30 ± 0.03	3.5 ± 0.5
LTNF 1	0.36 ± 0.07	0.72 ± 0.05	5.8 ± 0.3

star binarity can be considered a strong indication that significant accretion has occurred (Miszalski et al., 2009b; Tocknell et al., 2014; Sowicka et al., 2017). The secondary in the Necklace nebula (which shows prominent jet-like structures) has been shown to be a carbon dwarf, presenting significant contamination of AGB processed material that must have been accreted from the evolved primary (Miszalski et al., 2013a; further discussion of chemical contamination in binary CSPNe can be found in Section 8.3.1). Additionally, the kinematical ages of the jets (including those of the Necklace) also tend to be older than those measured for the central regions of these PNe, suggesting that the jets are launched prior to entering the CE (Corradi et al., 2011). This chronology is supported by observations of Fleming 1, where the precession period of the jets indicates that the orbital period of the binary at the time of ejection was much larger than the current post-CE period (i.e. before the CE in-spiral; Boffin et al., 2012). In a few cases, there are suggestions that the jets are formed after the ejection of the CE (or, in the case of some proto-PNe, contemporaneously), perhaps indicative of multiple mass transfer episodes or even fallback (Tocknell et al., 2014; Jones et al., 2016; Jones and Boffin, 2017a). This idea will be revisited in Section 8.4.2.

Many binary CSPNe with MS companions have been observed to be X-ray point sources (Kastner et al., 2012; Freeman et al., 2014), with most displaying relatively hard spectra that are unlikely to arise due to photospheric emission from the host CSPNe themselves (Montez Jr. et al., 2015). Instead, the emission is thought to originate from coronal emission from the spun-up main-sequence companion (Montez Jr. et al., 2010). Given the evolved nature of these systems, this activity is almost certainly a consequence of binary interaction (magnetic activity in late-type main-sequence stars decreases with age; Soker and Kastner, 2002), but it is unclear whether the spin-up is due to accretion or merely tidal locking in these short-period systems (Montez Jr. et al., 2015).

8.3 Non-Post–Common-Envelope Systems

Section 8.2 focussed on post-CE PNe, however even binaries that avoid the CE might be expected to have a significant impact on the formation of a PN. Indeed, some authors have even gone so far as to claim that the entire PN population may derive from binary stars, once wider binaries are included, and that most single stars may be precluded from forming an observable PN (e.g., Moe and De Marco, 2006). This idea is somewhat supported by the high binary fractions derived from methodologies sensitive to such wider binaries (see Section 8.2.1); however, more recent models of post-AGB evolution strongly indicate that single stars should be capable of producing observable PNe (Miller Bertolami, 2016).

In the following subsections, we will discuss some special cases amongst the PNe known to host wide-binary central stars, and their implications for the importance of such systems in the PN population as a whole.

8.3.1 Barium Stars

Section 8.2.4 discussed mass transfer as part of the CE, but several binary configurations will experience stable mass transfer and avoid the orbital shrinkage associated with a CE. One such class of objects are the barium stars, consisting of a WD and a cool (spectral type G to K), giant companion that has been chemically enriched with barium and other s-process elements (McClure et al., 1980). Given that the companions are not evolved enough to have produced such significant quantities of s-process elements, the chemical enrichment is believed to be the consequence of accreting s-process enriched material from the primary, most likely via wind Roche-lobe overflow (Boffin and Jorissen, 1988; Theuns et al., 1996). The wind Roche-lobe overflow mechanism – whereby, while the primary never fills its Roche lobe, its stellar wind does – allows for greatly increased accretion rates compared to the classical Bondi–Hoyle–Lyttleton accretion prescription. If the acceleration radius of the primary's wind is comparable or greater than its Roche-lobe radius, the wind will be strongly influenced by the binary potential and be accreted onto the secondary via the first Langrangian point.

Several barium star CSPNe have been discovered (e.g., Miszalski et al., 2012a, 2013b) though, to date, only one has a measured orbital period (LoTr 5 at a somewhat typical period for this class of object, $\sim$7.5 years; Aller et al., 2018). In some cases, the rotation rate of the giant secondary has been measured via photometric variability, indicating that they are highly spun up by the mass transfer (rotation periods of the order of a few days; Tyndall et al., 2013). Furthermore, analyses of the PNe surrounding these barium stars shows a tendency towards toroidal structures (Tyndall et al., 2013), indicative that mass loss through the system is also enhanced through the second and third Langrangian points, as predicted by smoothed particle hydrodynamics simulations of wind Roche-lobe overflow (Mohamed and Podsialowski, 2012). In this regard, Ba CSPNe offer a unique window into this poorly understood mass transfer process, given that the observed PN represents the remnant of the wind that escaped the binary potential.

8.3.2 Long-Period Radial Velocity Variables

A small number of long-period binary CSPNe have been discovered through long-term radial velocity monitoring campaigns. These sorts of campaigns have only been made possible by the new generation of high-stability, high-resolution spectrographs that permit the long-term monitoring of stellar sources with a precision of a few metres per second. Given the need for continued access to such facilities over long time periods, surveys for such long-period CSPNe have generally been limited to smaller telescopes, greatly restricting the number of sufficiently bright candidates. However, in spite of these limitations, one survey with the HERMES spectrograph mounted on the 1.2 m Mercator Telescope has revealed three long-period binary CSPNe with periods in the range 3 to 10 years (Van Winckel et al., 2014; Jones et al., 2017). In all three cases, the brightest component of the binary is the secondary star (in most cases a giant, though there is some doubt over the nature of the secondary of NGC 1514; Jones et al., 2017), providing some indication that these systems represent only the 'tip of the iceberg' when it comes to such long-period CSPNe, and that similar studies with larger telescopes (thus probing fainter binaries) should unearth many more such systems. Furthermore, the CSPN of the longest period of these systems, that of NGC 1514, was previously the subject of a long-term monitoring campaign lasting almost one year but failed to reveal its binary nature due to the its extreme eccentricity (the one-year campaign was unfortunate enough to observe the CSPN at phases when the variability was at a minimum; Jones et al., 2017). In light of this, previous monitoring campaigns that have found few variables cannot be considered conclusive, as many long-period systems would evade detection (either due to the velocity precision of the instrumentation used or the duration of the monitoring programmes). This, compounded with other surveys that have indicated a large number of radial velocity variable CSPNe but which failed to find periodicities (e.g., De Marco et al., 2004), is strongly indicative of a large number of long-period binary CSPNe that have, to date, evaded detection.

Given that there clearly should be an appreciable number of long-period binary CSPNe, the question becomes: does the evolution of these binaries vary significantly from isolated field stars, and what does that mean for the formation and/or evolution of any PNe resulting from theses systems? In the orbital period range of the three known systems, it is unlikely that any has experienced direct Roche-lobe overflow (i.e., one component filling its Roche lobe). However, they are clearly in the regime where wind Roche-lobe overflow could have an appreciable effect. The observed morphologies of the host PNe support the idea that their binary nature has significantly influenced the mass-loss evolution of the systems, with two of the three showing clearly axisymmetric, bipolar morphologies (NGC 1514 and LoTr 5; Jones et al., 2017, and references therein), while the morphology of the third (PN G052.7+50.7) is unclear due to its rather small extent (5″, Napiwotzki et al., 1994). NGC 1514 in particular, which hosts the longest period CSPN at $\sim$9 years, presents with a remarkable pair of dust rings (see Figure 8.4) somewhat similar to those observed in nebulae surrounding other binary phenomena such as symbiotic stars (e.g., Hen 2-104, the Southern Crab; Santander-García et al., 2004) and even the type II SN 1987A (Morris and Podsialowski, 2007). Additionally, two of the three long-period CSPNe show stellar chemical abnormalities clearly associated with a binary evolution – the CSPN of LoTr 5 is a barium star (see Section 8.3.1) while the CSPN of PN G052.7+50.7 shows a depletion

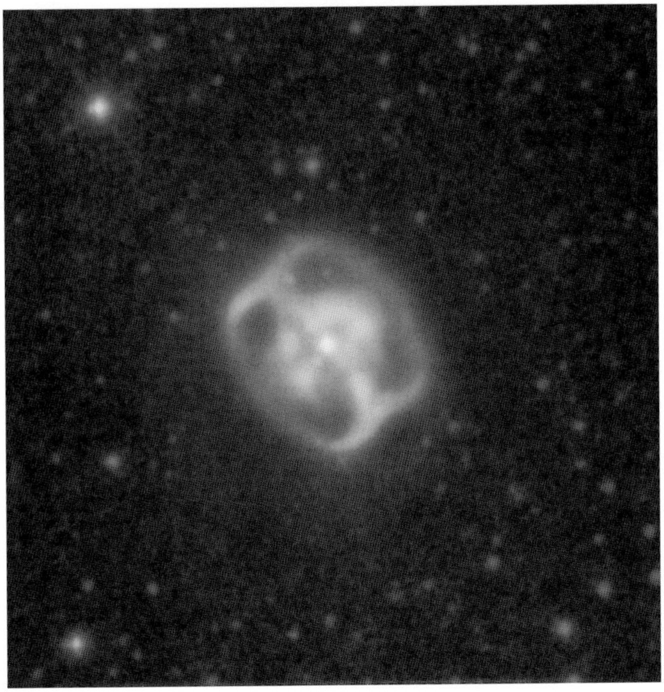

Figure 8.4 Image of the infrared emission from PN NGC 1514 highlighting the prominent rings, believed to represent the ends of a bipolar, hourglass-like structure – typical of the influence of a central binary star. Image credit: NASA/JPL-Caltech/Wise team.

in refractory elements similar to that found in many post-AGB binaries with circumbinary discs (Napiwotzki et al., 1994; Van Winckel et al., 2014, and references therein).

8.3.3 Resolved Companions

Beyond those few wide binaries for which orbital periods have been derived directly, there are also several CSPNe for which distant companions have been resolved. Ciardullo et al. (1999) observed 113 PNe as part of a Hubble Space Telescope 'snapshot' survey, finding 10 CSPNe with stars so close to their nuclei that their associations are very likely while a further six have possible associations (i.e., a resolved binary fraction of ~10%). The measured separations of these candidates equate to orbital periods roughly in the range of several hundreds to hundreds of thousands of years, a regime in which most are likely to have experienced little to no interaction even during the nebular progenitors' AGB ascent. In some cases, this hypothesis is supported by the observed morphologies of the host nebulae, with several presenting (near-)spherical morphologies (e.g., Abell 33, K1-14) and observable haloes as predicted by single-star models (e.g., NGC1535; Corradi et al., 2003). The majority, however, present aspherical morphologies, some clearly as a result of interaction with the interstellar medium (e.g., Abell 31; Wareing et al., 2007), but most are better explained by a binary interaction. It is perhaps possible that those with the shortest periods may be close enough to have experienced somewhat enhanced mass-loss rates via the so-called companion-reinforced attrition process (CRAP; Tout and Eggleton, 1988) or perhaps even some wind Roche-lobe overflow (see Section 8.3.1), but most would be considered too wide. This raises the intriguing possibility that these systems are (or were), in fact, triple stars wherein the PN progenitor forms a closer binary with an unresolved companion while the resolved companion is the third component of a hierarchical triple (Bear and Soker, 2017). Interestingly, in the only known triple CSPN, all three components are resolvable from the ground using adaptive optics imagery and, as such, the progenitor is unlikely to have experienced any strong binary interaction (Adam and Mugrauer, 2014). To date, no system where the nebular progenitor forms part of a close binary inside a hierarchical triple has been clearly detected (Jones and Boffin, 2017b) – perhaps the best remaining candidate is Abell 63, which is known to host a post-CE binary central star with a possible third companion at around 3,500 au separation; however, high-contrast imaging and astrometry are required to confirm the association (Ciardullo et al., 1999; Adam and Mugrauer, 2014).

8.4 Chemistry

8.4.1 Dual-Dust Chemistry

The dust chemistry of PNe is dominated by the interplay between carbon and oxygen, with the less abundant species locked away in the form of carbon monoxide leaving the more abundant to drive the nebular dust chemistry. Therefore, the majority of PNe are found to display either carbon- or oxygen-rich chemistry in the form of polycyclic aromatic hydrocarbons (PAHs, in the case of carbon-rich dust) or silicates (in the case of oxygen-rich dust). However, a significant fraction of PNe are found to display so-called dual-dust chemistry wherein both PAH and silicate emission features are observed in their spectra. It had been suggested that these objects may be transition objects, in a phase where the star is transitioning from O-rich to C-rich (following a thermal pulse, for example; Guzman-Ramirez et al., 2015), however the discovery of several dual-dust PNe in the Galactic bulge (where the population of old, low mass stars is not expected to have experienced significant third dredge-up and therefore should not present with the necessary enhanced C/O ratios) throws this hypothesis into doubt (Perea-Calderón et al., 2009; García-Rojas et al., 2018). Guzman-Ramirez et al. (2011) demonstrated that PAHs could be formed in oxygen-rich environments via the photodissociation of carbon monoxide in a UV-irradiated, dense torus, such as those associated with post-CE binary CSPNe (see Section 8.2.2). Mid-infrared imaging of dual-dust PNe supports this hypothesis, with PAH emission being restricted to the denser equatorial regions (Guzman-Ramirez et al., 2014).

The strong association between post-CE nebular morphologies (equatorial rings) and dual-dust chemistry provides a clear indication that complex molecular material can be formed almost immediately following the CE phase. This is particularly interesting in the context of the discovery of circumbinary exoplanets around post-CE systems (e.g., NN Ser; Marsh et al., 2014), demonstrating that it is not necessary for these exoplanets to have survived the CE but rather that they may have formed from the remnant of the ejected envelope (Schleicher et al., 2015). Bearing this in mind, the name PN may not be such a misnomer given that the nebular material may go on to form second-generation planets around post-CE cores.

8.4.2 The Abundance Discrepancy Problem

The spectra of PNe present a plethora of emission lines, both as a result of recombination and collisional excitation, from a wide range of elements. The intensities of these emission lines vary as a function of both the abundance of that particular chemical species and the physical conditions (electron density and temperature) of the emitting gas. However, for more than 70 years there has been a known discrepancy between the chemical abundances calculated using lines of the same species but with different emission mechanisms (Wyse, 1942). This discrepancy has generally been characterised by an abundance discrepancy factor (ADF) defined as

$$\mathrm{ADF}(X^{+i}) = \frac{X^{+i}_{\mathrm{RLs}}}{X^{+i}_{\mathrm{CELs}}} \tag{8.1}$$

where $\mathrm{ADF}(X^{+i})$ is the ADF of a given ion X^{+i}, and X^{+i}_{RLs} and X^{+i}_{CELs} are the ionic abundances of that species derived using recombination lines (RLs) and collisionally excited lines (CELs), respectively. The value of the ADF is almost universally, across all forms of astrophysical nebulae and irrelevant of the chosen ion, determined to be greater than unity

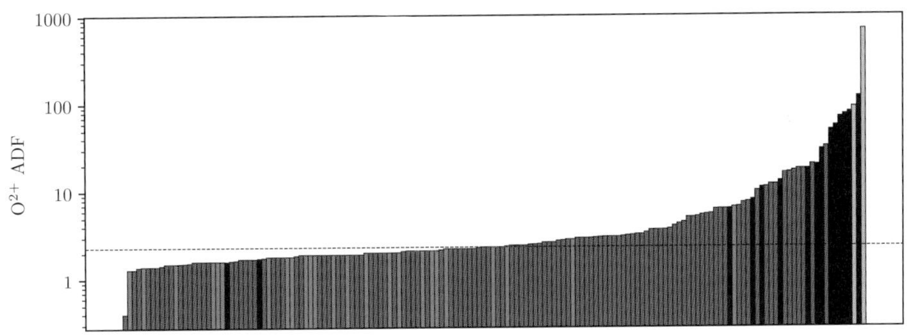

Figure 8.5 A compilation of O^{2+} ADFs from the literature including HII regions (light gray), PNe (gray) PNe with close binary CSPNe (dark gray) and PNe with born-again CSPNe (lightest gray). The horizontal dashed line marks the median value of all measured ADFs at 2.3, this value becomes 1.9 when considering only HII regions and rises to 2.6 for PNe alone. Based on data compiled by Roger Wesson (www.nebulousresearch.org/adfs/).

(see Figure 8.5; and McNabb et al., 2013), representing one of the greatest unresolved problems in nebular astrophysics today (Peimbert et al., 2017). Recently, the issue has been further complicated by the discovery that the ADFs measured in PNe with post-CE CSPNe far exceed those of the general population of ionised nebulae (although there are a few exceptions, e.g., IC 4776; Sowicka et al., 2017), with values reaching up to 800 in the central regions of some objects (Liu et al., 2006; Corradi et al., 2015; Wesson et al., 2018).

Several explanations have been put forward to explain the ADF problem, including various forms of inhomogeneity (temperature, chemical and density), non-Maxwellian electron velocity distributions and time-dependent photoionisation variations, with the true cause probably comprising some combination of all these factors (see, for example, the review of Peimbert et al., 2017). While the root cause across all ionised nebulae cannot have a binary origin (HII regions do not have an associated binary), it is clear that the impact of at least one of the proposed mechanisms driving the discrepancy is greatly amplified by a close-binary evolution in the CSPN. Spatially resolved ADF studies have indicated that in post-CE PNe there does appear to be a strong chemical inhomogeneity, with centrally peaked ADFs being associated with the presence of a second, higher metallicity (and, therefore, lower temperature) gas phase at the core of these nebulae (García-Rojas et al., 2016; Jones et al., 2016). Alongside the post-CE PNe, the highest ADF values are found in born-again PNe – those systems where it is believed that the CSPN underwent a very late thermal pulse, ejecting chemically enriched material into the old nebula (Wesson et al., 2008). This evolutionary scenario offers a clear explanation for the presence of a second gas phase in the inner regions of these nebulae, and perhaps offers insight into the origins of the high ADFs in post-CE PNe given that these born-again PNe are also strongly suspected to host binary CSPNe (Wesson et al., 2008). Indeed, it has been suggested that chemical reprocessing of CE material that has fallen-back onto the CSPN could prompt a form of born-again evolution in the post-CE PNe, offering a formation mechanism for the chemically enriched, second gas phase in these nebulae (Jones et al., 2016). Further exploration of this hypothesis is essential to understand the evolution of these systems through and beyond the CE, with spatiokinematical observations in both optical recombination lines (ORLs) and CELs combined with hydrodynamic simulations essential to evaluate the possible formation scenarios of these post-CE PNe and the origin of their high ADFs.

8.5 Summary

Central star binarity has long been thought to play a role in the shaping of PNe, but it is only now that the magnitude of its importance is becoming clear. Some 20% of all PNe host post-CE CSPNe that are detectable from the ground via photometric monitoring, with those nebulae showing marked morphological signatures of their binary evolution (bipolar structures, jets and equatorial rings). In *all* cases, where both are known, the nebular symmetry axes are found to lie perpendicular to the orbital plane of the binary, just as predicted by models of PN formation by binary interaction – the probability of finding such a correlation by chance is less than one in a million. Surveys employing different methodologies (radial velocity monitoring, infrared excesses, space-based photometry) indicate that this 20% binary fraction may well be a dramatic underestimate, with some determinations (that will include contributions from non-post-CE CSPNe) consistent with a near-100% fraction. It is also clear that binary systems that avoid a CE evolution can still have a dramatic impact on the resulting PNe, with those hosting long-period (years to tens of years) CSPNe also showing marked departures from sphericity, likewise those few PNe suspected to host the products of CE mergers. Even PNe hosting resolved binaries, with separations so wide that interaction should have been minimal, seem to show morphologies that are difficult to explain in a single-star scenario. There are still many open questions but it is now clear that central star binarity plays a crucial role in the evolution of a significant fraction of, and perhaps even all, PNe.

PNe with binary central stars offer an important window into a multitude of astrophysical processes, including the common-envelope phase, wind Roche-lobe overflow, launching of jets around stellar mass objects and the chemical contamination of stellar atmospheres, to name just a few. They also represent the progenitors of a wide range of astronomical phenomena, including chemically peculiar stars (such as barium stars and carbon stars), cataclysmic variables, novae, SNe Ia (irrelevant of their formation mechanism), R Coronae Borealis stars, low-mass X-ray binaries and, most likely, many other astronomical transients. Studies of PNe clearly have much to inform on the formation and evolution of these phenomena. PNe may yet have the final word in the debate over the formation of circumbinary exoplanets around post-CE systems, with first-generation planets having to survive the PN phase, and second-generation planets perhaps being formed from material associated with the PN itself.

Perhaps the most obvious role that PNe can play in a wider context is in forwarding our understanding of the CE phase. Studies have already revealed several aspects to the phase that will be critical in moving forward. The presence of jets, as well as inflated and chemically polluted main-sequence secondaries, strongly points to a significant amount of material accreted onto the companion, most likely just prior to entering the CE phase. The broad-scale morphologies of post-CE PNe themselves offer a unique opportunity to study the structure of the ejected envelope as the nebula itself is formed from this material. Post-CE CSPNe represent the immediate products of the CE phase, with the PN ensuring that the binary systems are 'fresh out of the oven', and, as such, their mass and period distributions are clearly those that CE population synthesis studies should be required to match. The abundance patterns in post-CE PNe, in particular the high ADFs and the distribution and kinematics of the CEL and RL emitting gas, have much to inform on the CE process, perhaps indicating that the final stages of the process result in the ejection of chemically enriched material into the already dissipating envelope. Understanding the CE and its products represent a critical issue in modern astrophysics, and a crucial aspect in our understanding of all binary evolution and particularly in the production of detectable gravitational wave sources as well as cosmologically important SNe Ia.

References

Adam, C., and Mugrauer, M. 2014. HIP 3678: a hierarchical triple stellar system in the centre of the planetary nebula NGC 246. *MNRAS*, **444**, 3459.

Afşar, M., and Ibanoğlu, C. 2008. Two-colour photometry of the binary planetary nebula nuclei UU Sagitte and V477 Lyrae: oversized secondaries in post-common-envelope binaries. *MNRAS*, **391**, 802.

Aller, A., Lillo-Box, J., Vučković, M., et al., 2018, *MNRAS*, **476**, 1140

Balick, B., and Frank, A. 2002. Shapes and shaping of planetary nebulae. *ARA&A*, **40**, 439.

Bear, E., and Soker, N. 2017. Planetary nebulae that cannot be explained by binary systems. *ApJ*, **837**, 10.

Bode, M. F., Roberts, J. A., Whittet, D. C. B., Seaquist, E. R., and Frail, D. A. 1987. An ancient planetary nebula surrounding the old nova GK Persei. *Nature*, **329**, 519.

Boffin, H. M. J., and Jorissen, A. 1988. Can a barium star be produced by wind accretion in a detached binary? *A&A*, **205**, 155.

Boffin, H. M. J., Miszalski, B., Rauch, T., et al. 2012. An interacting binary system powers precessing outflows of an evolved star. *Science*, **338**, 773.

Bond, H. E., 1976. Objects common to the catalogue of galactic planetary nebulae and the general catalogue of variable stars. *PASP*, **88**, 192.

Bond, H .E. 2000. Binary central stars of planetary nebulae. *ASPC*, **199**, 115.

Brown, A.J., Jones, D., Boffin, H.M.J., and Van Winckel, H., 2019, *MNRAS*, **482**, 4951.

Ciardullo, R., Bond, H. E., Sipior, M. S., et al. 1999. A Hubble Space Telescope survey for resolved companions of planetary nebula nuclei. *AJ*, **118**, 488.

Clayton, G. C., Sugerman, B. E. K., Stanford, S. A., et al. 2011. The circumstellar environment of R Coronae Borealis: white dwarf merger or final-helium shell flash? *ApJ*, **743**, 44.

Corradi, R. L. M., García-Rojas, J., Jones, D., and Rodríguez-Gil, P. 2015. Binarity and the abundance discrepancy problem in planetary nebulae. *ApJ*, **803**, 99.

Corradi, R. L. M., Rodríguez-Gil, P., Jones, D., et al. 2014. The planetary nebula IPHASXJ211420.0+434136 (Ou5): insights into common-envelope dynamical and chemical evolution. *MNRAS*, **441**, 2799.

Corradi, R. L. M., Sabin, L., Miszalski, B., et al. 2011. The Necklace: equatorial and polar outflows from the binary central star of the new planetary nebula IPHASX J194359.5+170901. *MNRAS*, **410**, 1349.

Corradi, R. L. M., Schönberner, D., Steffen, M., and Perinotto, M. 2003. Ionized haloes in planetary nebulae: new discoveries, literature compilation and basic statistical properties. *MNRAS*, **340**, 417.

De Marco, O. 2009. The origin and shaping of planetary nebulae: putting the binary hypothesis to the rest. *PASP*, **121**, 316.

De Marco, O., Bond, H. E., Harmer, D., and Fleming, A. J. 2004. Indications of a large fraction of spectroscopic binaries among nuclei of planetary nebulae. *ApJ*, **602**, 93.

De Marco, O., Hillwig, T. C., and Smith, A. J. 2008. Binary central stars of planetary nebulae discovered through photometric variability. I. What we know and what we would like to find out. *AJ*, **136**, 323.

De Marco, O., Long, J., Jacoby, G. H., et al. 2015. Identifying close binary central stars of PN with Kepler. *MNRAS*, **448**, 3587.

De Marco, Passy, J.-C., Moe, M., et al. 2011. On the α formalism for the common envelope interaction. *MNRAS*, **411**, 2277.

Douchin, D., De Marco, O., Frew, D. J., et al. 2015. The binary fraction of planetary nebula central stars. II. A larger sample and improved technique for the infrared excess search. *MNRAS*, **1448**, 3132.

Exter, K. M., Pollacco, D. L., and Bell, S. A. 2003. The planetary nebula K 1-2 and its binary central star VW Pyx. *MNRAS*, **341**, 1349.

Exter, K. M., Pollacco, D. L., Maxted, P. F. L., Napiwotzki, R., and Bell, S. A. 2005. A study of two post–common envelope binary systems. *MNRAS*, **359**, 315.

Ferguson, D. H., Liebert, J., Haas, S., Napiwotzki, R., and James, T. A. 1999. Masses and other parameters of the post–common envelope binary BE Ursae Majoris. *ApJ*, **518**, 866.

Freeman, M., Montez Jr., R., Kastner, J. H., et al. 2014. The CHANDRA planetary nebula survey (ChanPlaNS). II. X-ray emission from compact planetary nebulae. *ApJ*, **794**, 99.

Frew, D. J., Parker, Q. A., and Bojičiíc, I. S. 2016. The Hα surface brightness-radius relation: a robust statistical distance indicator for planetary nebulae. *MNRAS*, **455**, 1459.

García-Rojas, J., Corradi, R. L. M., Monteiro, H., et al. 2016. Imaging the elusive H-poor gas in the high adf planetary nebula NGC 6778. *ApJ*, **824**, 27.

García-Rojas, J., Delgado-Inglada, G., García-Hernandez, D. A., et al. 2018. C/O ratios in planetary nebulae with dual-dust chemistry from faint optical recombination lines. *MNRAS*, arXiv:1709.07958.

García-Segura, G., Villaver, E., Langer, N., Yoon, S. C., and Manchado, A. 2014. Single rotating stars and the formation of bipolar planetary nebula. *ApJ*, **783**, 74.

Guzman-Ramirez, L., Lagadec, E., Jones, D., Zijlstra, A. A., and Gesicki, K. 2014. PAH formation in O-rich planetary nebulae. *MNRAS*, **441**, 364.

Guzman-Ramirez, L., Lagadec, E., Wesson, R., et al. 2015. Witnessing the emergence of a carbon star. *MNRAS*, **451**, 1.

Guzman-Ramirez, L., Zijlstra, A. A., Níchuimín, R., et al. 2011. Carbon chemistry in Galactic bulge planetary nebulae. *MNRAS*, **414**, 1667.

Hilditch, R. W., Harries, T. J., and Hill, G. 1996. On the reflection effect in three sdOB binary stars. *MNRAS*, **279**, 1380.

Hillwig, T. C., Bond, H. E., Afşar, M., De Marco, O. 2015a. Binary central stars of planetary nebulae discovered through photometric variability. II. Modeling the central stars of NGC 6026 and NGC 6337. *AJ*, **140**, 319.

Hillwig, T. C., Bond, H. E., Frew, D. J., et al. 2016a. Binary central stars of planetary nebulae discovered through photometric variability. IV. The central stars of HaTr 4 and Hf 2-2. *AJ*, **152**, 34.

Hillwig, T. C., Frew, D. J., Louie, M., et al. 2015b. Binary central stars of planetary nebulae discovered through photometric variability. III. The central star of Abell 65. *AJ*, **150**, 30.

Hillwig, T. C., Frew, D. J., Reindl, N., et al. 2016c. Binary central stars of planetary nebulae discovered through photometric variability. V. The central stars of HaTr 7 and ESO 330-9. *AJ*, **153**, 24.

Hillwig, T. C., Jones, D., De Marco, O., et al. 2016b. Observational confirmation of a link between common envelope binary interaction and planetary nebula shaping. *ApJ*, **832**, 125.

Iben, I. and Tutukov, A. V. 1984. Supernovae of type I as end products of the evolution of binaries with components of moderate initial mass (M not greater than about 9 solar masses). *ApJS*, **54**, 335.

Jones, D., and Boffin, H. M. J. 2017a. Binary stars as the key to understanding planetary nebulae. *Nat. Astron.*, **1**, 117.

Jones, D., and Boffin, H. M. J. 2017b. On the possible triple central star system of PN SuWt 2: no ménage à trois at the heart of the Wedding Ring. *MNRAS*, **466**, 2034.

Jones, D., Boffin, H. M. J., Miszalski, B., et al. 2014. The post–common-envelope, binary central star of the planetary nebula Hen 2-11. *A&A*, **562**, 89.

Jones, D., Boffin, H. M. J., Rodríguez-Gil, P., et al. 2015. The post–common envelope central stars of the planetary nebulae Henize 2-155 and Henize 2-161. *A&A*, **580**, 19.

Jones, D., Boffin, H.M.J., Sowicka, P., et al., 2019, MNRAS, **482**, L75.

Jones, D., Lloyd, M., Santander-García, M., et al. 2010. Abell 41: shaping of a planetary nebula by a binary central star. *MNRAS*, **408**, 2312.

Jones, D., Mitchel, D. L., Lloyd, M., et al. 2012. The morphology and kinematics of the Fine Ring Nebula, planetary nebula Sp 1, and the shaping influence of its binary central star. *MNRAS*, **420**, 2271.

Jones, D., Van Winckel, H., Aller, A., Exter, K., and De Marco, O. 2017. The long-period binary central stars of the planetary nebulae NGC 1514 and LoTr 5. *A&A*, **600**, 9.

Jones, D., Wesson, R., García-Rojas, J., Corradi, R. L. M., and Boffin, H. M. J. 2016. NGC 6778: strengthening the link between extreme abundance discrepancy factors and central star binarity in planetary nebulae. *MNRAS*, **455**, 3263.

Kastner, J. H., Montez Jr., R., Balick, B., et al. 2012. The CHANDRA X-ray survey of planetary nebulae (ChanPlaNS): probing binarity, magnetic fields, and wind collisions. *ApJ*, **144**, 58.

Kwok, S., Purton, C. R., and Fitzgerald, P. M. 1978. On the origin of planetary nebulae. *ApJ*, **219**, 125.

Liu, X.-W., Barlow, M. J., Zhang, Y., Bastin, R. J., and Storey, P. J. 2016. Chemical abundances for Hf 2-2, a planetary nebula with the strongest-known heavy-element recombination lines. *MNRAS*, **368**, 1959.

MacLeod, M., and Ramirez-Ruiz, E. 2015. Asymmetric accretion flows within a common envelope. *ApJ*, **803**, 41.

Manick, R., Miszalski, B., and McBride, V. 2015. A radial velocity survey for post–common-envelope Wolf-Rayet central stars of planetary nebulae: first results and discovery of the close binary nucleus of NGC 5189. *MNRAS*, **448**, 1789.

Maoz, D., and Badenes, C. 2010. The supernova rate and delay time distribution in the Magellanic Clouds. *MNRAS*, **407**, 1314.

Maoz, D., and Hallakoun, N. 2017. The binary fraction, separation distribution, and merger rate of white dwarfs from SPY. *MNRAS*, **467**, 1414.

Maoz, D., Mannucci, F., and Nelemans, G. 2014. Observational clues to the progenitors of Type Ia supernovae. *ARA&A*, **52**, 107.

Marsh, T. R., Parsons, S. G., Bours, M. C. P., et al. 2014. The planets around NN Serpentis: still there. *MNRAS*, **437**, 475.

McClure, R. D., Fletcher, J. M., and Nemec, J. M. 1980. The binary nature of the barium stars. *ApJ*, **238**, 35.

McNabb, I. A., Fang, X., Liu, X.-W., Bastin, R. J., and Storey, P. J. 2013. Plasma diagnostics for planetary nebulae and H II regions using the N II and O II optical recombination lines. *MNRAS*, **428**, 3443.

Mendez, R. H. and Niemela, V. S. 1981. The binary central star of NGC 2346 and the extinction puzzle. *ApJ*, **250**, 240.

Miller Bertolami, M. M. 2016. New models for the evolution of post-asymptotic giant branch stars and central stars of planetary nebulae. *A&A*, **588**, 25.

Miszalski, B., Acker, A., Moffat, A. F. J., Parker, Q. A., and Udalski, A. 2009a. Binary planetary nebulae nuclei towards the Galactic bulge. I. Sample discovery, period distribution, and binary fraction. *A&A*, **496**, 813.

Miszalski, B., Acker, A., Parker, Q. A., and Moffat, A. F. J. 2009b. Binary planetary nebulae nuclei towards the Galactic bulge. II. A penchant for bipolarity and low-ionisation structures. *A&A*, **505**, 249.

Miszalski, B., Boffin, H. M. J., and Corradi, R. L. M. 2013a. A carbon dwarf wearing a Necklace: first proof of accretion in a post–common-envelope binary central star of a planetary nebula with jets. *MNRAS*, **428**, 39.

Miszalski, B., Boffin, H. M. J., Frew, D. J., et al. 2012a. A barium central star binary in the Type I diamond ring planetary nebula Abell 70. *MNRAS*, **419**, 39.

Miszalski, B., Boffin, H. M. J., Jones, D., et al. 2013b. SALT reveals the barium central star of the planetary nebula Hen 2-39. *MNRAS*, **436**, 3068.

Miszalski, B., Corradi, R. L. M., Boffin, H. M. J., et al. 2011a. ETHOS 1: a high-latitude planetary nebula with jets forged by a post–common-envelope binary central star. *MNRAS*, **413**, 1264.

Miszalski, B., Crowther, P. A., De Marco, O., et al. 2012b. IC 4663: the first unambiguous [WN] Wolf–Rayet central star of a planetary nebula. *MNRAS*, **423**, 934.

Miszalski, B., Jones, D., Rodríguez-Gil, P., et al. 2011b. Discovery of close binary central stars in the planetary nebulae NGC 6326 and NGC 6778. *A&A*, **531**, 158.

Miszalski, B., Manick, R., Mikołajewska, J., et al. 2018. SALT HRS discovery of a long period double-degenerate binary in the planetary nebula NGC 1360. *MNRAS*, **473**, 2275.

Miszalski, B., Manick, R., Mikołajewska, J., Van Winckel, H., and Iłkiewicz, K., 2018b. SALT HRS Discovery of the Binary Nucleus of the Etched Hourglass Nebula MyCn 18. *PASA*, **35**, 27.

Moe, M., and De Marco, O. 2006. Do most planetary nebulae derive from binaries? I. Population synthesis model of the galactic planetary nebula population produced by single stars and binaries. *ApJ*, **650**, 916.

Mohamed, S., and Podsialowski, Ph. 2012. Mass transfer in Mira-type binaries. *BaltA*, **21**, 88.

Montez Jr., R., De Marco, O., Kastner, J. H., and Chu, Y. H. 2010. X-ray emission from the binary central stars of the planetary nebulae HFG 1, DS 1, and LoTr 5. *ApJ*, **721**, 1820.

Montez Jr., R., Kastner, J. H., Balick, B., et al. 2015. The CHANDRA planetary nebula survey (ChanPlaNS). III. X-ray emission from the central stars of planetary nebulae. *ApJ*, **721**, 1820.

Morris, T., and Podsialowski, Ph. 2007. The triple-ring nebula around SN 1987A: fingerprint of a binary merger. *Sci*, **315**, 1103.

Nandez, J. L. A., and Ivanova, N. 2016. Common envelope events with low-mass giants: understanding the energy budget. *MNRAS*, **460**, 3992.

Napiwotzki, R., Heber, U., and Köppen, J. 1994. Analysis of BD+33°2642: a newly detected planetary nebula in the Galactic halo and its central star. *A&A*, **292**, 239.

Nordhaus, J., and Blackman, E. G. 2006. Low-mass binary-induced outflows from asymptotic giant branch stars. *MNRAS*, **370**, 2004.

Nordhaus, J., Blackman, E. G., and Frank, A. 2007. Isolated versus common envelope dynamos in planetary nebula progenitors. *MNRAS*, **376**, 599.

Paczynski, B. 1976. Common envelope binaries. *IAUS*, **73**, 75.

Peimbert, M., Peimbert, A., and Delgado-Inglada, G. 2017. Nebular spectroscopy: a guide on H II regions and planetary nebulae. *PASP*, **129**, 082001.
Perea-Calderón, J. V., García-Hernández, D. A., García-Lario, P., Szczerba, R., and Bobrowsky, M. 2009. The mixed chemistry phenomenon in Galactic Bulge PNe. *A&A*, **495**, 5.
Perlmutter, S., Aldering, G., Goldhaber, G., et al. 1999. Measurements of Ω and Γ from 42 high-redshift supernovae. *ApJ*, **517**, 565.
Phillips, M. M. 1993. The absolute magnitudes of Type IA supernovae. *ApJ*, **413**, 105.
Reiss, A. G., Filippenko, A. V., Challis, P., et al. 1998. Observational evidence from supernovae for an accelerating universe and a cosmological constant. *AJ*, **116**, 1009.
Ricker, P. M., and Taam, R. E. 2012. An AMR study of the common-envelope phase of binary evolution. *ApJ*, **746**, 74.
Rodríguez-Gil, P., Santander-García, M., Knigge, C., et al. 2010. The orbital period of V458 Vulpeculae, a post-double common-envelope nova. *MNRAS*, **407**, 21.
Santander-García, M., Corradi, R. L. M., Mampaso, A., et al. 2004. Hen 2-104: a close-up look at the Southern Crab. *A&A*, **485**, 117.
Santander-García, M., Rodríguez-Gil, P., Corradi, R. L. M., et al. 2015. The double-degenerate, super-Chandrasekhar nucleus of the planetary nebula Henize 2-428. *Nature*, **519**, 63.
Schleicher, D. R. G., Dreizler, S., Völschow, M., Banerjee, R., and Hessman, F. V. 2015. Planet formation in post–common-envelope binaries. *AN*, **336**, 458.
Schmidt, B. P., Suntzeff, N. B., Phillips, M. M., et al. 1998. The high-Z supernova search: measuring cosmic deceleration and global curvature of the Universe using Type IA supernovae. *ApJ*, **507**, 46.
Soker, N. 2015. Close stellar binary systems by grazing envelope evolution. *ApJ*, **800**, 114.
Soker, N., and Kastner, J. H. 2002. X-ray emission from central binary systems of planetary nebulae. *ApJ*, **570**, 245.
Soszyński, I., Stępień, K., Pilecki, B., et al. 2015. Ultra-short-period binary systems in the OGLE fields toward the Galactic bulge. *AcA*, **65**, 39.
Sowicka, P., Jones, D., Corradi, R. L. M., et al. 2017. The planetary nebula IC 4776 and its post–common-envelope binary central star. *MNRAS*, **471**, 3529.
Steffen, W., Koning, N., Esquivel, A., et al. 2013. A wind-shell interaction model for multipolar planetary nebulae. *MNRAS*, **436**, 470.
Theuns, T., Boffin, H. M. J., and Jorissen, A., 1996. Wind accretion in binary stars. II. Accretion rates. *MNRAS*, **280**, 1264.
Tocknell, J., De Marco, O., and Wardle, M. 2014. Constraints on common envelope magnetic fields from observations of jets in planetary nebulae. *MNRAS*, **439**, 2014.
Tout, C. A., and Eggleton, P. P. 1988. Tidal enhancement by a binary companion of stellar winds from cool giants. *MNRAS*, **231**, 823.
Tovmassian, G., Yungelson, L., Rauch, T., et al. 2010. The double-degenerate nucleus of the planetary nebula TS 01: a close binary evolution showcase. *ApJ*, **714**, 178.
Tsebrenko, D., and Soker, N. 2015. The fraction of type Ia supernovae exploding inside planetary nebulae (SNIPs). *MNRAS*, **447**, 2568.
Tyndall, A. A., Jones, D., Boffin, H. M. J., et al. 2013. Two rings but no fellowship: LoTr 1 and its relation to planetary nebulae possessing barium central stars. *MNRAS*, **436**, 2082.
Van Winckel, H., Jorissen, A., Exter, K., et al. 2014. Binary central stars of planetary nebulae with long orbits: the radial velocity orbit of BD+33°2642 (PN G052.7+50.7) and the orbital motion of HD 112313 (PN LoTr5). *A&A*, **563**, 10.

Wareing, C. J., Zijlstra, A. A., and O'Brien, T. J. 2007. The interaction of planetary nebulae and their asymptotic giant branch progenitors with the interstellar medium. *MNRAS*, **382**, 1233.

Wesson, R., Barlow, M. J., Liu, X. W., et al. 2008. The hydrogen-deficient knot of the 'born-again' planetary nebula Abell 58 (V605 Aql). *MNRAS*, **383**, 1639.

Wesson, R., Jones, D., García-Rojas, J., Boffin, H.M.J., and Corradi, R.L.M., 2018, MNRAS, **480**, 4589.

Woods, T. E., Ivanova, N., van der Sluys, M. V., and Chaichenets, S. 2012. On the formation of double white dwarfs through stable mass transfer and a common envelope. *ApJ*, **744**, 12.

Wyse, A. B. 1942. The spectra of ten gaseos nebula. *ApJ*, **95**, 356.

9

Massive Star Evolution: Binaries as Two Single Stars

CYRIL GEORGY AND SYLVIA EKSTRÖM[1]

9.1 The Basics

Massive stars are stars more massive than about 8 to $10 M_{\odot}$, that are bound to die in the explosion of a supernova or collapsing directly into a black hole (Heger et al., 2003; Jones et al., 2013). Their life is marked by the succession of all burning stages up to silicon burning.

The modelling of massive stars is done by the mean of one-dimension (1D) stellar evolution codes (see Section 9.2), which solve the four equations of stellar structure (see Kippenhahn and Weigert, 1990, for a classical textbook), coupled with a network of nuclear reactions providing the energy generation rate inside the stellar model, and the time evolution of the chemical content of the star. On top of that, various effects can be accounted for, such as rotation or mass loss, particularly important in the context of the evolution of massive stars.

An important aspect of massive star evolution that became prominent during the past decade is that a significant fraction of massive stars are found in multiple systems. Large surveys in different environments suggest that 50% to 70% of all massive stars (Sana et al., 2012, 2013) are in binary (or multiple) systems where components are sufficiently close to interact at least once during their lifetime. Evolution of massive stars in multiple systems can be very different from single-star evolution (van den Heuvel, 1975; Vanbeveren, 1991; Podsiadlowski et al., 1992; Vanbeveren et al., 1998; Eldridge et al., 2008; de Mink et al., 2013; among others). Taking multiplicity into account is thus extremely important when dealing with massive star populations.

It is, however, important to keep in mind that it is utopian to model correctly multiple stellar systems without being able to accurately describe the evolution of a single star. In this chapter, we discuss uncertainties of key physical processes taking place in single-star models and their consequences. Observational constraints of the multiplicity properties of massive stars are given in Chapter 10. These shortcomings should be kept in mind by anybody using results from stellar evolution computations or from population synthesis.

[1] The authors acknowledge support from the Swiss National Science Foundation (project number 200020-172505). CG acknowledge the Partnership for Research in Advanced Computing in Europe (PRACE) for awarding access to resource MareNostrum 4 based in Spain at Barcelona Supercomputing Center. The support of David Vicente and Janko Strassburg from Barcelona Supercomputing Center, Spain, to the technical work is gratefully acknowledged.

9.2 Mass Loss

Mass loss is a very important ingredient to be included in stellar evolution code when dealing with massive stars. At solar metallicity, a $15\,M_\odot$ star loses 2 to $3\,M_\odot$ during its entire life (but only a few 10ths of a solar mass during its main sequence). At the other extremity of the mass spectrum, a $120\,M_\odot$ star loses up to $100\,M_\odot$ during its lifetime, including several 10ths during the main sequence (see, e.g., Ekström et al., 2012). The situation is even more extreme in the case of very massive stars, which lose most of their mass in the course of their life (Yusof et al., 2013).

During the main sequence phase, winds from massive stars are in general described in the framework of the radiation-driven wind theory ('CAK' theory; see Lucy and Solomon, 1970; Castor et al., 1975). It has been improved over the years to accommodate supplementary processes such as multiple scattering, wind clumping, etc. (see the review by Puls et al., 2008). It has led to the computation of theoretical mass-loss rates that are extensively used in stellar evolution calculations (Vink et al., 2000, 2001). However, the face values of the obtained mass-loss rates computed with this recipe seems to disagree by a factor of two to three with mass-loss rates obtained observationally at different wavelengths (e.g., Najarro et al., 2011; Šurlan et al., 2013; Rauw et al., 2015). This uncertainty on the mass-loss rates that should be adopted in stellar evolution models has a major impact on the results of such simulations. As an illustration, Figure 9.1 shows tracks in the Hertzsprung–Russell diagram for two $60\,M_\odot$ models with different prescriptions for the computation of the mass-loss rates on the main sequence: a model using Vink et al. (2000, solid line) and one using Kudritzki and Puls (2000, dotted line). Mass-loss rates during the advanced stages are kept the same (see also Keszthelyi et al., 2017). It shows that a difference of a factor of two in the mass-loss rates during the main sequence modifies completely the late evolution and the endpoint just before the supernova.

For the most massive O and Wolf–Rayet (WR) stars, a dependence of the mass-loss rates on the Eddington factor $\Gamma_{\rm Edd} = \frac{\kappa L}{4\pi cGM}$ seems to appear when they reach large $\Gamma_{\rm Edd}$ (Vink, 2006; Gräfener et al., 2011; Vink et al., 2011; Bestenlehner et al., 2014). This dependence is around ${\Gamma_{\rm Edd}}^2$ for O stars with $\Gamma_{\rm Edd} < 0.70$, and around ${\Gamma_{\rm Edd}}^4$ for WR stars with $\Gamma_{\rm Edd} > 0.70$.

Mass-loss rates usually scale with the metallicity, but the scaling to adopt is not yet completely clear. The Z dependence has been estimated initially to be 0.4 by Abbott (1982), and then between 0.5 (Kudritzki, 2002) and 0.85 (Vink et al., 2001). Mokiem et al. (2007) find the empirical scaling deduced from observations of early B-type and O-type stars to be 0.83 for a smooth wind solution, or 0.72 when a correction for clumping is applied.

During the red supergiant phase, the mass-loss rates are even less well known. Observational determinations exhibit a very strong scatter, with variations over more than two orders of magnitude at a given luminosity (van Loon et al., 2005; Mauron and Josselin, 2011; Beasor and Davies, 2016, 2018; Georgy, 2017). As the winds from red supergiants are still not well understood from a theoretical point of view, the current prescriptions used in stellar evolution code are fits to these observational data. Questions such as, 'in what kind of star, when and how long do the heavy mass-loss episodes occur? What is the average

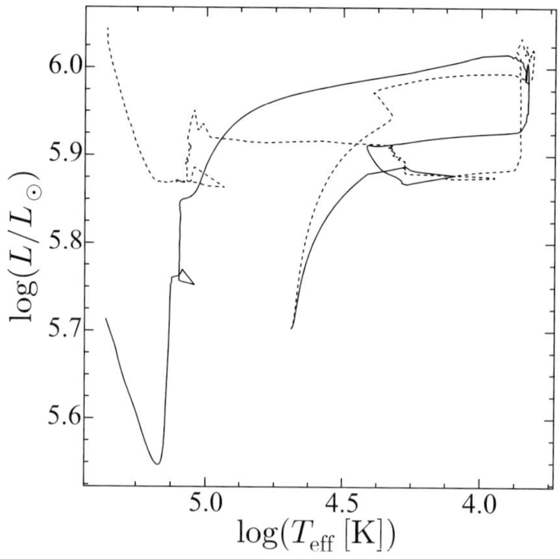

Figure 9.1 Tracks in the Hertzsprung–Russell diagram of a $60\,M_\odot$ model computed with two different prescriptions for the mass-loss rates during the main sequence: Vink et al. (2000, solid line) and Kudritzki and Puls (2000, dotted line). The impact on the tracks is clearly visible, due to the much higher rates with the Vink et al. (2000) prescription.

mass-loss rate at a given evolutionary stage?' need to be answered in order to correctly model the late stages of stellar evolution.

In cases in which the mass-loss rates are best described by the prescriptions leading to rates in the low range of the observations, stars up to about $30\,M_\odot$ end their life as a red supergiant. In such cases, explanations to the so-called 'red supergiant problem' (Smartt et al., 2009) have to be found. Strong extinction of the supernova event due to the dust in the strong winds of the more massive red supergiants (Walmswell and Eldridge, 2012) or direct collapse into a black hole (Adams et al., 2017) have been suggested. In case average mass-loss rates during the red supergiant phases are underestimated in usual prescriptions, it allows the star to evolve back towards the hot side of the Hertzsprung–Russell diagram (Vanbeveren et al., 1998; Georgy, 2012; Meynet et al., 2015). In this context, it is also interesting to note that a theoretical explanation of the pulsating properties of a class of variable blue supergiants, the 'α-Cygni' variables, seems to require that these stars have lost their hydrogen-rich envelope through a strong mass-loss episode, possibly during a previous red supergiant stage (Saio et al., 2013).

9.3 Internal Mixing Processes

Mixing processes are any physical process able to transport chemical elements or angular momentum from one place to another inside a star. They are important as they can considerably affect the way a star evolves, as well as its surface (and observable) properties.

Historically, the first of such processes included in the modelling of stars was convection (even before people started considering heat transfer by radiation; see Eddington, 1926). Rotation is another such process and was studied very early during the twentieth century (von Zeipel, 1924; Eddington, 1925; Endal and Sofia, 1976). In this chapter, we focus only on these two processes (each of them would still require a whole book for an in depth description, such as Maeder, 2009, for rotation), and we will not consider other mixing processes such as radiative diffusion, gravitational settling, internal waves, magnetic field or tides, among others.

9.3.1 Convection

The treatment of convection in classical 1D stellar evolution codes is a longstanding problem of stellar evolution (e.g., Arnett et al., 2015). In most stellar evolution codes (e.g., the Geneva Stellar Evolution code, Eggenberger et al., 2008; MESA, Paxton et al., 2011, 2013, 2015, 2017; STERN, Petrovic et al., 2005; KEPLER, Heger et al., 2000; FRANEC, Chieffi and Limongi, 2013 for the most widely used code in the massive star community), it is a two-step process: (1) determine the boundaries of the convective zones, often using the 'Schwarzschild' or 'Ledoux' criterion (e.g., Kippenhahn and Weigert, 1990), and possibly extend them empirically to account for the 'overshoot' (Maeder, 1975), which cannot be predicted by the linear perturbation theory used to elaborate the stability criteria; (2) compute the thermal gradient inside the convective zone, often using the 'mixing-length theory' (MLT) or one of its derivative (Böhm-Vitense, 1958). This was devised before the publication in the west of Kolmogorov's work on turbulent cascade (Kolmogorov, 1941, 1962), and may be inconsistent with that now well-established theory (see also Arnett et al., 2009).

There are a variety of ways of implementing convection in 1D stellar evolution codes, following theoretical works or guided by hydrodynamics simulations (Zahn, 1991; Freytag et al., 1996; Meakin and Arnett, 2007). Recent works have shown how sensitive the outputs of 1D stellar modelling are to the detailed implementation of convection (Martins and Palacios, 2013; Georgy et al., 2014b).

In this context, hydrodynamics simulations of convection in stellar interiors are very important, as they allow one to study the behaviour of the flow inside the convective regions from first principles. With the rapid increase of computational resources during the past years and the development of efficient parallelised hydrodynamics code, it has become possible to perform such simulations in different regimes: convective envelopes of cool stars (e.g., Freytag and Höfner, 2008; Chiavassa et al., 2009; Magic et al., 2013; Viallet et al., 2013), intermediate helium-burning shell in intermediate mass stars (Herwig et al., 2006; Woodward et al., 2015), or late burning stages of massive star life, such as carbon-burning shell (Cristini et al., 2017), oxygen-burning shell (Meakin and Arnett, 2007; Müller et al., 2016; Jones et al., 2017), or silicon-burning shell (e.g., Couch et al., 2015). These simulations provide very important information about the physics of the convective boundary (shape, location, motion, boundary mixing) and the interaction between the

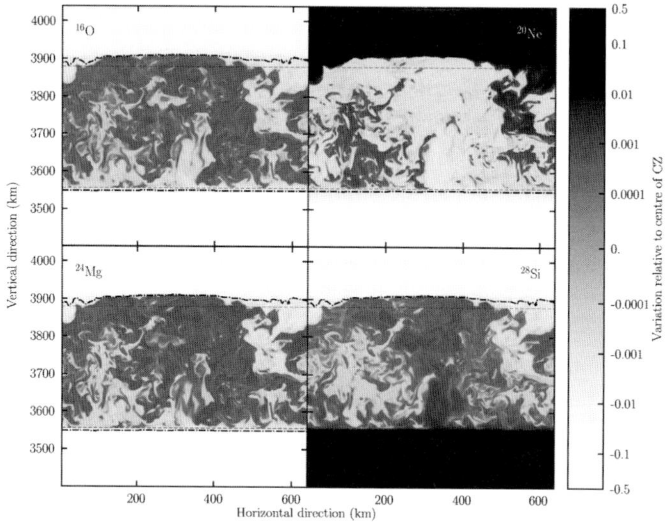

Figure 9.2 Abundance of ^{16}O (top-left panel), ^{20}Ne (top-right panel), ^{24}Mg (bottom-left panel) and ^{28}Si (bottom-right panel) relative to the average value in the middle of the convective zone. Black and dark greys indicate overabundances, while white and light greys indicate underabundances. The figure shows a 2D vertical slice through the 3D computation domain. The thin dashed line indicates the position of the convective zone in the corresponding 1D model, and the thick dot-dashed line shows the convective boundary in the 3D simulation.

nuclear burning and the convection (see Figure 9.2, where the chemical abundances of ^{16}O, ^{20}Ne, ^{24}Mg and ^{28}Si with respect to the centre of the convective zone are shown in the case of the simulation of a neon-burning shell). The results of such simulations can be used to deduce prescriptions for the convective boundary mixing (Freytag et al., 1996; Herwig, 2000; Jones et al., 2017). Using specific averaging techniques (the 'RANS' approach, see, e.g., Viallet et al., 2013; Mocák et al., 2014), such simulations are also useful in providing a new theoretical framework and new algorithms for the treatment of convection in 1D stellar evolution code (Arnett et al., 2015).

The need for a new treatment of convection and convective boundary mixing is observationally motivated by the asteroseismic data available for main-sequence B-type stars (Moravveji et al., 2015), or from nucleosynthesis considerations at different production sites (Battino et al., 2016; Choplin et al., 2017). The modelling of convection has also a deep impact on the final structure of the star, and can considerably affect its final fate by changing its compactness (supernova explosion or not, and the type of remnant left behind; see O'Connor and Ott, 2011; Sukhbold and Woosley, 2014; Davis et al., 2017).

9.3.2 Rotation

Rotation has two different ways of influencing stellar evolution: (1) It creates a deformation of the star's shape such that the stellar characteristics become dependent on the colatitude considered, affecting the stellar parameters deduced from observations. It also induces an anisotropy in the mass lost by the star. (2) It triggers an internal mixing that transports chemicals and angular momentum, and induces many deviations from the standard non-rotating evolution. We detail both effects in this section.

Surface Deformation

The deviation from spherical symmetry is proportional to the rotation ratio $\omega = \Omega_{\rm surf}/\Omega_{\rm crit}{}^2$. In the frame of the Roche model, the equatorial radius is at most $R_{\rm eq,crit} = \frac{3}{2} R_{\rm pol,crit}$, with the reasonable assumption that $R_{\rm pol,crit} = R_{{\rm pol},\omega=0}$. An oblateness has been observed by interferometry in the case of the rapidly rotating star Achernar (α Eridani), for example, in agreement with the value of 1.5 (Domiciano de Souza et al., 2003; Vinicius et al., 2006; Carciofi et al., 2008).

Because of the deformation, the effective gravity in the Roche approximation becomes dependent on the rotation rate and the colatitude:

$$\vec{g}_{\rm eff} = \vec{g}_{\rm eff}(\Omega,\theta) = \left(-\frac{GM}{r^2} + \Omega^2 r \,\sin^2\theta\right)\,\vec{e}_r + \Omega^2 r\,\sin\theta\,\cos\theta\,\vec{e}_\theta.$$

Subsequently, the flux inherits from this $\Omega - \theta$ dependence (e.g. Maeder, 1999):

$$\vec{F} = \vec{F}(\Omega,\theta) = -\frac{L}{4\pi\,GM^\star}\,\vec{g}_{\rm eff}(\Omega,\theta),$$

where $M^\star = M\left(1 - \frac{\Omega^2}{2\pi\,G\rho_M}\right)$. According to the Stefan–Boltzman law, we have $F = \sigma T^4$, so the effective temperature also becomes dependent on Ω and θ (von Zeipel, 1924):

$$T_{\rm eff} = \left[\frac{L}{4\pi\sigma\,GM^\star}g_{\rm eff}(\Omega,\theta)\right]^{1/4}.$$

Espinosa Lara and Rieutord (2011) have proposed a more general expression, valid also in the case of rapid rotation:

$$T_{\rm eff} = \left(\frac{L}{4\pi\sigma GM}\right)^{1/4}\sqrt{\frac{\tan\vartheta}{\tan\theta}}\;g_{\rm eff}^{1/4},$$

where ϑ is the solution of the equation:

$$\cos\vartheta + \ln\tan\frac{\vartheta}{2} = \frac{1}{3}\omega^2\tilde{r}^3\cos^3\theta + \cos\theta + \ln\tan\frac{\theta}{2}.$$

The latitude dependence of the temperature has been observed in interferometry for the star Altair (α Aquilae; see Monnier et al., 2007) and the stars Alderamin and Rasalhague (α Ceph and α Oph; see Zhao et al., 2009), with values compatible with the models.

Below $\omega = 0.7$, the deformation remains insignificant (see Figure 9.3), but above this value, the angle with which the stars are observed becomes an important parameter. A star

[2] We use here $\Omega_{\rm crit} = \sqrt{\frac{8}{27}\frac{GM}{R^3_{\rm pol,crit}}}$, where $R_{\rm pol,crit}$ is the polar radius at the critical velocity.

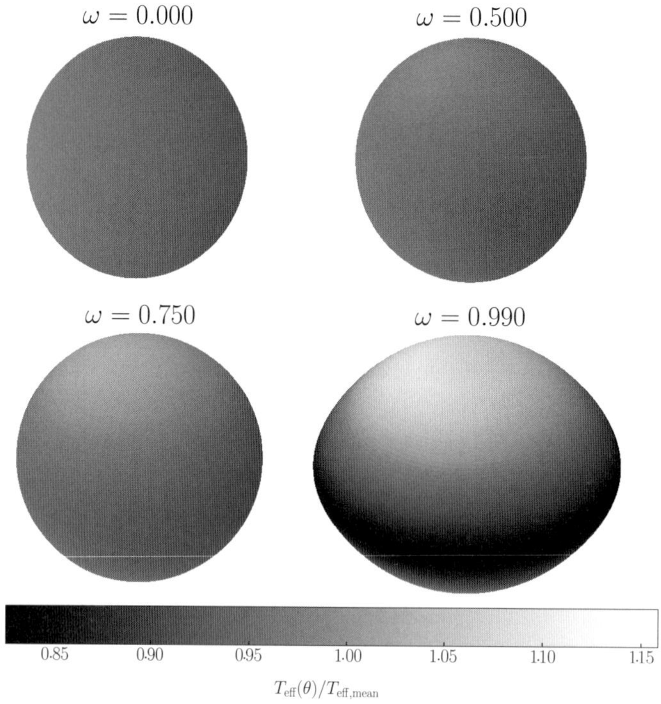

Figure 9.3 Effective temperature of rotating stars relative to the mean $T_{\rm eff}$ for four different rotation rates $\omega = \Omega_{\rm surf}/\Omega_{\rm crit}$. The $T_{\rm eff}(\theta)$ law is from Espinosa Lara and Rieutord (2011).

observed pole-on will present a higher L and $T_{\rm eff}$ than the same star seen with an average inclination. In contrast, a star seen equator-on will appear dimmer and cooler (e.g., Georgy et al., 2014a). The stellar parameters deduced for such stars will be affected.

Hotter poles and cooler equators modify the mass flux ejected by the star, which gets a latitudinal dependence (Maeder and Meynet, 2000):

$$\frac{\mathrm{d}\dot{M}(\theta)}{\mathrm{d}\sigma} \sim A(\alpha,k)\left(\frac{L}{4\pi\, G\, M^{\star}}\right)^{\frac{1}{\alpha}-\frac{1}{8}} \frac{g_{\rm eff}^{1-\frac{1}{8}}}{(1-\Gamma_{\Omega}(\theta))^{\frac{1}{\alpha}-1}},$$

where $\mathrm{d}\sigma$ is the surface unit, $A(\alpha,k)$ is an empirically determined function of the force multiplier parameters α and k (Lamers et al., 1995), and $\Gamma_{\Omega}(\theta)$ is the local Eddington factor taking into account the effects of rotation:

$$\Gamma_{\Omega}(\theta) = \frac{\kappa_{\rm es} L}{4\pi cGM\left(1-\frac{\Omega^2}{2\pi G\rho_{\rm m}}\right)},$$

with $\kappa_{\rm es}$ the electron-scattering opacity, and $\rho_{\rm m}$ the internal average density. According to Georgy et al. (2011), for a nearly critically rotating star, we have $\dot{M}(\mathrm{pol}) = 3.25\,\dot{M}(\mathrm{eq})$. This anisotropy modifies the angular momentum budget, since the mass loss occurs preferentially in the region of the poles, which removes less angular momentum than what the equatorial regions would do (see also Müller and Vink, 2014).

Transport Processes

In 1D, the transport of angular momentum is, according to Zahn (1992) and Maeder and Zahn (1998):

$$\rho \frac{\partial}{\partial t}\left(r^2\bar{\Omega}\right)_{M_r} = \underbrace{\frac{1}{5r^2}\frac{\partial}{\partial r}\left(\rho r^4 \bar{\Omega} U(r)\right)}_{\text{advective term}} + \underbrace{\frac{1}{r^2}\frac{\partial}{\partial r}\left(\rho D_{\text{v}} r^4 \frac{\partial \bar{\Omega}}{\partial r}\right)}_{\text{diffusive term}},$$

where $U(r)$ is the function such that the radial component of the meridional circulation is $U(r)\ P_2 \cos(\theta)$, and D_{v} the diffusion coefficient due to various mechanisms such as convection or shear. The transport of chemical species has been shown to be satisfactorily approximated by a diffusion-only process (Chaboyer and Zahn, 1992) and can be expressed as:

$$\rho \frac{\partial X_i}{\partial t} = \frac{1}{r^2}\frac{\partial}{\partial r}\left(\rho r^2 \left(D_{\text{v}} + D_{\text{eff}}\right)\frac{\partial X_i}{\partial r}\right),$$

with D_{eff} the diffusion coefficient due to meridional circulation and horizontal turbulence.

The implementation of rotation in the various codes existing on the market varies a lot from one to another. Some of them use a diffusion-only approximation to express the transport of angular momentum, as for the species. When the advective term is taken into account, three different expressions for the horizontal turbulence can be used (Zahn, 1992; Maeder, 2003; Mathis et al., 2004), and two for the shear turbulence (Maeder, 1997; Talon and Zahn, 1997). Maeder et al. (2013) proposed also a global diffusion coefficient covering all the various instabilities triggered by rotation (GSF and Solberg-Høiland instabilities, thermohaline mixing, etc.) and the interplay they have with each other. The various ways of expressing the mixing process induced by rotation leads to large differences between the outputs of different codes, as highlighted by Chieffi and Limongi (2013). A study of the impact the different implementations have on the stellar evolution is presented in Meynet et al. (2013).

Effects on the Evolution

Rotational mixing brings fresh fuel into the core, allowing it to grow further, hence resulting in longer lifetimes and a higher luminosity. However, the behaviour of a star is not monotonic with increasing rotation. At the beginning of the main sequence, the hydrostatic effects dominate and the star presents a slightly lower L and T_{eff}. As the evolution proceeds, the mixing induces a slower shrinking of the core, increasing the luminosity, while the T_{eff} decreases because of the modified chemical structure. Note that the effects of mixing are dominant for low and average rotation, while the rapid rotators show a composite behaviour between strong hydrostatic effects and mixing effects. The brightest stars are therefore not the fastest ones, but the slightly above-average rotators (see Figure 9.4). In

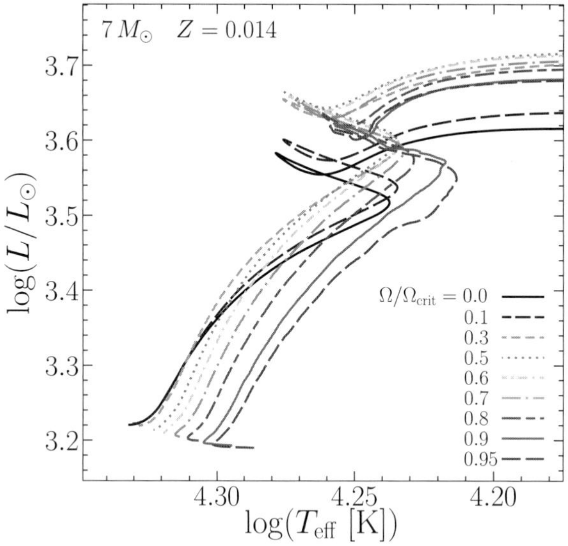

Figure 9.4 Main sequence tracks in the Hertzsprung–Russell diagram for models of $7\,M_\odot$ with various initial rotation rates. (Models from Georgy et al., 2013.)

some extreme cases, the mixing during the main sequence is so strong that the star evolves quasichemically homogeneously (usually at low metallicity; see Yoon et al., 2006; Szécsi et al., 2015). In that case, the star becomes extremely hot and compact, becoming a possible progenitor for long, soft gamma-ray bursts (and gravitational waves; de Mink and Mandel, 2016).

Evolution of the Surface Velocity

If only the local conservation of angular momentum is considered, the surface velocity is expected to decrease during the main sequence, because of the natural expansion of the surface (see the short-dash-dotted curve in Figure 9.5 for a $20\,M_\odot$ model). On top of this natural evolution, two competing processes come into play: the mass loss (that removes angular momentum from the surface and triggers meridional currents), and the internal transport (that generally brings angular momentum from the centre to the surface). The strength of both the transport and the winds scales with the mass and metallicity, higher mass stars having stronger winds and more efficient transport, as do more metallic stars compared to low-Z ones. The net result depends on their relative strength, as shown in Figure 9.5. In the $3\,M_\odot$ and the $9\,M_\odot$ models (black solid and grey long-dashed curves respectively), the winds are negligible; the only difference is the strength of the transport, which is a little larger inside the $9\,M_\odot$ model. The $20\,M_\odot$ model (grey short-dashed curve) has a stronger transport, but the winds also become nonnegligible, so the net result is a slight decrease of V_{surf} during the main sequence. The case with no transport at all (local conservation of the angular momentum) is shown (dark grey, short-dashed-dotted

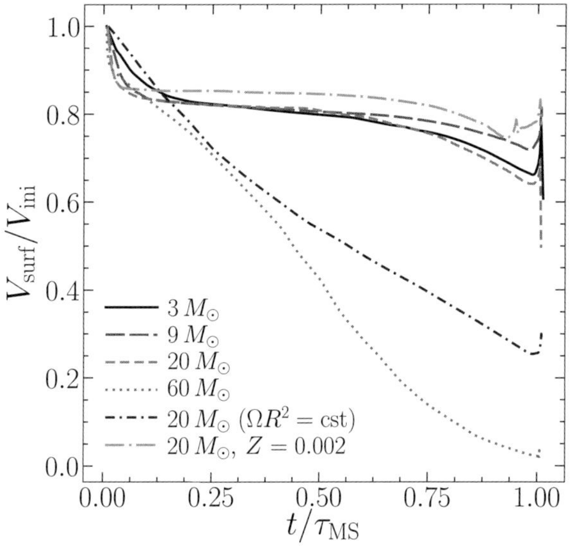

Figure 9.5 Evolution of the equatorial velocity (normalised to the initial one) during the main sequence for various models at $Z = 0.014$ computed with an initial rotation ratio $\omega = 0.50$.

curve) for comparison. In the $60\,M_\odot$ model, the transport is very strong but the wind effect largely dominates and the star is very efficiently braked. Actually, in that mass range, any comparison of rotation rates with observations should be considered as a check for the mass-loss rate used rather than a check for the treatment of rotation itself. A low metallicity weakens the internal transport but also the winds, as illustrated by the grey long-dashed-dotted curve in Figure 9.5 ($20\,M_\odot$ model at $Z = 0.002$), to be compared with the grey short-dashed one (same mass, $Z = 0.014$).

Note that the surface velocity and its ratio to the critical velocity V/V_{crit} evolve differently: $V_{crit} = \sqrt{\frac{2}{3}\frac{GM}{R_{pol}}}$ decreases always during the main sequence, because the mass decreases and the radius inflates. Therefore, even if the surface velocity decreases with the evolution, the ratio to the critical velocity usually increases during the main sequence.

9.4 Conclusion

It is hence now clear that there are important uncertainties than must be taken into account when modelling single massive stars. For example, it has been discussed how the inclusion of various mass-loss rates prescription can in fact affect the evolution of massive stars. In particular, the mass loss during the red supergiant phase, which is still largely unknown, plays a key role for the subsequent evolution of the star and its final location in the Hertzsprung–Russell diagram.

Another key ingredient of stellar modelling is the treatment of convection. Here again, the difficulties of a correct treatment in classical 1D stellar evolution codes are a source of major shortcomings in our understanding of massive star life. The hopes for a better understanding and an improved modelling of convection come nowadays from multidimensional hydrodynamics simulations.

Finally, the implementation of rotation, which is suspected to play an important role as a mixing process inside massive stars, also suffers from uncertainties. The various possible ways of implementing rotation in stellar evolution codes make the prediction of simulations largely uncertain.

All of the uncertainties discussed in this chapter, related to single-star evolution, also appear in the modelling of multiple systems, and they should be kept in mind when considering the results of any simulations involving stars.

References

Abbott, D. C. 1982. The theory of radiatively driven stellar winds. II. The line acceleration. *ApJ*, **259**, 282.

Adams, S. M., Kochanek, C. S., Gerke, J. R., et al., 2017. The search for failed supernovae with the Large Binocular Telescope: confirmation of a disappearing star. *MNRAS*, **468**, 4968–4981.

Arnett, D., Meakin, C., and Young, P. A. 2009. Turbulent convection in stellar interiors. II. The velocity field. *ApJ*, **690**, 1715–1729.

Arnett, W. D., Meakin, C., Viallet, M., et al., 2015. Beyond mixing-length theory: a step toward 321D. *ApJ*, **809**, 30.

Battino, U., Pignatari, M., Ritter, C., et al., 2016. Application of a theory and simulation-based convective boundary mixing model for AGB star evolution and nucleosynthesis. *ApJ*, **827**, 30.

Beasor, E. R., and Davies, B. 2016. The evolution of red supergiants to supernova in NGC 2100. *MNRAS*, **463**, 1269–1283.

Beasor, E. R., and Davies, B. 2018. The evolution of red supergiant mass-loss rates. *MNRAS*, **475**, 55–62.

Bestenlehner, J. M., Gräfener, G., Vink, J. S., et al., 2014. The VLT-FLAMES Tarantula Survey. XVII. Physical and wind properties of massive stars at the top of the main sequence. *A&A*, **570**, A38.

Böhm-Vitense, E. 1958. Über die Wasserstoffkonvektionszone in Sternen verschiedener Effektivtemperaturen und Leuchtkräfte. Mit 5 Textabbildungen. *ZAp*, **46**, 108.

Carciofi, A. C., Domiciano de Souza, A., Magalhães, A. M., et al., 2008. On the determination of the rotational oblateness of Achernar. *ApJ*, **676**, L41.

Castor, J. I., Abbott, D. C., and Klein, R. I. 1975. Radiation-driven winds in Of stars. *ApJ*, **195**, 157–174.

Chaboyer, B., and Zahn, J.-P. 1992. Effect of horizontal turbulent diffusion on transport by meridional circulation. *A&A*, **253**, 173–177.

Chiavassa, A., Plez, B., Josselin, E., et al., 2009. Radiative hydrodynamics simulations of red supergiant stars. I. Interpretation of interferometric observations. *A&A*, **506**, 1351–1365.

Chieffi, A., and Limongi, M. 2013. Pre-supernova evolution of rotating solar metallicity stars in the mass range 13-120 M and their explosive yields. *ApJ*, **764**, 21.

Choplin, A., Ekström, S., Meynet, G., et al., R. 2017. Pre-supernova mixing in CEMP-no source stars. *A&A*, **605**, A63.

Couch, S. M., Chatzopoulos, E., Arnett, W. D., et al., 2015. The three-dimensional evolution to core collapse of a massive star. *ApJ L*, **808**, L21.

Cristini, A., Meakin, C., Hirschi, R., et al., 2017. 3D hydrodynamic simulations of carbon burning in massive stars. *MNRAS*, **471**, 279–300.

Davis, A., Jones, S., and Herwig, F. 2017. Convective boundary mixing in a post-He core burning massive star model. *ArXiv e-prints*, 1712.00114.

de Mink, S. E., and Mandel, I. 2016. The chemically homogeneous evolutionary channel for binary black hole mergers: rates and properties of gravitational-wave events detectable by advanced LIGO. *MNRAS*, **460**, 3545–3553.

de Mink, S. E., Langer, N., Izzard, R. G., et al., 2013. The rotation rates of massive stars: the role of binary interaction through tides, mass transfer, and mergers. *ApJ*, **764**, 166.

Domiciano de Souza, A., Kervella, P., Jankov, S., et al., 2003. The spinning-top Be star Achernar from VLTI-VINCI. *A&A*, **407**, L47.

Eddington, A. S. 1925. Circulating currents in rotating stars. *The Observatory*, **48**, 73–75.

Eddington, A. S. 1926. *The Internal Constitution of the Stars*. Dover Publications Inc.

Eggenberger, P., Meynet, G., Maeder, A., et al., 2008. The Geneva stellar evolution code. *Astrophys.&Sp.Sci.*, **316**, 43–54.

Ekström, S., Georgy, C., Eggenberger, P., et al., 2012. Grids of stellar models with rotation. I. Models from 0.8 to 120 $M_\odot$ at solar metallicity (Z = 0.014). *A&A*, **537**, A146.

Eldridge, J. J., Izzard, R. G., and Tout, C. A. 2008. The effect of massive binaries on stellar populations and supernova progenitors. *MNRAS*, **384**, 1109–1118.

Endal, A. S., and Sofia, S. 1976. The evolution of rotating stars. I. Method and exploratory calculations for a 7-solar-mass star. *ApJ*, **210**, 184–198.

Espinosa Lara, F., and Rieutord, M. 2011. Gravity darkening in rotating stars. *A&A*, **533**, A43.

Freytag, B., and Höfner, S. 2008. Three-dimensional simulations of the atmosphere of an AGB star. *A&A*, **483**, 571–583.

Freytag, B., Ludwig, H.-G., and Steffen, M. 1996. Hydrodynamical models of stellar convection: the role of overshoot in DA white dwarfs, A-type stars, and the Sun. *A&A*, **313**, 497–516.

Georgy, C. 2012. Yellow supergiants as supernova progenitors: an indication of strong mass loss for red supergiants? *A&A*, **538**, L8.

Georgy, C. 2017. Evolution models of red supergiants. Pages 193–198 of: Eldridge, J. J., Bray, J. C., McClelland, L. A. S., and Xiao, L. (eds), *The Lives and Death-Throes of Massive Stars*. IAU Symposium, vol. 329.

Georgy, C., Ekström, S., Granada, A., et al., 2013. Populations of rotating stars. I. Models from 1.7 to 15 $M_\odot$ at Z = 0.014, 0.006, and 0.002 with Ω/Ω_{crit} between 0 and 1. *A&A*, **553**, A24.

Georgy, C., Granada, A., Ekström, S., et al., 2014a. Populations of rotating stars. III. SYCLIST, the new Geneva population synthesis code. *A&A*, **566**, A21.

Georgy, C., Meynet, G., and Maeder, A. 2011. Effects of anisotropic winds on massive star evolution. *A&A*, **527**, A52.

Georgy, C., Saio, H., and Meynet, G. 2014b. The puzzle of the CNO abundances of α Cygni variables resolved by the Ledoux criterion. *MNRAS*, **439**, L6–L10.

Gräfener, G., Vink, J. S., de Koter, A., et al., 2011. The Eddington factor as the key to understand the winds of the most massive stars: evidence for a Γ-dependence of Wolf–Rayet type mass loss. A&A, **535**, A56.

Heger, A., Fryer, C. L., Woosley, S. E., et al., 2003. How massive single stars end their life. *ApJ*, **591**, 288.

Heger, A., Langer, N., and Woosley, S. E. 2000. Presupernova evolution of rotating massive stars. I. Numerical method and evolution of the internal stellar structure. *ApJ*, **528**, 368–396.

Herwig, F. 2000. The evolution of AGB stars with convective overshoot. *A&A*, **360**, 952–968.

Herwig, F., Freytag, B., Hueckstaedt, R. M., et al., 2006. Hydrodynamic simulations of He shell flash convection. *ApJ*, **642**, 1057–1074.

Jones, S., Andrassy, R., Sandalski, S., et al., 2017. Idealized hydrodynamic simulations of turbulent oxygen-burning shell convection in 4π geometry. *MNRAS*, **465**, 2991–3010.

Jones, S., Hirschi, R., Nomoto, K., et al., 2013. Advanced burning stages and fate of 8-10 M stars. *ApJ*, **772**, 150.

Keszthelyi, Z., Puls, J., and Wade, G. A. 2017. Modeling the early evolution of massive OB stars with an experimental wind routine: the first bi-stability jump and the angular momentum loss problem. *A&A*, **598**, A4.

Kippenhahn, R., and Weigert, A. 1990. *Stellar Structure and Evolution*, XVI, Springer-Verlag Berlin Heidelberg New York. Also Astronomy and Astrophysics Library.

Kolmogorov, A. 1941. The local structure of turbulence in incompressible viscous fluid for very large Reynolds' numbers. *Akademiia Nauk SSSR Doklady*, **30**, 301–305.

Kolmogorov, A. N. 1962. A refinement of previous hypotheses concerning the local structure of turbulence in a viscous incompressible fluid at high Reynolds number. *Journal of Fluid Mechanics*, **13**, 82–85.

Kudritzki, R. P. 2002. Line-driven winds, ionizing fluxes, and ultraviolet spectra of hot stars at extremely low metallicity. I. Very massive O stars. *ApJ*, **577**, 389.

Kudritzki, R.-P., and Puls, J. 2000. Winds from hot stars. *ARA&A*, **38**, 613–666.

Lamers, H. J. G. L. M., Snow, T. P., and Lindholm, D. M. 1995. Terminal velocities and the bistability of stellar winds. *ApJ*, **455**, 269.

Lucy, L. B., and Solomon, P. M. 1970. Mass loss by hot stars. *ApJ*, **159**, 879.

Maeder, A. 1975. Stellar evolution. IV. Evolution of a star of 1.5 solar masses from the main-sequence to the red-giant branch with and without overshooting from convective core. *A&A*, **43**, 61–69.

Maeder, A. 1997. Stellar evolution with rotation. II. A new approach for shear mixing. *A&A*, **321**, 134–144.

Maeder, A. 1999. Stellar evolution with rotation IV: von Zeipel's theorem and anisotropic losses of mass and angular momentum. *A&A*, **347**, 185–193.

Maeder, A. 2003. Stellar rotation: evidence for a large horizontal turbulence and its effects on evolution. *A&A*, **399**, 263–269.

Maeder, A. 2009. *Physics, Formation and Evolution of Rotating Stars*. Berlin, Heidelberg: Springer Verlag.

Maeder, A., and Meynet, G. 2000. Stellar evolution with rotation. VI. The Eddington and Omega -limits, the rotational mass loss for OB and LBV stars. *A&A*, **361**, 159–166.

Maeder, A., and Zahn, J.-P. 1998. Stellar evolution with rotation. III. Meridional circulation with MU -gradients and non-stationarity. *A&A*, **334**, 1000–1006.

Maeder, A., Meynet, G., Lagarde, N., and Charbonnel, C. 2013. The thermohaline, Richardson, Rayleigh-Taylor, Solberg-Høiland, and GSF criteria in rotating stars. *A&A*, **553**, A1.

Magic, Z., Collet, R., Asplund, M., et al., 2013. The Stagger-grid: a grid of 3D stellar atmosphere models. I. Methods and general properties. *A&A*, **557**, A26.

Martins, F., and Palacios, A. 2013. A comparison of evolutionary tracks for single Galactic massive stars. *A&A*, **560**, A16.

Mathis, S., Palacios, A., and Zahn, J.-P. 2004. On shear-induced turbulence in rotating stars. *A&A*, **425**, 243–247.

Mauron, N., and Josselin, E. 2011. The mass-loss rates of red supergiants and the de Jager prescription. *A&A*, **526**, A156.

Meakin, C. A., and Arnett, D. 2007. Turbulent convection in stellar interiors. I. Hydrodynamic simulation. *ApJ*, **667**, 448–475.

Meynet, G., Chomienne, V., Ekström, S., et al., 2015. Impact of mass-loss on the evolution and pre-supernova properties of red supergiants. *A&A*, **575**, A60.

Meynet, G., Ekström, S., Maeder, A., et al., 2013. Models of rotating massive stars: impacts of various prescriptions. Page 3 of: Goupil, M., Belkacem, K., Neiner, C., Lignières, F., and Green, J. J. (eds), *Studying Stellar Rotation and Convection*. Lecture Notes in Physics, vol. 865. Berlin: Springer Verlag.

Mocák, M., Meakin, C., Viallet, M., et al., 2014. Compressible hydrodynamic mean-field equations in spherical geometry and their application to turbulent stellar convection data. *ArXiv e-prints*, 1401.5176.

Mokiem, M. R., de Koter, A., Vink, J. S., et al., 2007. The empirical metallicity dependence of the mass-loss rate of O- and early B-type stars. *A&A*, **473**, 603–614.

Monnier, J. D., Zhao, M., Pedretti, E., et al., 2007. Imaging the surface of Altair. *Science*, **317**, 342.

Moravveji, E., Aerts, C., Pápics, P. I., et al., 2015. Tight asteroseismic constraints on core overshooting and diffusive mixing in the slowly rotating pulsating B8.3V star KIC 10526294. *A&A*, **580**, A27.

Müller, B., Viallet, M., Heger, A., et al., 2016. The last minutes of oxygen shell burning in a Massive Star. *ApJ*, **833**, 124.

Müller, P. E., and Vink, J. S. 2014. Rotating massive O stars with non-spherical 2D winds. *A&A*, **564**, A57.

Najarro, F., Hanson, M. M., and Puls, J. 2011. L-band spectroscopy of Galactic OB-stars. *A&A*, **535**, A32.

O'Connor, E., and Ott, C. D. 2011. Black hole formation in failing core-collapse supernovae. *ApJ*, **730**, 70.

Paxton, B., Bildsten, L., Dotter, A., et al., 2011. Modules for experiments in stellar astrophysics (MESA). *ApJ S*, **192**, 3.

Paxton, B., Cantiello, M., Arras, P., et al., 2013. Modules for experiments in stellar astrophysics (MESA): planets, oscillations, rotation, and massive stars. *ApJ S*, **208**, 4.

Paxton, B., Marchant, P., Schwab, J., et al., 2015. Modules for experiments in stellar astrophysics (MESA): binaries, pulsations, and explosions. *ApJ S*, **220**, 15.

Paxton, B., Schwab, J., Bauer, E. B., et al., 2018. Modules for experiments in stellar astrophysics (MESA): convective boundaries, element diffusion, and massive star explosions. *ApJ S*, **234**, 34.

Petrovic, J., Langer, N., Yoon, S.-C., et al., 2005. Which massive stars are gamma-ray burst progenitors? *A&A*, **435**, 247–259.

Podsiadlowski, P., Joss, P. C., and Hsu, J. J. L. 1992. Presupernova evolution in massive interacting binaries. *ApJ*, **391**, 246–264.

Puls, J., Vink, J. S., and Najarro, F. 2008. Mass loss from hot massive stars. *A&A Rev.*, **16**, 209–325.

Rauw, G., Hervé, A., Nazé, Y., et al., 2015. Simultaneous X-ray and optical spectroscopy of the Oef supergiant λ Cephei. *A&A*, **580**, A59.

Saio, H., Georgy, C., and Meynet, G. 2013. Evolution of blue supergiants and α Cygni variables: puzzling CNO surface abundances. *MNRAS*, **433**, 1246–1257.

Sana, H., de Mink, S. E., de Koter, A., et al., 2012. Binary interaction dominates the evolution of massive stars. *Science*, **337**, 444.

Sana, H., de Koter, A., de Mink, S. E., et al., 2013. The VLT-FLAMES Tarantula Survey. VIII. Multiplicity properties of the O-type star population. *A&A*, **550**, A107.

Smartt, S. J., Eldridge, J. J., Crockett, R. M., et al., 2009. The death of massive stars. I. Observational constraints on the progenitors of Type II-P supernovae. *MNRAS*, **395**, 1409–1437.

Sukhbold, T., and Woosley, S. E. 2014. The compactness of presupernova stellar cores. *ApJ*, **783**, 10.

Šurlan, B., Hamann, W.-R., Aret, A., et al., 2013. Macroclumping as solution of the discrepancy between Hα and P v mass loss diagnostics for O-type stars. *A&A*, **559**, A130.

Szécsi, D., Langer, N., Yoon, S.-C., et al., 2015. Low-metallicity massive single stars with rotation: evolutionary models applicable to I Zwicky 18. *A&A*, **581**, A15.

Talon, S., and Zahn, J.-P. 1997. Anisotropic diffusion and shear instabilities. *A&A*, **317**, 749–751.

van den Heuvel, E. P. J. 1975. Modes of mass transfer and classes of binary X-ray sources. *ApJ L*, **198**, L109.

van Loon, J. T., Cioni, M.-R. L., Zijlstra, A. A., et al., 2005. An empirical formula for the mass-loss rates of dust-enshrouded red supergiants and oxygen-rich Asymptotic Giant Branch stars. *A&A*, **438**, 273–289.

Vanbeveren, D. 1991. Massive close binaries: observations and evolutionary computations. *Space Sci. Rev.*, **56**, 249–311.

Vanbeveren, D., De Donder, E., van Bever, J., et al., 1998. The WR and O-type star population predicted by massive star evolutionary synthesis. *New Astronomy*, **3**, 443–492.

Viallet, M., Meakin, C., Arnett, D., et al., 2013. Turbulent convection in stellar interiors. III. Mean-field analysis and stratification effects. *ApJ*, **769**, 1.

Vinicius, M. M. F., Zorec, J., Leister, N. V., et al., 2006. α Eridani: rotational distortion, stellar and circumstellar activity. *A&A*, **446**, 643–660.

Vink, J. S. 2006 . Massive star feedback – from the first stars to the present. Page 113 of: Lamers, H. J. G. L. M., Langer, N., Nugis, T., and Annuk, K. (eds), *Stellar Evolution at Low Metallicity: Mass Loss, Explosions, Cosmology*. Astronomical Society of the Pacific Conference Series, vol. 353. San Francisco: Astonomical Society of the Pacific.

Vink, J. S., de Koter, A., and Lamers, H. J. G. L. M. 2000. New theoretical mass-loss rates of O and B stars. *A&A*, **362**, 295–309.

Vink, J. S., de Koter, A., and Lamers, H. J. G. L. M. 2001. Mass-loss predictions for O and B stars as a function of metallicity. *A&A*, **369**, 574–588.

Vink, J. S., Muijres, L. E., Anthonisse, B., et al., 2011. Wind modelling of very massive stars up to 300 solar masses. *A&A*, **531**, A132.

von Zeipel, H. 1924. The radiative equilibrium of a rotating system of gaseous masses. *MNRAS*, **84**, 665–683.

Walmswell, J. J., and Eldridge, J. J. 2012. Circumstellar dust as a solution to the red supergiant supernova progenitor problem. *MNRAS*, **419**, 2054–2062.

Woodward, P. R., Herwig, F., and Lin, P.-H. 2015. Hydrodynamic simulations of H entrainment at the top of He-shell flash convection. *ApJ*, **798**, 49.

Yoon, S.-C., Langer, N., and Norman, C. 2006. Single star progenitors of long gamma-ray bursts. I. Model grids and redshift dependent GRB rate. *A&A*, **460**, 199–208.
Yusof, N., Hirschi, R., Meynet, G., et al., 2013. Evolution and fate of very massive stars. *MNRAS*, **433**, 1114–1132.
Zahn, J.-P. 1991. Convective penetration in stellar interiors. *A&A*, **252**, 179–188.
Zahn, J.-P. 1992. Circulation and turbulence in rotating stars. *A&A*, **265**, 115–132.
Zhao, M., Monnier, J. D., Pedretti, E., et al., 2009. Imaging and modeling rapidly rotating stars: α Cephei and α Ophiuchi. *ApJ*, **701**, 209.

10
Binarity at High Masses

HUGUES SANA[1]

10.1 Massive Stars in the Grand Scheme of Things

Stars with birth masses larger than about eight times the mass of our Sun end their life in powerful explosions, as core-collapse supernovae (CCSNe) and gamma-ray bursts (GRBs). They are the progenitors of neutrons stars and black holes, the coalescence of which produces observable gravitational wave (GW) events (Abbott et al., 2016). With large luminosities, strong stellar winds and spectacular explosions, these massive stars heat and enrich the surrounding gas clouds, where new generations of stars and planets form, and they drive the chemical evolution of galaxies (Bromm et al., 2009; de Rossi et al., 2010). Entire populations of massive stars, and their final explosions, are so bright that they can be seen throughout the Universe and allow astronomers to study its properties across cosmic times (Tanvir et al., 2009; Stark, 2016). Furthermore, the first generation of massive stars is thought to have played a pivotal role in the assembly of the first galaxies (Bromm and Larson, 2004; Robertson et al., 2010) and to have contributed to the reionisation of the early Universe. Given the decisive role of massive stars for a wide range of astrophysical problems, it is of paramount importance to understand how massive stars form, live and die, both in our Galaxy and in the Universe near and far. Unfortunately, too many uncertainties remain, including their formation mechanism, multiplicity, internal mixing and stellar winds (Langer, 2012). In this chapter, we focus on recent observational constraints on the stellar multiplicity at high masses.

10.1.1 The Issue of Multiplicity

The abundance of massive stars recently found in close multiple systems has called for a paradigm shift in our thinking: massive stars rarely, if at all, form and live in isolation (Mason et al., 2009; Sana et al., 2012; Kobulnicky et al., 2014; Sana et al., 2014; Maíz Apellániz et al., 2016). Recent observational constraints have revealed that the majority of massive, O-type stars are part of a binary system so tight that the two stars will exchange mass, or even merge, before any of the components explode as supernova (Sana et al., 2012). This interaction has severe consequences for their further evolution and final fate,

[1] Supported by the Research Foundation–Flanders FWO, Odysseus program under project G0F8H6N.

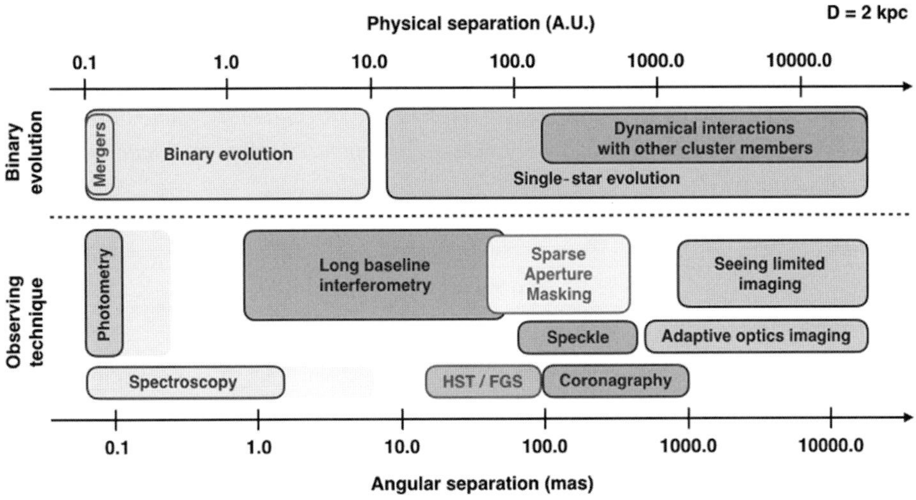

Figure 10.1 Sketch of typical separation ranges relevant for massive star evolution and respective binary detection domains of a variety of observational techniques. A distance of 2 kpc has been assumed, which is typical for nearby O-type stars.

including the properties of their final explosions and the nature of the compact objects that are left behind.

10.1.2 The Parameter Space

The nature of the binary interactions, their timing and outcome critically depend on the initial properties of the binaries, including the components initial masses, orbital period ($P_{\rm orb}$) or separation, eccentricity (e) and mass ratio (M_2/M_1). Massive star formation theories cannot yet reliably predict the birth multiplicity properties of massive stars. Observational constraints on the actual binary fraction and on the distribution of their orbital properties are thus needed to constrain the final stages of massive star formation and to set up the initial conditions on their future evolution. Constraining these quantities, however, requires one to map a range of orbital separation spanning at least four orders of magnitude, from about 0.1 au to $10^3 - 10^4$ au. This can only be achieved by combining a variety of observational techniques, each with its own range of sensitivity (Figure 10.1). The following section will emphasise recent progress obtained through spectroscopy and high-angular resolution techniques as they are ideally suited to cover the separation range relevant to binary evolution.

10.2 The Spectroscopic Regime

Multi-epoch spectroscopy efficiently detects massive binaries with orbital periods up to a few years through their radial velocity (RV) variations. Modern spectroscopic surveys have

Table 10.1 *Overview of the results of recent spectroscopic surveys (column 1), including the sample spectral type (column 2), the sample size N (column 3), the numbers of detected binaries N_{SB} (column 4) and of binaries with an orbital solution N_{SBO} (column 5). The observed (f_{SB}^{obs}) and bias-corrected ($f_{SB}^{corrected}$) binary fraction are provided in columns 6 and 7. The associated error bars were computed following binomial statistics. Adapted from Sana (2017).*

Region/survey	**SpT**	N	N_{SB}	N_{SBO}	f_{SB}^{obs}	$f_{SB}^{corrected}$	**Reference**
				Milky Way: O-type stars[a]			
BESO Survey	O	243	166	–	0.68 ± 0.03	–	Chini et al. (2012)
Young Clusters	O	71	40	34	0.56 ± 0.06	0.69 ± 0.09	Sana et al. (2012)
GOSSS	O	194	97 (low)	–	0.50 ± 0.03	–	Sota et al. (2014)
	O		117 (high)	–	0.60 ± 0.03	–	
IACOB[b]	O	141	66	–	0.47 ± 0.04	–	Simón-Díaz (unpubl.)
Clusters/Assoc	O	161	68 (low)	68	0.42 ± 0.04	–	Aldoretta et al. (2015)
	O		96 (high)	68	0.60 ± 0.04	–	
OWN	O	205	102	85	0.50 ± 0.03	–	Barbá et al. (2017)
Cyg OB2	O	45	23	23	0.51 ± 0.07	–[c]	Kobulnicky et al. (2014)
	OB	*128*	*48*	*48*	0.38 ± 0.04	*0.55*[c]	
				Milky Way: B-type stars			
Cyg OB2	B0-2	83	25	25	0.30 ± 0.05	–[c]	Kobulnicky et al. (2014)
BESO Survey	B0-3	226	105	–	0.46 ± 0.03	–	Chini et al. (2012)
	B4-9	353	67	–	0.19 ± 0.02	–	
				Large Magellanic Cloud (LMC)			
30Dor/VFTS	O	360	124	–	0.35 ± 0.03	0.51 ± 0.03	Sana et al. (2013)
30Dor/TMBM	O			79	–	≈ 0.58	Almeida et al. (2017)
30Dor/VFTS	B0-3	408	102	–	0.25 ± 0.02	0.58 ± 0.11	Dunstall et al. (2015)

Notes: [a]The O-type samples in the Milky Way have strong overlap, so that the quoted measurements should not be viewed as being independent. [b]Numbers only concerns O stars with more than three observational epochs. The IACOB survey covers another ~ 40 Northern O stars, but with fewer epochs so far. [c]Kobulnicky et al. (2014) do not provide separate completeness correction for the O- and B-type subsamples in Cyg OB2.

sample sizes from several tens to a few hundreds objects, allowing them to reach statistical uncertainties of the order of 3% to 5% on their binary detection rates. The detection probability of a given binary depends strongly on its orbital properties. For example, the amplitudes of the RV variations and the chance to sample the phases of large variations both decrease largely for long-period, highly eccentric systems and for systems with low-mass ratios (typically $M_2/M_1 < 0.25$). These binaries are thus more difficult to detect than their shorter period counterparts. The observational results and, for the studies providing a bias correction, the intrinsic binary fractions are listed in Table 10.1.

For O-type stars, these surveys consistently find an observed spectroscopic binary fraction (f_{SB}^{obs}) larger than 50%. These numbers are not corrected for observational biases and

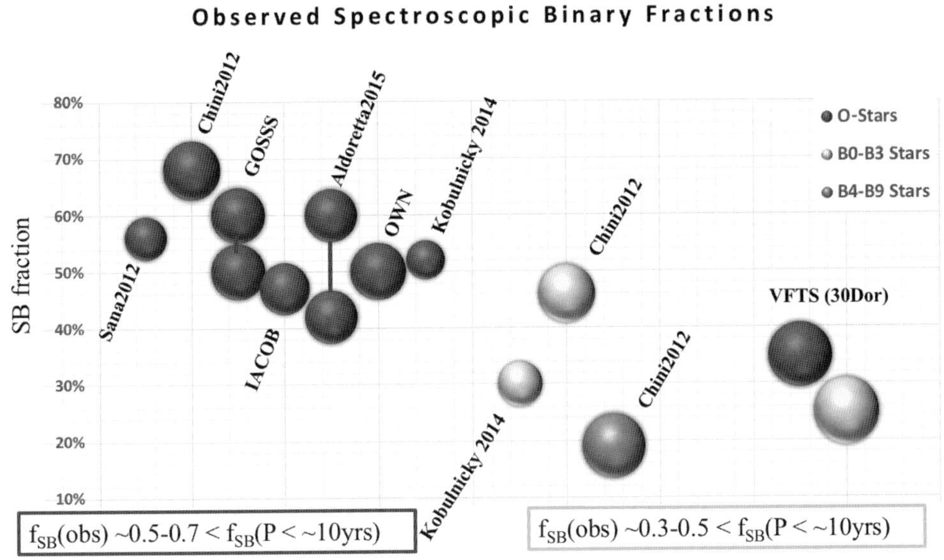

Figure 10.2 Detected spectroscopic binary fraction $f_{\rm SB}^{\rm obs}$ for O- and B-type stars. The symbols areas are proportional to the sample sizes N. Based on figures given in Table 10.1.

set thus a strict lower limit on the overall multiplicity fraction of O stars (Figure 10.2). An accurate bias correction requires the knowledge of both the observations sensitivity and of the intrinsic distributions of orbital parameters, and is thus far from trivial. The number of studies that have attempted such a correction are limited. Those that have indicate that, as far as O-type stars are concerned, one can expect a correction factor of the order of 1.25 to 1.5 between the observed ($f_{\rm SB}^{\rm obs}$) and bias-corrected ($f_{\rm SB}^{\rm corrected}$) binary fractions (e.g., Sana et al., 2012; Kobulnicky et al., 2014). Because the vast majority of the binaries in the spectroscopic regimes have separations that are of the order of, or less than, the radius of a red supergiant, one can further expect that most of the detected binary systems will interact, leaving no doubt on the fact that binary interaction through mass transfer or coalescence is a widespread phenomena among O-type stars. Aside from the spectroscopic binary fraction, the observed period distributions that have been recorded so far only show small differences despite the different metallicity and age distributions of the samples involved (Figure 10.3). This points towards a relative universality of the multiplicity properties within the environments explored so far.

For early B-type stars, observational constraints are scarcer and less stringent. This results from the overall fainter luminosities and smaller RV amplitudes at play owing to the smaller stellar masses of B stars, and an overall smaller detection rate. While the uncertainties remain large, results from Kobulnicky et al. (2014) and Dunstall et al. (2015) seem to indicate bias-corrected binary fractions $\gtrsim 0.5$. Unfortunately, the current data do not yet allow us to firmly answer the question of whether the multiplicity properties of

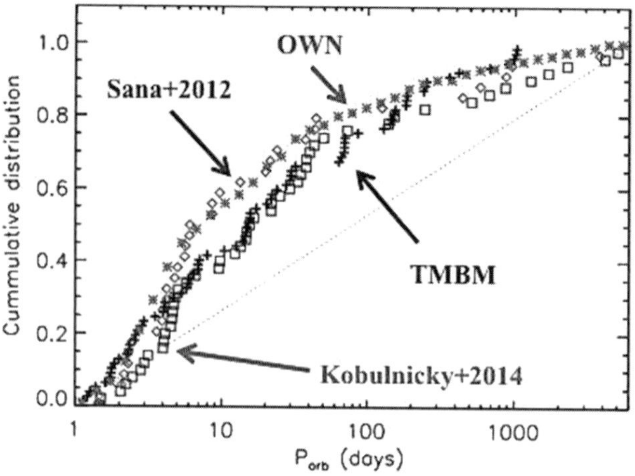

Figure 10.3 Observed orbital period distributions in four recent spectroscopic surveys (see Table 10.1).

early B stars are significantly different from those of O-type stars. The answers to these questions are important, as progenitors of core-collapse supernovae are dominated by stars born as B-type on the zero-age main sequence (ZAMS).

10.3 The High-Angular Resolution Regime

Recent progress in the efficiency and sensitivity of high-angular resolution techniques – in particular, in long-baseline optical-infrared interferometry – has allowed researchers to cross the interferometric gap that was separating the spectroscopic regime and the classical and AO regimes (Mason et al., 1998; see also Figure 10.1). Results of two recent surveys now allow for a complete coverage of the separation range down to $\sim$0.001″. These surveys are the Southern Massive Stars at high-angular resolutions (SMaSH+; Sana et al., 2014) at the ESO Very Large Telescope (VLT) and Interferometer (VLTI), and the High-Angular Resolution Survey by Aldoretta et al. (2015), which used the Fine Guidance Sensor (FGS) onboard the Hubble Space Telescope (HST). Small-angular separation pairs detected by SMaSH+ and by the FGS are displayed in Figure 10.4 and reach a contrast of 5-magnitude or more over the entire interferometric gap.

With more than 200 new companions detected, these surveys have yielded a record number of companions and multiplicity fractions. Accounting for known but unresolved spectroscopic or eclipsing companions – but excluding runaway stars – the SMaSH+ obtained a fraction of multiple systems of $f_{\rm m} = 0.91 \pm 0.03$ before any detection-bias correction. The fraction of triples and higher-order multiples reaches >35% when limiting to magnitude contrast of ΔH <5. It reaches over 60% when including fainter companions, although a number of these are probably due to spurious alignment along the line of sight.

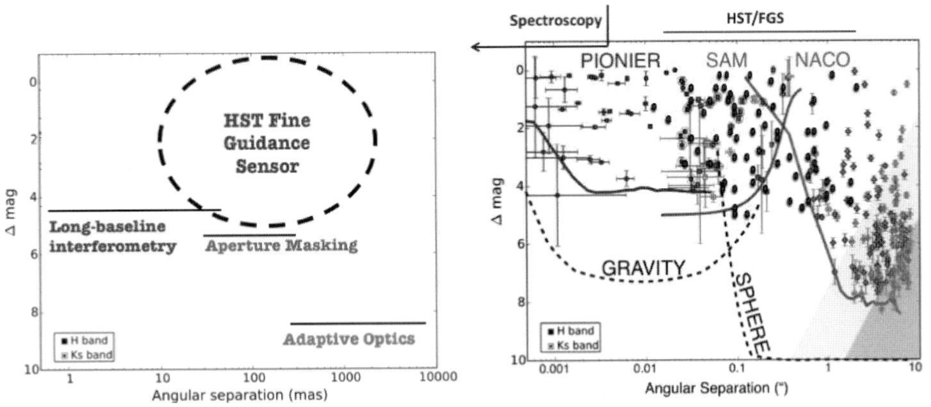

Figure 10.4 Parameter space (left) and resolved O-type binaries (right) of the SMaSH+ (PIONIER, Sparse Aperture Interferometric Mask [SAM], NACO) and the FGS surveys. Light- and dark-grey shade areas indicate regions were the chance of spurious alignements is above 1% and 50%, respectively. The sensitivity range of the second-generation of the VLT(I) instruments GRAVITY and SPHERE are also indicated.

These results show again that massive stars are formed and live almost exclusively as part of a multiple system.

The second generation of VLT(I) instruments as well as recent instrumental developments at the Gemini and the Center for High-Angular Resolution Astronomy (CHARA) facilities are currently allowing us to push the contrast boundaries even further. As an illustration, the companion discovery space opened by the extreme AO and coronagraphic system Spectro-Polarimetric High-Contrast Exoplanet Research (SPHERE) and by the four-beam combiner GRAVITY at the VLT and VLTI, respectively, is sketched in Figure 10.4. Figure 10.5 reveals the detection of a previously unknown companion around the high-order multiple system QZ Car, with a contrast of about 6.5 to 7 mag and at a separation of only 0.6″. Performed on a significant sample, that kind of observations will allow us to investigate whether the large number of faint companions seen, e.g., in Figure 10.4, at separation over 1″, continues at smaller separations or whether one can identify a physical scale length at which no such companions exist.

10.4 Future Work

While it is clear that multiplicity plays a fundamental role in the formation and evolution of massive stars, its exact implication on the evolution and on the properties of entire populations of massive stars, and of the distribution of their end-of-life products, remains hardly quantifiable. This is because the multiplicity properties and the binary physics are still insufficiently constrained and because accurate observational constraints have so far mostly focused on O-type stars at solar and LMC metallicity. On the one hand, obtaining accurate physical and atmospheric properties of large samples of massive O, B and

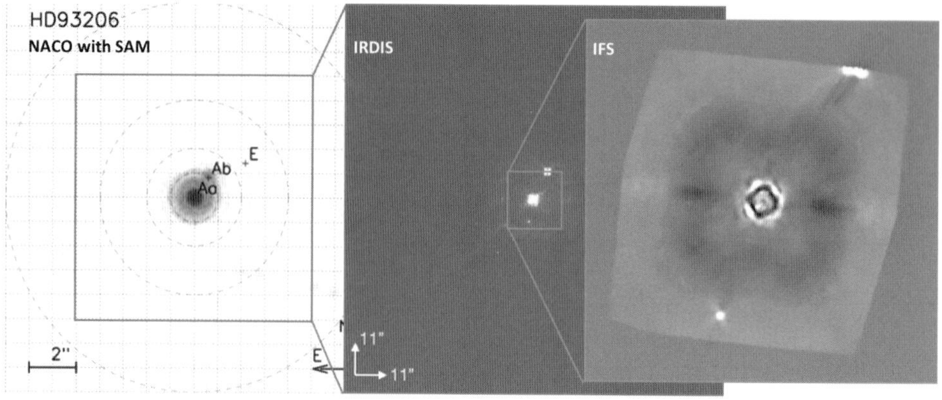

Figure 10.5 Comparison of the field of view of NACO (left), SPHERE/IRDIS (middle) and SPHERE/IFS (right) for the massive high-order multiple system QZ Car. The NACO observations were done with the SAM mask inserted into the light path, hence the shape of the point-spread function, while the SPHERE observations were performed with the YJH coronagraph. The bright companion on the upper edge of the IFS field of view is Ab, while the less bright companion to the South is a new companion that was not visible in the NACO image obtained by the SMaSH+ collaboration. The SPHERE images are courtesy of Rainot et al. (in prep.).

Wolf–Rayet binaries is a necessary task if we are to confront in any meaningful way modern binary evolution computation. More details on the uncertainties in modelling massive binaries can be found in Chapter 9. On the other hand, it is important to expand the region of the parameter space where high-quality constraints have been obtained. While offering an important window, the sole O-type mass domain (corresponding to the 15-60 $M_\odot$ range) is indeed insufficient to fully explore the role and implication of the multiplicity of mass stars. For example, the distribution of core-collapse supernovae is actually dominated by progenitors with initial masses in the 8 to 16 $M_\odot$ range (i.e., B0–B3 stars on the ZAMS). Similarly, the massive double black hole binaries such as those involved in many of the gravitational wave events detected so far are likely to come from more massive stars and at a lower metallicity regime (Belczynski et al., 2016; de Mink and Mandel, 2016; Marchant et al., 2016). Obtaining high-quality constraints on the multiplicity properties at lower- and higher-mass stars and at sub-LMC metallicity is thus one important task for our community in the year to come.

References

Abbott, B. P., Abbott, R., Abbott, T. D., et al. 2016. GW151226: Observation of Gravitational Waves from a 22-Solar-Mass Binary Black Hole Coalescence. *Physical Review Letters*, **116**(24), 241103.

Aldoretta, E. J., Caballero-Nieves, S. M., Gies, D. R., et al., 2015. The Multiplicity of Massive Stars: A High Angular Resolution Survey with the Guidance Sensor. *AJ*, **149**, 26.

Almeida, L. A., Sana, H., Taylor, W., et al., 2017. The Tarantula Massive Binary Monitoring. I. Observational Campaign and OB-type Spectroscopic Binaries. *A&A*, **598**, A84.

Barbá, R. H., Gamen, R., Arias, J. I., et al., 2017. OWN Survey: A Spectroscopic Monitoring of Southern Galactic O and WN-Type Stars. Pages 89–96 of: Eldridge, J. J., Bray, J. C., McClelland, L. A. S., and Xiao, L. (eds), *The Lives and Death-Throes of Massive Stars*. IAU Symposium, vol. 329.

Belczynski, K., Holz, D. E., Bulik, T., and O'Shaughnessy, R. 2016. The First Gravitational-Wave Source from the Isolated Evolution of Two Stars in the 40–100 Solar Mass Range. *Nature*, **534**, 512–515.

Bromm, V., and Larson, R. B. 2004. The First Stars. *ARA&A*, **42**, 79–118.

Bromm, V., Yoshida, N., Hernquist, L., et al., 2009. The formation of the First Stars and Galaxies. *Nature*, **459**, 49–54.

Chini, R., Hoffmeister, V. H., Nasseri, A., et al., 2012. A Spectroscopic Survey on the Multiplicity of High-Mass Stars. *MNRAS*, **424**, 1925–1929.

de Mink, S. E., and Mandel, I. 2016. The Chemically Homogeneous Evolutionary Channel for Binary Black Hole Mergers: Rates and Properties of Gravitational-Wave Events Detectable by Advanced LIGO. *MNRAS*, **460**, 3545–3553.

de Rossi, M. E., Tissera, P. B., and Pedrosa, S. E. 2010. Impact of Supernova Feedback on the Tully-Fisher Relation. *A&A*, **519**, A89.

Dunstall, P. R., Dufton, P. L., Sana, H., et al., 2015. The VLT-FLAMES Tarantula Survey. XXII. Multiplicity Properties of the B-Type Stars. *A&A*, **580**, A93.

Kobulnicky, H. A., Kiminki, D. C., Lundquist, M. J., et al., 2014. Toward Complete Statistics of Massive Binary Stars: Penultimate Results from the Cygnus OB2 Radial Velocity Survey. *ApJ S*, **213**, 34.

Langer, N. 2012. Presupernova Evolution of Massive Single and Binary Stars. *ARA&A*, **50**, 107–164.

Maíz Apellániz, J., Sota, A., Arias, J. I., et al., 2016. The Galactic O-Star Spectroscopic Survey (GOSSS). III. 142 Additional O-type Systems. *ApJ S*, **224**, 4.

Marchant, P., Langer, N., Podsiadlowski, P., Tauris, T. M., and Moriya, T. J. 2016. A New Route Towards Merging Massive Black Holes. *A&A*, **588**, A50.

Mason, B. D., Gies, D. R., Hartkopf, W. I., et al., 1998. ICCD Speckle Observations of Binary Stars. XIX. An Astrometric/Spectroscopic Survey of O Stars. *AJ*, **115**, 821–+.

Mason, B. D., Hartkopf, W. I., Gies, D. R., et al., 2009. The High Angular Resolution Multiplicity of Massive Stars. *AJ*, **137**, 3358–3377.

Robertson, B. E., Ellis, R. S., Dunlop, J. S., et al., 2010. Early Star-Forming Galaxies and the Reionization of the Universe. *Nature*, **468**, 49–55.

Sana, H. 2017. The Multiplicity of Massive Stars: A 2016 View. Pages 110–117 of: Eldridge, J. J., Bray, J. C., McClelland, L. A. S., and Xiao, L. (eds), *The Lives and Death-Throes of Massive Stars*. IAU Symposium, vol. 329.

Sana, H., de Koter, A., de Mink, S. E., et al., 2013. The VLT-FLAMES Tarantula Survey. VIII. Multiplicity Properties of the O-Type Star Population. *A&A*, **550**, A107.

Sana, H., de Mink, S. E., de Koter, A., et al., 2012. Binary Interaction Dominates the Evolution of Massive Stars. *Science*, **337**, 444.

Sana, H., Le Bouquin, J.-B., Lacour, et al., 2014. Southern Massive Stars at High Angular Resolution: Observational Campaign and Companion Detection. *ApJ S*, **215**, 15.

Sota, A., Maíz Apellániz, J., et al., 2014. The Galactic O-Star Spectroscopic Survey (GOSSS). II. Bright Southern Stars. *ApJ S*, **211**, 10.

Stark, D. P. 2016. Galaxies in the First Billion Years after the Big Bang. *ARA&A*, **54**, 761–803.

Tanvir, N. R., Fox, D. B., Levan, A. J., et al., C. 2009. A γ-ray Burst at a Redshift of z˜8.2. *Nature*, **461**, 1254–1257.

11

Luminous Blue Variables: Their Formation and Instability in the Context of Binary Interactions

ANDREA MEHNER

11.1 Introduction

Massive stars are found primarily in higher-order multiple systems, and binary interactions are expected to have a significant effect on their evolution (see Chapter 10). Sana et al. (2012) show that 71% of O-type stars experience binary interactions during their lives. About 8% of early-type stars could be merger products (de Mink et al., 2014). A close companion may also dominate the evolution of a star through episodes of mass accretion or by triggering eruptive mass-loss events (Kashi and Soker, 2010; Smith, 2011).

Understanding the physics of mass loss is critical for stellar evolution theory. Progress has been made for mass loss via steady line-driven winds in the most massive stars (Puls et al., 2008). However, the impact of eruptive mass-loss episodes on stellar evolution remains uncertain (Smith, 2014). One stellar class that experiences large mass-loss episodes are Luminous Blue Variables (LBVs). They are luminous, evolved massive stars that are subject to irregular and eruptive variability accompanied by mass ejection (Conti, 1984). The physical mechanism of their outbursts are poorly understood. Binary interactions may be required, which will be explored in this chapter.

11.1.1 Luminous Blue Variables

LBVs have masses from tens to more than 100 $M_{\odot}$ and their luminosities of $log(L/L_{\odot}) = 5 - 7$ can reach or exceed the Humphreys–Davidson limit (Humphreys and Davidson, 1979). Figure 11.1 shows a schematic upper Hertzsprung–Russell (H-R) diagram displaying all LBVs and LBV candidates in the Galaxy, Large Magellanic Cloud (LMC) and Small Magellanic Cloud (SMC). LBVs are characterised by extreme mass loss and by remarkable photometric and spectroscopic variability (Humphreys and Davidson, 1994; Vink, 2012). They are surrounded by circumstellar envelopes of 0.1 to 40 $M_{\odot}$, created during episodic, irregular ejections.

LBVs experience two types of eruptions with timescales from several months to decades during which they can lose a significant fraction of their mass (Humphreys and Davidson, 1994): S Doradus outbursts with visual magnitude changes of 1 to 2 mag and giant eruptions with visual magnitude changes of >2 mag. During S Doradus outbursts, LBVs transit

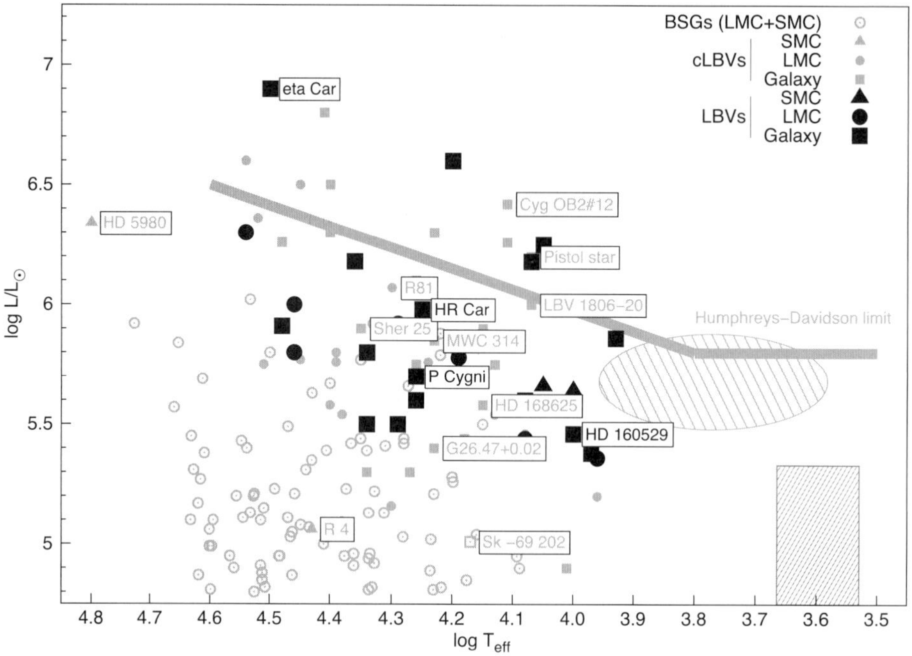

Figure 11.1 Schematic upper H-R diagram showing (candidate) LBVs in the Galaxy, LMC and SMC. The objects mentioned in this chapter are labelled. The solid curve is the upper luminosity boundary for the LMC (Humphreys and Davidson, 1979). Confirmed LBVs (black) and candidate LBVs (grey) are displayed at their quiescent phase. BSGs are indicated with dark grey–open circles (Trundle et al., 2004; Trundle and Lennon, 2005; Hunter et al., 2008). RSGs and Yellow Hypergiants are indicated by hatched regions.

in the H-R diagram from hot to cooler temperatures. This transit occurs over several months to several years. In the quiescent state, LBVs have effective temperatures in the range of $T_{\rm eff} \sim 12{,}000$ to $30{,}000$ K and spectra that would classify them as Blue Supergiants (BSGs) or late Of-type/WN stars. Mass-loss rates are on the order of $\dot{M} \sim 10^{-7}$ to 10^{-6} $M_\odot$ yr^{-1} in the quiescent phase. During an S Doradus outburst, their effective temperature lowers to $T_{\rm eff} \sim 8000$ K and their spectra resemble F-type supergiants. Mass-loss rates of $\dot{M} \sim 10^{-4} - 10^{-5}$ $M_\odot$ yr^{-1} are observed, which is atypical for F-type supergiants. Giant eruptions may give rise to mass loss of the order of $\dot{M} \sim 10^{-1}$ to 1 $M_\odot$ yr^{-1} with resulting atmospheres that resemble F-type or G-type supergiants.

About 40 LBVs are known in the Local Group (Richardson and Mehner, 2018). Stars that resemble LBVs in their photometry and spectra and have nebulae indicative of past eruptive mass loss, but have not yet been observed to undergo S Doradus outbursts, are candidate LBVs. Classification of various stars as an LBV or LBV candidate is controversial on account of the long and irregular timescales of the S Doradus variability.

11.1.2 Evolutionary Stage of the Luminous Blue Variables

How the LBVs fit into the evolutionary sequence for massive stars is as yet unclear. LBVs are nitrogen- and helium-rich compared to O-type stars, but have more hydrogen in their atmospheres than Wolf–Rayet (WR) stars. In the traditional view, LBVs represent a transitionary phase between main sequence O-type stars and WR stars (Langer, 2012). However, observational evidence suggests that the progenitors of some core-collapse supernovae (SNe) are LBVs, and thus did not reach the WR phase. Stellar modelling also shows that rotating stars with initial masses of 20 to 25 $M_\odot$ can appear as LBVs at the time of explosion, following significant mass loss in the Red Supergiant (RSG) phase (Groh et al., 2013).

The majority of type IIn SNe display pre-explosion outbursts of which some may be LBV outbursts (Ofek et al., 2014). Kotak and Vink (2006) first proposed an LBV-SN connection, suggesting that the radio light curves of some SNe indicated an interaction with circumstellar material consistent with previous S Doradus outbursts. The progenitor of SN 2005gl was identified as a very luminous star, consistent with an LBV (Gal-Yam and Leonard, 2009). Superluminous SNe such as SN 2006gy are associated with LBV progenitors (Smith et al., 2007). The strongest evidence for LBVs as SN progenitors comes from SN 2009ip and SN 2010mc. Both showed pre-SN photometric outbursts and were classified as type IIn events (Mauerhan et al., 2013; Ofek et al., 2013; Pastorello et al., 2013). However, no SN progenitor has been classified as a genuine LBV due to the lack of spectroscopic confirmation, and the connection remains uncertain.

A constraint on LBV evolution may come from their spatial distribution. Smith and Tombleson (2015) find that LBVs are more isolated from O-type stars than WR stars. They conclude that the LBV phenomenon is inconsistent with a single-star scenario in which O-type stars undergo an LBV phase on their way to becoming WR stars. Instead, LBVs could be the more massive analogs of blue stragglers. Proposed binary scenarios for the formation of LBVs are discussed in Section 11.3.

Various types of massive stars could eventually become LBVs, e.g., fast-rotating single stars or stars in binary systems, following a period as mass gainers or as the product of binary mergers. Until the phenomenon is better understood, the term ‘LBV’ should be used to describe a broad phenomenon that may not be restricted to a specific evolutionary stage of massive stars.

11.1.3 The Cause of the Luminous Blue Variable Instability

Originally, S Doradus outbursts were thought to occur at constant bolometric luminosity. This implies that a change in temperature corresponds to a change in the photospheric radius. Some authors suggest that the eruption leads to a greatly enhanced mass loss, which is so optically thick that a pseudophotosphere forms in the wind (Humphreys and Davidson, 1994). However, photometric and spectroscopic variations predicted by atmosphere models find that the observed mass-loss rates are too low to form a pseudophotosphere (de Koter et al., 1996). More recent studies show that the bolometric luminosity is not constant

(Groh et al., 2009b). It has been suggested that the observed radius change of the photosphere could be a pulsation or envelope inflation driven by the iron opacity bump (Gräfener et al., 2012). It is doubtful that the giant eruptions can be explained by a normal line-driven stellar wind (Smith and Owocki, 2006; cf. Davidson, 1987). Instead, these may be continuum-driven super-Eddington winds or hydrodynamic explosions.

The underlying physical trigger of S Doradus outbursts and giant eruptions is still under debate. The high stellar luminosities near the Eddington limit probably enable instabilities (Glatzel and Kiriakidis, 1993; Shaviv, 2001; Vink and de Koter, 2002; Smith and Owocki, 2006; Gräfener et al., 2012; Guzik and Lovekin, 2014; Owocki, 2015). Different types of instabilities are proposed, among which are radiation and turbulent pressure instabilities, vibrations and dynamical instabilities (see Humphreys and Davidson, 1994, for an overview). In addition to explanations from stellar interior and atmospheric instabilities, an external trigger by binarity is getting more attention in the past years (Gallagher, 1989; Kashi, 2010; Kashi and Soker, 2010; Smith, 2011).

11.2 Luminous Blue Variable Outbursts and Binary Interaction

Few LBV binaries are known to date. However, detecting companions of LBVs is difficult due to their high luminosities and broad, variable wind emission spectra. It is worth considering if and how a close companion may assist the S Doradus outbursts and giant eruptions. Several scenarios can be envisioned in observational and theoretical work: (1) During an S Doradus outburst, the primary may exceed its Roche-lobe radius and become large enough for the companion to pass through the primary's outer layers. While the companion is not the cause of the instability, it may amplify the event. (2) A close companion in an eccentric orbit could periodically destabilise the upper atmosphere of the LBV, leading to episodes of large mass loss. Repeated envelope grazing could escalate in a catastrophic eruption. (3) In addition to these tidal interactions, mass transfer can significantly affect the LBV instability and evolution.

In the following, I provide a summary about the known LBV binarity and review the two Galactic LBV binary systems, η Car and HR Car. The literature on η Car is extensive, and this section is kept short. The detection of a companion to HR Car is recent, and some of the results presented here are work in progress.

11.2.1 LBV Binarity

In the Milky Way, there are only two confirmed close binary systems: η Car (Davidson and Humphreys, 2012; and references therein) and HR Car (Boffin et al., 2016). The direct detection of close companions to Galactic LBVs requires milliarcsecond resolution, which is only reachable with optical interferometry. Radial velocity studies require time series of years to decades and even when existent may be inconclusive because of the complex and intrinsic wind variability observed in the broad emission line spectra.

Nazé et al. (2012) performed an X-ray survey of about half of the known Galactic (candidate) LBVs. They find that the X-ray emission of three objects (the LBV η Car, and the candidate LBVs Cyg OB2 #12 and GAL 026.47+00.02) could be explained by wind–wind collisions in a binary system. Aldoretta et al. (2015) conducted an all-sky survey with the Fine Guidance Sensor on the Hubble Space Telescope to search for resolved binary systems among massive stars. They find signatures for companions to six out of 13 (candidate) LBVs. In the Galaxy, they identify only the LBV HD 160529 as potential binary. Their detections in the LMC are questionable because of source crowding. Martayan et al. (2016) presented an imaging study of several Galactic (candidate) LBVs and find that $\sim$30% may have a companion in a wide orbit, with unknown eccentricity. Their results contain no significant detection of a wide companion to a genuine LBV.

Several LBV candidates are known or are suspected binaries. MWC 314 is a Galactic semi-detached binary with a bipolar nebula. The primary star is losing mass to a circumbinary disk. It is debated if the primary is an LBV candidate or if binary interactions between two lower mass stars match the observations (cf. Lobel et al. 2013 and Richardson et al. 2016). It may have a third, distant companion (Martayan et al., 2016). Figer et al. (2004) suggest that the Galactic candidate LBV 1806-20 is a double-lined binary, but confirmation through spectroscopic follow-up is needed. Interferometric observations and radial velocity variations indicate that the Pistol Star could be a binary (Martayan et al., 2016). HD 168625, known for its circumstellar rings similar to SN 1987A, is suspected to have several companions (Aldoretta et al., 2015; Martayan et al., 2016). The debated LBV candidate R 81 is an eclipsing binary in the LMC (Tubbesing et al., 2002). The candidate LBV R 4 is a binary in the SMC (Zickgraf et al., 1996). HD 5980 is allegedly an eclipsing LBV+WR multiple system in the SMC (Koenigsberger et al., 2010, 2014). While the observed eruptions in the later system are LBV-like, the two emission-line stars have temperatures of $T_{\mathrm{eff}} > 60{,}000$ K, much hotter than any other LBV, exemplifying the concept of the LBV phenomenon (Figure 11.1). The eruptions could be attributed to binary interactions between two WR stars. This system is too complex to be further discussed in this contribution.

Additional observational evidence for LBV binarity is provided by the often bipolar nebulae, such as the Homunculus nebula around η Car. About 50% of LBVs are surrounded by bipolar nebulae (Weis, 2011). Some theoretical models can produce bipolar nebulae through strong polar winds (Maeder and Desjacques, 2001; Owocki, 2011), but others argue that binary-star models are required to form these nebulae (Soker, 2004). Kashi and Soker (2010) even suggest that LBV outbursts and giant eruptions are an indication of mass transfer and accretion, and binarity is thus a prerequisite. The candidate LBVs HD 168625 and Sher 25 show similar nebular morphologies as the triple-ring nebula around SN 1987A, which indicates the imprint of a binary merger (Morris and Podsiadlowski 2007, 2009; the progenitor of SN 1987A was Sk -69 202, a BSG).

11.2.2 η Car

Eta Car is one of the most massive, evolved stars known (see Davidson and Humphreys, 2012, for a review). The star is famous for its Great Eruption in the first half of the nineteenth century during which 10 to 40 $M_\odot$ were ejected, creating the bipolar Homunculus nebula. A smaller (lesser) eruption took place around 1890 with 0.1 to 1 $M_\odot$ ejected. Eta Car is often referred to as an LBV, yet its giant eruption could have been caused by a different mechanism than the one in play for standard S Doradus outbursts. Its lesser eruption resembled an S Doradus outburst.

It has been established that a companion in a 5.5-year orbit triggers changes in the photometry and spectra of η Car during each periastron passage (e.g., Mehner et al. 2011, 2015). Particularly revealing and in support of binarity are X-ray results (Pittard and Corcoran, 2002; Corcoran et al., 2010, 2017) and hydrodynamic wind modelling (Okazaki et al., 2008; Madura et al., 2012). Models suggest that the current orbital parameters are $e = 0.9$, $P = 5.54$ yr, $a = 15.9$ au, $M_1 = 100\ M_\odot$ and $M_2 = 30\ M_\odot$ (Pittard and Corcoran, 2002; Okazaki et al., 2008; Parkin et al., 2009). The periastron separation between the two stars is 1.6 au.

Kashi and Soker (2010) advocate binary interactions to explain the giant and lesser eruptions of η Car. They argue that mass transfer from the primary to the companion during periastron passage can lead to a bright transient event on timescales of weeks to months. Accounting for a change in the orbital parameters due to the huge amount of mass ejected during the giant eruption, they find that the peaks in the light curve during the eruption coincide with periastron passages. They speculate that all giant LBV eruptions like η Car's or P Cygni's are triggered in some way by stellar companions. Smith (2011) proposes that η Car's giant eruption was an explosion triggered by a violent collision of the two stars in the system. They outline a scenario in which the evolved, unstable primary increases its radius to larger than periastron separation. As the companion then plunges into the wind of the primary, an eruption or explosion is induced, leading to substantial mass ejection and a luminous transient.

11.2.3 HR Car

HR Car is only the second Galactic LBV that has been confirmed to be a close binary system. It is surrounded by a bipolar nebula (Weis et al., 1997) and is highly variable, photometrically and spectroscopically. It had two S Doradus outbursts in the recent past, with maxima around 1991 and 2001 (Szeifert et al., 2003). Visual monitoring indicates that the star has been in quiescent phase for the last 10 years with $V \sim 8.7 - 9$ mag (Figure 11.2).

Interferometric observations revealed that HR Car is a binary star with an orbital period of several years (Boffin et al., 2016). The best-fit orbit of the companion has a period of about 6.5 years with an eccentricity of $e = 0.1$. The total mass of the system is about 50 $M_\odot$. The Brγ emission line forming region is more extended than the continuum emitting region, which indicates wind–wind interaction. The separation at periastron is around 2 mas ($\sim$10 au at a distance of 5 kpc). During quiescent state, the stars are well separated. In the maximum of an S Doradus outburst, the primary may become large enough for the companion to pass through its outer layers at periastron. Preliminary analysis suggests that HR Car is the outcome of a mass transfer event (Boffin et al., in prep.).

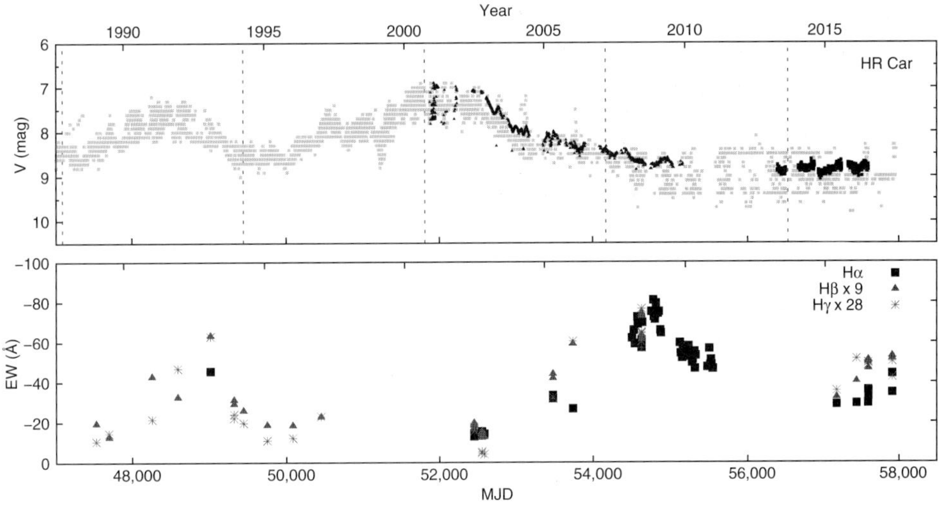

Figure 11.2 Visual light curve of HR Car and equivalent widths of Balmer emission lines. The light curve shows All Sky Automated Survey (ASAS) data (black triangles; Pojmanski 1997), American Association of Variable Star Observers (AAVSO) CCD observations (black circles; www.aavso.org), and AAVSO Vis data (grey squares). Periastron passages are marked with grey dashed lines for an orbital period of 2,362 days. The strengthening of the Balmer emission lines in 2008 indicates an increase in the mass-loss rate, maybe triggered at periastron passage.

Spectroscopic and photometric data of HR Car cover the last three decades. This data set was analysed to identify signatures of the companion and its potential impact on the S Doradus cycle (Boffin et al., in prep.). Radial velocity variations in absorption and emission features from the wind of HR Car show a periodicity of about 6.5 years, confirming spectroscopically the binary nature of the object.

The Hα equivalent width shows no signs of periodicity and no strong correlation with the light curve (Figure 11.2). However, an increase in the Hα equivalent width is observed in 2008, shortly after a periastron passage. Buemi et al. (2017) already noted that HR Car's mass-loss rate had increased from 6.1 $\times 10^{-6}$ $M_{\odot}\mathrm{yr}^{-1}$ during its quiescent phase in 1994–1995 to 1.17×10^{-5} $M_{\odot}\mathrm{yr}^{-1}$ in mid-2013, during its current quiescent phase. Boffin et al. (in prep.) estimate a mass-loss rate of 1.3×10^{-5} $M_{\odot}\mathrm{yr}^{-1}$ in 2017. This increase is unexpected for the star in quiescence. It originated shortly after a periastron passage.

In addition, the light curve of the two observed S Doradus outbursts suggestively follows periastron passages (see the upper panel of Figure 11.2). At close approach, the companion may be altering the structure of HR Car's upper atmosphere, leading to changes in opacity, temperature and density, and may either drive the star into higher mass-loss episodes or into outburst phases. However, it is clear that this does not occur periodically at each periastron.

The HR Car binary system has a similar orbital period as the η Car system, but a much lower system mass and lower eccentricity. No giant eruption has been recorded for HR Car, and estimates of the ejecta mass are of the order of 1 $M_{\odot}$ (White, 2000), much less than for η Car. Nevertheless, the discovery of a close companion to HR Car highlights the possible role of binarity in the generation and evolution of LBVs. Continued interferometric and spectroscopic monitoring of HR Car is ongoing to constrain the orbital period. Combined with a precise Gaia distance, this will provide better mass estimates and thus improved insights into their evolutionary path. The next periastron in 2020 will offer an excellent opportunity for a multiwavelength, multitechnique campaign to provide constraints for further theoretical modeling.

11.3 Binary Scenarios for the Formation of LBVs

The past few years have seen a heated exchange on the topic of LBV single-star or binary-star evolution. Smith and Tombleson (2015) find that LBVs have a larger distance from nearby OB clusters than WR stars. They argue that LBVs either have lower initial masses and thus longer lifetimes than inferred from their luminosities or that binary interaction must play a role. Disputing this claim, Humphreys et al. (2016) note that the sample in Smith and Tombleson (2015) is a mixture of less luminous LBVs more luminous classical LBVs, unconfirmed LBVs and unconfirmed candidate LBVs. They argue that separating them into the appropriate groups alleviates the conflict with single-star models. Humphreys et al. (2016) also state that the observed LBV velocities are too small to be consistent with the proposed mass gainer and kick scenario. Smith (2016) shows that by subdividing the LBV sample as Humphreys et al. (2016) suggest, the result of relative LBV isolation holds up. Aghakhanloo et al. (2017) confirm with models of young stellar clusters and their passive dissolution that if LBVs evolve as single stars, then their evolutionary track ages are incompatible with their observed spatial distribution, i.e., their kinematical ages. They also find that a wide range of kick velocities up to 100 km s^{-1} is consistent with the spatial distribution found for LBVs. The debate is ongoing. Aadland et al. (2018) used spatially complete photometric catalogs of the LMC, M31, and M33, and find that LBVs are no more isolated than luminous blue stars or WR stars.

We discuss two binary-related scenarios for the formation of LBVs. They are not mutually exclusive: (1) LBVs are mass gainers and received a kick when the primary exploded; and (2) LBVs are the product of mergers.

11.3.1 Rejuvenated Mass Gainer and Kick Scenario

Smith and Tombleson (2015) suggest that the observed spatial distribution of LBVs can be explained with them being rejuvenated stars. They propose a binary evolution scenario in which LBVs are mass gainers in stable mass transfer. The less massive star gains mass from the primary star and increases its mass, luminosity and rotation rate. The primary star is stripped off its hydrogen envelope and eventually explodes as type Ib/c SN. The mass gainer (LBV) receives a kick by the SN.

This scenario is supported by theoretical work (Justham et al. 2014; Podsiadlowski 2017; and references therein). Because the radius of a star expands when leaving the main sequence, a star is much more likely to encounter mass transfer during its post–main-sequence phase. Post–main-sequence accretion increases the envelope mass relative to the core mass. The result is that the star may never become an RSG (or if it was in the RSG region when the accretion started, the star will leave this region), but spend its remaining life in the BSG region. Some of these post-accretion stars could become LBVs. In the case that the secondary star lives through its main-sequence phase before accreting material, the age of the rejuvenated star of initial mass of 15 to 30 $M_\odot$ could be twice as inferred from its post-accretion luminosity. This would alleviate the discrepancy between the kinematic age and the single-star evolution nuclear age. In some cases, the close companion may not have exploded yet. Also, initial triple systems could explain that some LBVs are found in binaries.

11.3.2 Merger Scenario

About 8% of early-type stars are estimated to be merger products (de Mink et al., 2014). The most famous example for a binary merger is SN 1987A (Podsiadlowski, 1992; Morris and Podsiadlowski, 2007). The likely scenario for its progenitor is that a massive close binary experienced dynamical mass transfer during the second RSG phase and merged, resulting in a BSG. Mergers of two massive O-type stars could result in LBVs and LBV SN progenitors. Premerger multiple-star systems could account for the fact that some LBVs are found in binaries. For example, triple-star merger models have been proposed for η Car (Portegies Zwart and van den Heuvel, 2016) and for the LBV candidate R 4 (Pasquali et al., 2000).

There is observational evidence for LBVs being merger products: (1) LBVs are fast rotators with velocities reaching the critical regime in their minimum phase (Groh et al., 2009a). Rapid rotation may promote the LBV instability. (2) LBVs appear to have a lower (albeit uncertain) binary fraction than similar massive stars. (3) A merger scenario could account for the observed spatial distribution of LBVs.

Justham et al. (2014) discuss merger scenarios that could produce LBVs. They examine models in which massive stars gain mass soon after the end of core hydrogen burning. As the primary star leaves the main-sequence phase and expands, it enters into a contact phase with its companion. Some of these massive contact binaries become unstable and merge to form BSGs. These may eventually become LBVs and explode in the region of the H-R diagram where LBVs reside during outbursts. Figure 11.3 presents a schematic of this scenario. Another merger scenario that could result in LBVs involves collisional mergers of massive stars in dense young clusters (Glebbeek et al., 2013).

11.4 Conclusion

The place of LBVs in massive star evolution and the cause of their instability remain under debate. Recent observational and theoretical work has shown that binary interactions could play a key role in the formation of LBVs and in the underlying instability mechanism. It is

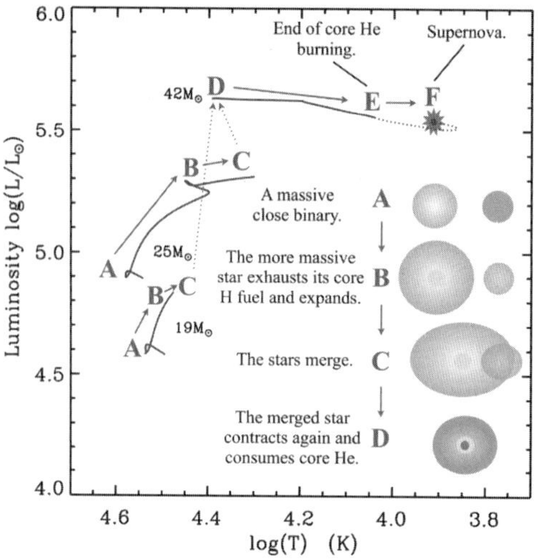

Figure 11.3 Schematic outline of a merger scenario example leading to an LBV SN progenitor. A binary (25 $M_{\odot}$ + 19 $M_{\odot}$), merges to form a 42 $M_{\odot}$ single star. The merger product becomes a BSG during core helium burning and explodes as an LBV. (Figure from Justham et al. 2014. Reproduced by permission of the American Astronomical Society (AAS).)

not clear yet if any of them play a driving role. These ideas are also not mutually exclusive. LBVs may be a diverse group of unstable evolved hot stars, originating from different evolutionary paths. Binary evolution involving mass transfer or mergers could achieve the observed LBV spatial distribution and apparent low multiplicity rate. A close companion in an eccentric orbit may periodically destabilise the upper atmosphere of the LBV, amplifying or triggering S Doradus outbursts and giant eruptions. Initial triple systems and even binary interactions in postmerger systems may account for the fact that some LBVs are found in close binary systems.

Continued interferometric and spectroscopic monitoring of LBVs will likely result in more discoveries of LBVs in close binary systems. The LBVs in the Milky Way are suitable for such studies, as the example of HR Car has shown. Radial velocity studies could not only reveal more spectroscopic binaries, but long-term spectroscopic monitoring is also required to study the driving mechanisms of the eruptive mass-loss episodes on timescales of years or decades. Imaging studies probe the presence of potential companions in wide orbits, which provide further constraints on models of LBV formation and evolution.

References

Aadland, E. and Massey, P. and Neugent, K. F. and Drout, M. R. 2018. Shedding Light on the Isolation of Luminous Blue Variables. *AJ*, **156**, 294.

Aghakhanloo, M., Murphy, J. W., Smith, N., and Hložek, R. 2017. Modelling luminous-blue-variable isolation. *MNRAS*, **472**, 591–603.

Aldoretta, E. J., Caballero-Nieves, S. M., Gies, D. R., et al. 2015. The multiplicity of massive stars: a high angular resolution survey with the guidance sensor. *AJ*, **149**, 26.

Boffin, H. M. J., Rivinius, T., Mérand, A., et al. 2016. The Luminous Blue Variable HR Carinae has a partner: discovery of a companion with the VLTI. *A&A*, **593**, A90.

Buemi, C. S., Trigilio, C., Leto, P., et al. 2017. Exploring the multifaceted circumstellar environment of the Luminous Blue Variable HR Carinae. *MNRAS*, **465**, 4147–4158.

Conti, P. S. 1984. Basic observational constraints on the evolution of massive stars. Maeder, A., and Renzini, A. (eds), *Observational Tests of the Stellar Evolution Theory*. IAU Symp., **105**, 233. Springer Netherlands.

Corcoran, M. F., Hamaguchi, K., Pittard, J. M., et al. 2010. Recent X-ray variability of η Carinae: the quick road to recovery. *ApJ*, **725**, 1528–1535.

Corcoran, M. F., Liburd, J., Morris, D., et al. 2017. The 2014 X-ray minimum of η Carinae as seen by Swift. *ApJ*, **838**, 45.

Davidson, K. 1987. The relation between apparent temperature and mass-loss rate in hypergiant eruptions. *ApJ*, **317**, 760–764.

Davidson, K., and Humphreys, R. M. (eds). 2012. *Eta Carinae and the Supernova Impostors*. ASSL, **384**. New York: Springer.

de Koter, A., Lamers, H. J. G. L. M., and Schmutz, W. 1996. Variability of Luminous Blue Variables. II. Parameter study of the typical LBV variations. *A&A*, **306**, 501.

de Mink, S. E., Sana, H., Langer, N., Izzard, R. G., and Schneider, F. R. N. 2014. The incidence of stellar mergers and mass gainers among massive stars. *ApJ*, **782**, 7.

Figer, D. F., Najarro, F., and Kudritzki, R. P. 2004. The double-lined spectrum of LBV 1806-20. *ApJ L*, **610**, L109–L112.

Gal-Yam, A., and Leonard, D. C. 2009. A massive hypergiant star as the progenitor of the supernova SN 2005gl. *Nature*, **458**, 865–867.

Gallagher, J. S. 1989. Close binary models for Luminous Blue Variables stars. Davidson, K., Moffat, A. F. J., and Lamers, H. J. G. L. M. (eds), *IAU Colloq. 113: Physics of Luminous Blue Variables*. ASSL, **157**, 185–192. Dordrecht: Springer.

Glatzel, W., and Kiriakidis, M. 1993. Stability of massive stars and the Humphreys/Davidson limit. *MNRAS*, **263**, 375.

Glebbeek, E., Gaburov, E., Portegies Zwart, S., and Pols, O. R. 2013. Structure and evolution of high-mass stellar mergers. *MNRAS*, **434**, 3497–3510.

Gräfener, G., Owocki, S. P., and Vink, J. S. 2012. Stellar envelope inflation near the Eddington limit: implications for the radii of Wolf–Rayet stars and Luminous Blue Variables. *A&A*, **538**, A40.

Groh, J. H., Damineli, A., Hillier, D. J., et al. 2009a. Bona fide, strong-variable galactic Luminous Blue Variable stars are fast rotators: detection of a high rotational velocity in HR Carinae. *ApJ L*, **705**, L25–L30.

Groh, J. H., Hillier, D. J., Damineli, A., et al. 2009b. On the nature of the prototype Luminous Blue Variable Ag Carinae. I. Fundamental parameters during visual minimum phases and changes in the bolometric luminosity during the S-Dor cycle. *ApJ*, **698**, 1698–1720.

Groh, J. H., Meynet, G., and Ekström, S. 2013. Massive star evolution: Luminous Blue Variables as unexpected supernova progenitors. *A&A*, **550**, L7.

Guzik, J. A., and Lovekin, C. C. 2014. Pulsations and hydrodynamics of Luminous Blue Variable stars. *ArXiv e-prints*, 1402.0257.

Humphreys, R. M., and Davidson, K. 1979. Studies of luminous stars in nearby galaxies. III. Comments on the evolution of the most massive stars in the Milky Way and the Large Magellanic Cloud. *ApJ*, **232**, 409–420.

Humphreys, R. M., and Davidson, K. 1994. The Luminous Blue Variables: astrophysical geysers. *PASP*, **106**, 1025–1051.

Humphreys, R. M., Weis, K., Davidson, K., and Gordon, M. S. 2016. On the social traits of Luminous Blue Variables. *ApJ*, **825**, 64.

Hunter, I., Lennon, D. J., Dufton, P. L., et al. 2008. The VLT-FLAMES survey of massive stars: atmospheric parameters and rotational velocity distributions for B-type stars in the Magellanic Clouds. *A&A*, **479**, 541–555.

Justham, S., Podsiadlowski, P., and Vink, J. S. 2014. Luminous Blue Variables and superluminous supernovae from binary mergers. *ApJ*, **796**, 121.

Kashi, A. 2010. Luminous Blue Variable eruptions triggered and powered by binary interaction. Kalogera, V., and van der Sluys, M. (eds), *American Institute of Physics Conference Series*. Am. Inst. of Phys. Conf. Ser., **1314**, 55–56. Melville: American Institute of Physics.

Kashi, A., and Soker, N. 2010. Periastron passage triggering of the 19th century eruptions of Eta Carinae. *ApJ*, **723**, 602–611.

Koenigsberger, G., Georgiev, L., Hillier, D. J., Morrell, N., Barbá, R., and Gamen, R. 2010. A $\sim$ 40 Year Variability Cycle in the Luminous Blue Variable/Wolf–Rayet Binary System HD 5980? *AJ*, **139**, 2600–2611.

Koenigsberger, G., Morrell, N., Hillier, D. J., et al. 2014. The HD 5980 multiple system: masses and evolutionary status. *AJ*, **148**, 62.

Kotak, R., and Vink, J. S. 2006. Luminous Blue Variables as the progenitors of supernovae with quasi-periodic radio modulations. *A&A*, **460**, L5–L8.

Langer, N. 2012. Presupernova evolution of massive single and binary stars. *ARA&A*, **50**, 107–164.

Lobel, A., Groh, J. H., Martayan, C., et al. 2013. Modelling the asymmetric wind of the Luminous Blue Variable binary MWC 314. *A&A*, **559**, A16.

Madura, T. I., Gull, T. R., Owocki, S. P., et al. 2012. Constraining the absolute orientation of η Carinae's binary orbit: a 3D dynamical model for the broad [Fe III] emission. *MNRAS*, **420**, 2064–2086.

Maeder, A., and Desjacques, V. 2001. The shape of Eta Carinae and LBV nebulae. *A&A*, **372**, L9–L12.

Martayan, C., Lobel, A., Baade, D., et al. 2016. Luminous Blue Variables: an imaging perspective on their binarity and near environment. *A&A*, **587**, A115.

Mauerhan, J. C., Smith, N., Filippenko, et al. 2013. The unprecedented 2012 outburst of SN 2009ip: a luminous blue variable star becomes a true supernova. *MNRAS*, **430**, 1801–1810.

Mehner, A., Davidson, K., Martin, J. C., et al. 2011. Critical differences and clues in Eta Car's 2009 event. *ApJ*, **740**, 80.

Mehner, A., Davidson, K., Humphreys, R. M., et al. 2015. Eta Carinae's 2014.6 spectroscopic event: clues to the long-term recovery from its Great Eruption. *A&A*, **578**, A122.

Morris, T., and Podsiadlowski, P. 2007. The triple-ring nebula around SN 1987A: fingerprint of a binary merger. *Science*, **315**, 1103.

Morris, T., and Podsiadlowski, P. 2009. A binary merger model for the formation of the Supernova 1987A triple-ring nebula. *MNRAS*, **399**, 515–538.

Nazé, Y., Rauw, G., and Hutsemékers, D. 2012. The first X-ray survey of Galactic Luminous Blue Variables. *A&A*, **538**, A47.

Ofek, E. O., Sullivan, M., Cenko, S. B., et al. 2013. An outburst from a massive star 40 days before a supernova explosion. *Nature*, **494**, 65–67.

Ofek, E. O., Sullivan, M., Shaviv, N. J., et al. 2014. Precursors prior to Type IIn supernova explosions are common: precursor rates, properties, and correlations. *ApJ*, **789**, 104.

Okazaki, A. T., Owocki, S. P., Russell, C. M. P., and Corcoran, M. F. 2008. Modelling the RXTE light curve of η Carinae from a 3D SPH simulation of its binary wind collision. *MNRAS*, **388**, L39–L43.

Owocki, S. 2011. Theory of winds from hot, luminous massive stars. *Bulletin de la Societe Royale des Sciences de Liege*, **80**, 16–28.

Owocki, S. P. 2015. Instabilities in the envelopes and winds of very massive stars. Page 113 of: Vink, J. S. (ed), *Very Massive Stars in the Local Universe*. ASSL, **412**. Springer.

Parkin, E. R., Pittard, J. M., Corcoran, M. F., Hamaguchi, K., and Stevens, I. R. 2009. 3D modelling of the colliding winds in η Carinae: evidence for radiative inhibition. *MNRAS*, **394**, 1758–1774.

Pasquali, A., Nota, A., Langer, N., Schulte-Ladbeck, R. E., and Clampin, M. 2000. R4 and its circumstellar nebula: evidence for a binary merger? *AJ*, **119**, 1352–1358.

Pastorello, A., Cappellaro, E., Inserra, C., et al. 2013. Interacting supernovae and supernova impostors: SN 2009ip, is this the end? *ApJ*, **767**, 1.

Pittard, J. M., and Corcoran, M. F. 2002. In hot pursuit of the hidden companion of eta Carinae: an X-ray determination of the wind parameters. *A&A*, **383**, 636–647.

Podsiadlowski, P. 1992. The progenitor of SN 1987 A. *PASP*, **104**, 717–729.

Podsiadlowski, P. 2017. Binary evolution and the progenitor of SN 1987A. *ArXiv e-prints*, 1702.03973.

Pojmanski, G. 1997. The All Sky Automated Survey. *Acta Astron.*, **47**, 467–481.

Portegies Zwart, S. F., and van den Heuvel, E. P. J. 2016. Was the nineteenth century giant eruption of Eta Carinae a merger event in a triple system? *MNRAS*, **456**, 3401–3412.

Puls, J., Vink, J. S., and Najarro, F. 2008. Mass loss from hot massive stars. *A&A Rev.*, **16**, 209–325.

Richardson, N. D., Moffat, A. F. J., Maltais-Tariant, R., et al. 2016. Spectroscopy, MOST photometry, and interferometry of MWC 314: is it an LBV or an interacting binary? *MNRAS*, **455**, 244–257.

Richardson, N. D., and Mehner, A. 2018. The 2018 Census of Luminous Blue Variables in the Local Group. *Research Notes of the American Astronomical Society*, **2**(3), 121.

Sana, H., de Mink, S. E., de Koter, A., et al. 2012. Binary interaction dominates the evolution of massive stars. *Science*, **337**, 444.

Shaviv, N. J. 2001. The theory of steady-state super-Eddington winds and its application to novae. *MNRAS*, **326**, 126–146.

Smith, N. 2011. Explosions triggered by violent binary-star collisions: application to Eta Carinae and other eruptive transients. *MNRAS*, **415**, 2020–2024.

Smith, N. 2014. Mass loss: its effect on the evolution and fate of high-mass stars. *ARA&A*, **52**, 487–528.

Smith, N. 2016. The isolation of Luminous Blue Variables: on subdividing the sample. *MNRAS*, **461**, 3353–3360.

Smith, N., and Owocki, S. P. 2006. On the role of continuum-driven eruptions in the evolution of very massive stars and population III stars. *ApJ L*, **645**, L45–L48.

Smith, N., and Tombleson, R. 2015. Luminous Blue Variables are antisocial: their isolation implies that they are kicked mass gainers in binary evolution. *MNRAS*, **447**, 598–617.

Smith, N., Li, W., Foley, R. J., et al. 2007. SN 2006gy: Discovery of the most luminous supernova ever recorded, powered by the death of an extremely massive star like η Carinae. *ApJ*, **666**, 1116–1128.

Soker, N. 2004. Why a single-star model cannot explain the bipolar nebula of η Carinae. *ApJ*, **612**, 1060–1064.

Szeifert, T., Kaufer, A., Crowther, P. A., Stahl, O., and Sterken, C. 2003. High-resolution spectroscopy of two LBV cycles of HR Car. Page 243 of: van der Hucht, K., Herrero, A., and Esteban, C. (eds), *A Massive Star Odyssey: From Main Sequence to Supernova*. IAU Symp., **212**, 243. San Francisco: Astronomical Society of the Pacific.

Trundle, C., and Lennon, D. J. 2005. Understanding B-type supergiants in the low metallicity environment of the SMC II. *A&A*, **434**, 677–689.

Trundle, C., Lennon, D. J., Puls, J., and Dufton, P. L. 2004. Understanding B-type supergiants in the low metallicity environment of the SMC. *A&A*, **417**, 217–234.

Tubbesing, S., Kaufer, A., Stahl, O., et al. 2002. The eclipsing hypergiant R 81 (B2.5Ia-O) in the Large Magellanic Cloud: system properties from spectroscopic and photometric monitoring. *A&A*, **389**, 931–944.

Vink, J. S. 2012. Eta Carinae and the Luminous Blue Variables. Davidson, K., and Humphreys, R. M. (eds), *Eta Carinae and the Supernova Impostors*. ASSL, **384**, 221. New York: Springer.

Vink, J. S., and de Koter, A. 2002. Predictions of variable mass loss for Luminous Blue Variables. *A&A*, **393**, 543–553.

Weis, K. 2011. Nebulae around Luminous Blue Variables: large bipolar variety. Neiner, C., Wade, G., Meynet, G., and Peters, G. (eds), *Active OB Stars: Structure, Evolution, Mass Loss, and Critical Limits*. IAU Symp., **272**, 372–377. Cambridge: Cambridge University Press.

Weis, K., Duschl, W. J., Bomans, D. J., Chu, Y.-H., and Joner, M. D. 1997. The bipolar structure of the LBV nebula around HR Carinae. *A&A*, **320**, 568–574.

White, S. M. 2000. The radio nebula around HR Carinae. *ApJ*, **539**, 851–857.

Zickgraf, F.-J., Kovacs, J., Wolf, B., Stahl, O., Kaufer, A., and Appenzeller, I. 1996. R4 in the Small Magellanic Cloud: a spectroscopic binary with a B[e]/LBV-type component. *A&A*, **309**, 505–514.

12

Type Ia Supernovae: Where Are They Coming From and Where Will They Lead Us?

FERDINANDO PATAT AND NA'AMA HALLAKOUN

12.1 Source of Embarrassment vs. Excitement

About 10 years ago, we opened a similar review (Patat, 2009) with the following quotation from Mario Livio: 'The fact that we do not know yet what are the progenitor systems of some of the most dramatic explosions in the universe has become a major embarrassment and one of the key unresolved problems in stellar evolution' (Livio, 2001). Today, almost 20 years after Livio had expressed his feelings on the matter, and although we have arguably learnt a lot on this topic, there still are good reasons for being embarrassed – or, if you are an enthusiast, to keep the excitement high.

In this review, which is explicitly meant for nonsupernova specialists, we will give a review of our understanding of Type Ia progenitors. For more details, the reader is referred to the exhaustive review by Maoz et al. (2014).

12.1.1 Why and How?

One may think that, once having established that Type Ia supernovae (hereafter SN Ia) are good standard candles, there is no need for understanding what they actually are. After all, a candle is a candle is a candle: why should it be important to know what wax it is made of?

We reckon there are a number of reasons, which can be summarised as follows:

- Using SNe Ia in cosmology requires an understanding of the evolution of the luminosity and the SN rate with cosmic epoch. Both depend on the nature of the progenitors.
- Galaxy evolution depends on the energetic and nucleosynthetic output of SNe Ia.
- A knowledge of the initial conditions and of matter distribution in the environment of the exploding star is essential for the understanding of the explosion itself.
- An unambiguous identification of the progenitors, coupled with observationally determined SN rates, will help placing constraints on the theory of binary star evolution.

When addressing the progenitor problem, there are a few fundamental facts that need to be kept in mind:

1. SNe Ia do not show hydrogen lines.

2. They are homogene-ous[isable] in terms of peak luminosity, spectra and light curve shape.
3. They appear in elliptical galaxies (as opposed to core-collapse SNe).
4. The preceding facts rule out core-collapse of massive (M>8 $M_\odot$), young stars.

From this comes the conundrum that has been bothering the SN community for a long time: (a) you need to make a low-mass, 'old' star explode; (b) you need to produce almost always about the same luminosity (i.e., the same amount of radioactive ^{56}Ni); and (c) you ought to hide the most abundant element in the Universe (which, in itself, is quite a remarkable fact).

The theoretical solution to this riddle was first put forward by Whelan and Iben (1973), who invoked 'mass transfer between members of a binary system'.

12.1.2 The Basics

Well before the preceding scenario was proposed, from the observations of both SNe and SN remnants it was clear that energetics and composition require the thermonuclear combustion of a degenerate object (Hoyle and Fowler, 1960). In addition, the absence of H and He indicates that the exploding object is most likely a CO white dwarf (for which the Chandrasekhar mass is $M_{CH} \sim 1.4$ $M_\odot$).

The general understanding is that the core of the star is burned into Fe-group elements plus some Intermediate Mass Elements (IME). The observed luminosity (typically peaking at $M_V \sim -19.3$) requires ∼0.6 $M_\odot$ of radioactive ^{56}Ni.

In terms of energetics, this can be illustrated with a few back-of-the envelope calculations. The nuclear energy of 0.6 $M_\odot$ of Ni is about 10^{51} erg. A similar amount is derived for the remaining 0.8 $M_\odot$ of iron-peak elements, so that the total nuclear energy is $E_N \sim 2\times10^{51}$ erg. Of this, about 1% is radiated as visible light ($\sim10^{49}$ erg). The gravitational binding energy of a 1.4 $M_\odot$ CO white dwarf (WD) is $E_G \sim 0.5\times10^{51}$ erg. Therefore, $E_N > E_G$, so that the thermonuclear explosion is indeed capable of completely disrupting the WD. Given that only a small fraction of the total energy is released in the form of radiation, we can estimate the expected ejecta velocity assuming that all energy is converted to kinetic energy: $v_e \approx \sqrt{2E_K/M_{WD}} \sim 10^4$ km s^{-1}, which is indeed what is observed. Therefore, the hypothesis holds in terms of energy budget.

All this led the community to the following consensus statement: SNe Ia represent thermonuclear disruptions of mass accreting CO WDs, when these approach the Chandrasekhar limit and ignite carbon.

Somehow, sometime and somewhere.

As it turns out, growing a WD to the critical limit is not that easy: if our understanding of stellar evolution is correct, there is no way a single star can produce a WD with a mass close to M_{CH}. Therefore, one way or another, the WD needs to accrete material, either from a nondegenerate companion (via some mass transfer mechanism, like Roche-lobe overflow or wind) or a degenerate companion (via merging or collision). These two scenarios are commonly dubbed single-degenerate (SD) and double-degenerate (DD).

12.2 The SNe Ia Progenitor Problem: An Observational Approach

More than 40 years have passed from the first suggestion for solving the conundrum and, despite the fact that SNe Ia are used for precision cosmology, the nature of the progenitor system(s) is still unknown.

Very schematically, the problem has been addressed through the following channels:

- Populations of potential progenitors
- Preexplosion imaging of (nearby) explosion sites
- Explosion properties carrying progenitors' imprints
- Circumstellar environment
- SN remnants
- Explosion rates as a function of time and location
- Binary population synthesis

We will briefly discuss each of them in the next subsections.

12.2.1 Candidate Populations

One thing is speculating about the existence of binary systems with $M \sim M_{CH}$ accreting WD, another is looking around for real examples and see what they tell us. Among the SD scenario, the following binary systems have been considered as possible candidates: recurrent novae (RNe), super-soft X-ray sources (SSS), rapidly accreting WDs and He-rich donors. On the DD side, the attention has obviously focused on binary WD systems.

A fair amount of work went into investigating RN systems like U Sco, RS Oph and T CrB, all containing WDs claimed to have masses close to the critical limit. However, a number of questions remain open. For instance: is the WD really a CO WD? If this is instead a ONeMg WD, then the star would not explode as a Type Ia. In these systems, it is not possible to directly observe the properties of the WD (for instance, because it is embedded in an accretion disk), and so this is a question likely doomed to remain unanswered. Another crucial point is related to the accretion: these systems may host a massive WD, but one cannot be sure its mass is increasing with time. There is the possibility that during the RN phase, the WD actually loses at each outburst more mass than it accretes during the inter-outburst phase. Finally, there is a fundamental issue about the number of these systems in the Milky Way (MW), which may well not be sufficient to account for the observed rate.

In general, SD systems are supposed to spend some Myr in a phase of steady nuclear burning while accreting material from the donor. In virtue of this fact, they should be detectable as supersoft X-ray sources (Di Stefano, 2010). Under certain conditions, also DD systems may undergo an SSS phase, with much lower luminosities, though. Therefore, the detection of an SSS in coincidence with an SN Ia explosion site would allow a direct study of at least some of the progenitor's properties. This has been first applied to SN 2007on, for which a preexplosion detection, based on X-ray imaging, was claimed, but turned out not to be fully convincing (Roelofs et al., 2008; Voss and Nelemans, 2008). The conclusion is that if the bulk of Type Ia is to be explained by SSS, only 1% of the WD mass accretion takes place in this phase to match the observed MW rate (Maoz et al., 2014).

Until a few years ago, binary WD systems, first suggested by Tutukov and Yungelson (1981), were considered as a disfavoured scenario. First of all, it was not clear whether there were enough suitable systems in nature, and second, the range of WD masses was seen as being in contradiction with the observed homogeneity of SNe Ia. In addition, from a theoretical point of view, the WD merger was suspected to result in an accretion induced core collapse (AICC), leading to the formation of a neutron star rather than to a thermonuclear explosions (Nomoto and Iben, 1985).

The first large-scale attempt to find suitable WD–WD candidates was the European Southern Observatory (ESO)–SN Ia Progenitor survey (SPY) search (Napiwotzki et al., 2004). This campaign included about 1,000 WDs and led to the detection of one system with a global mass larger than M_{CH} and that would merge in less than one Hubble time. Other similar studies are presented by Badenes et al. (2009) and Maoz and Hallakoun (2017). Of course, like for the SD candidates, the crucial question is whether there are enough such systems to sustain the observed rate. We will come back on this in greater detail in Section 12.3.

12.2.2 Pre-explosion Sites

The idea is very simple: one determines the SN position as accurately as possible and goes back to preexplosion archival data and searches for traces of the precursor system. While this had led to a number of positive detections for core-collapse systems, this has never been the case for thermonuclear SNe, and only not really stringent constraints on the progenitor's luminosity could be placed, until the nearby SN 2011fe exploded in M101. This was a very close-by (~6 Mpc) and very early (discovered most likely a few hours after explosion) standard Type Ia, which provided a unique opportunity to probe the earliest phases in great detail, across a wide wavelength range. The host being a Messier object, rich pre-explosion, HST data were available, and this allowed Weidong Li and collaborators to draw some firm conclusions (Li et al., 2011): a red giant at the location of the SN is ruled out in the decade before the explosion, hence removing RS Oph and T CrB systems as viable progenitors. Stars with $M > 3.5 M_{\odot}$ are also excluded, while a main sequence or sub-giant donor is allowed (i.e., systems such as U Sco in quiescence). A red giant was also ruled out for another, even closer (3.5 Mpc) event: SN 2014J (Kelly et al., 2014).

12.2.3 An Interlude on WD Spin-Up/Down

The common understanding of the physics of SNe Ia is that the exploding star has to reach some critical condition while approaching M_{CH}. Nevertheless, WD rotation can have a significant influence, as it has been shown in the so-called spin-up/spin-down scenario in the context of SD progenitors: a WD that has grown in mass, even beyond the critical limit, could be rotation-supported against collapse and ignition during the accretion. This was first introduced by Yoon and Langer (2004), and it has received quite some interest in recent years. If this mechanism is at work, the traces of the process (including the donor) could be completely wiped out by the time of the explosion. Interestingly, DD mergers explosions delayed by rotational support have also been proposed (Tornambé and Piersanti, 2013).

The bottom line is that there may be a population of rapidly spinning WDs, waiting to spin down and eventually explode. This is of course a rather frustrating thought, very effectively phrased by Maoz et al. (2014): ‘Observationally, a spin-up/spin-down scenario could potentially “erase” many of the clues we discuss in this review’.

12.2.4 Explosion Properties

A number of aspects seen during the explosion relate (unfortunately often ambiguously) to the progenitor properties:

- Early light curve and spectral evolution
- Shocks on (possible) companion star
- Ejecta asymmetries (polarimetry)
- Emission from hydrogen
- Radio, X-ray circumstellar medium (CSM) emission
- Narrow time-variable absorption
- Circumstellar (CS) dust and light echoes (spectrophotometry and polarimetry)

Each of these channels has been explored in depth. Here we will focus on the first two, as these were fundamental for studying the case of SN 2011fe, which provided the most stringent constraints ever. The rationale is very simple: in the phases immediately following the thermonuclear runaway, the interaction between the SN ejecta and a stellar object in the vicinity of the explosion is expected to produce some effects that depend, for instance, on the size of the star and the extent of its atmosphere (Kasen, 2010).

For SN 2011fe, Nugent et al. (2011), based on very early optical and UV observations, were able to exclude the presence of shock effects from SN ejecta hitting an alleged companion, definitely ruling out a red giant (RG) companion, in line with the independent conclusions reached by Li et al. (2011). Bloom et al. (2012) went one step further, ruling out even a main-sequence (MS) star, hence, concluding that the DD scenario is the most probable for this one object. In addition, for the first time, their observations provided a direct upper limit to the size of the exploding object, $R<0.02\ R_{\odot}$, so that the exploding star can be either a WD or a neutron star. Ongoing and planned surveys have the potential of leading to much more numerous very early discoveries, so that we believe this channel of study will gain importance in the future.

The study of overall asymmetries in the ejecta has also been used to set constraints on the progenitor. In general, the fact that normal Type Ia show a continuum polarisation consistent with zero has normally been interpreted as a sign of spherical symmetry (see, for instance, Patat et al., 2009), which, in turn, is probably inconsistent with explosion scenarios that imply a high degree of asymmetry, such as the case of violent mergers or head-on collisions (Bulla et al., 2016). Interestingly, different polarisation properties were derived for subluminous events, possibly pointing to an explosion arising within a fast-rotating WD (Patat et al., 2010).

So far, no emission coming from the interaction between SN ejecta and circumstellar material has been detected, hence indicating a clean environment, in turn favouring a DD scenario. At least at a first glance. Spin-up/down can of course delay the explosion so much that the material lost by the progenitor system before the explosion has a sufficient time to diffuse into the interstellar medium. This would make that material undetectable, at least through emission (optical or X-ray). But it may be detectable in absorption, using trace elements such as Na or K, which enable the observability of very tiny column densities. In addition, if the gas is far enough to be 'untouched' by the ejecta, but close enough to be affected by the SN radiation field, one may be able to observe time dependencies. This was the idea behind a series of studies following Patat et al. (2007), who detected a marked variability in the well-studied SN 2006X. As the variability is expected to take place only for a limited range of distances and physical conditions, hence taking very long to accumulate a sufficient sample, other searches have moved towards a more statistical approach, looking for an excess of blue-shifted Na absorption features signalling the presence of material lost by the progenitor (Sternberg et al., 2011; Maguire et al., 2013). These studies led to the conclusion that $\sim$20% of Type Ia show these features. Unfortunately, the interpretation is not univocal in terms of the progenitor's nature, as this gas may both be material lost by the non-degenerate companion, but also the relic of the common-envelope phase of a DD system.

We finally mention the polarisation studies along the lines of sight to reddened Type Ia, as they represent a recent addition to the toolbox. Very intriguingly, the interstellar polarisation derived for these objects shows a polarisation dependence that diverges in a remarkable way from what is seen in our own Galaxy (Patat et al., 2015). Excluding the anticopernican conclusion that there is something special with the MW dust, the alternative explanations rely on the special properties of the explosion environment. A further study has shown that, as a matter of fact, it is not true that there are no stars in the Galaxy showing these peculiar properties. In fact, very similar polarisation laws are observed in a subset of objects embedded in protoplanetary nebulae (Cikota et al., 2017).

Although there still are some unclear aspects, this may provide some support to the alternative core-degenerate (CD) scenario proposed by Kashi and Soker (2011). In this model, the WD merges with an asymptotic giant branch (AGB) star during the common-envelope phase. The envelope is ejected, the explosion of the merged core is delayed by rotation, it spins down (via magnetic dipole radiation) and eventually explodes. This model, which broadly belongs to the DD family, has the advantage of clearing the immediate surroundings, and still leaves material to be seen in absorption, which would reconcile it with the previously reported results.

12.2.5 Surviving Companions

Obviously, one substantial difference between DDs and SDs is the presence of a companion that becomes unbound after the explosion and hence runs away from the explosion site.

The models predict that the companion star survives the explosion, it gains an anomalous transversal and rotational speed and it probably gains a peculiar chemical composition in the atmospheric layers caused by the pollution by SN ejecta. With this concept in mind (no star $\Longleftrightarrow$ no SD), a few teams set out with the following agenda: search historical SN remnants (SNRs) for weird-composition, fast-moving, rapidly-rotating stars not far from the estimated explosion centre. A positive detection was first claimed by Ruiz-Lapuente et al. (2004) for the so-called star G in Tycho's remnant. This was not confirmed by Kerzendorf et al. (2009, 2013) (but see the rebuttal by Bedin et al., 2014). No detection was reported for SN 1006 (Kerzendorf et al., 2012), for SNR 0509-67.5 (Schaefer and Pagnotta, 2012) and for Kepler (Kerzendorf et al., 2014). We believe it is fair to state that, so far, there is no convincing evidence for direct companion detection.

12.2.6 SN Remnants

The idea is to look for possible interactions between the remnant and preexplosion material, and to compare with the predictions of hydrodynamical modelling. In the SD scenario, one expects a cavity to be blown in the interstellar medium (ISM) (3–30 pc) during the fast-wind phase of the accretion (Hachisu et al., 1996). Such cavity was not detected in 7 Type Ia SNR (Badenes et al., 2007). In addition, X-ray observations are consistent with a uniform ISM density. Therefore, either the fast-wind phase does not take place, or the accretion stopped well before the explosion.

12.2.7 SN Rates and Binary Population Synthesis

The Galactic rate of Type Ia explosion is known: $\sim 5 \times 10^{-3}$ yr^{-1}. This can be used to constrain the progenitor scenarios using Binary Population Synthesis (BPS). The models show that, in principle, DDs can produce this value, while SDs can only explain some fraction of the observed rate, with WD+MS being the most productive, and WD+RG the less productive. The observed rate (as a function of time) results from the combination of the star formation rate (SFR) and the so-called delay time distribution (DTD), which provides the time between the birth of the progenitor system and the explosion. Using different DTDs produced by BPS can be directly compared to observed rates as a function of redshift. The match provided by DDs is definitely better than that deduced for SDs (see Maoz et al., 2014, for a comprehensive discussion).

From the above it is clear that, although there still is some room for the SD channel, the DD scenario has gained quite some interest, as it may indeed explain the bulk of Type Ia. For this reason in the next section, we discuss in some detail its viability.

12.3 Are There Enough Double White Dwarf Mergers to Explain the Milky Way's SN Ia Rate?

If most SNe Ia share the same formation channel, then any candidate scenario competing for the title should have enough progenitor systems to account for the observed SN Ia rate in the Galaxy. The situation of the SD scenario in this sense is rather poor: even the most optimistic estimates for the number of recurrent novae in the MW (Pagnotta and Schaefer, 2014) cannot account for the majority of SN Ia explosions. But are there enough double-WD systems to explain the Galactic SN Ia rate through the DD channel? In order to answer this question, we need to characterise the local double-WD population.

The first systematic searches for detached double WDs were carried out in the 1980s, mostly with no detections (Robinson and Shafter, 1987; Foss et al., 1991), or with a low rate of success (Bragaglia et al., 1990). Marsh et al. (1995) decided upon a different approach, and focused their attempts on low-mass ($<0.45\ M_\odot$) WDs that have not had enough time to evolve from single stars during the lifetime of the Universe. Their method proved fruitful, with five out of seven cases turning out to be double WDs. However, the eventual mergers of these low-mass systems are not likely to result in an SN Ia explosion. Further studying a sample of 46 WDs, Maxted and Marsh (1999) estimated a binary fraction of 1.7 to 19%. Combined with a few other surveys that were carried out in the 1990s, 18 close double WDs were found among a total of $\sim$200 WDs surveyed (see Napiwotzki et al., 2004, and references therein). An important step forward was initiated in the early 2000s, with the ESO SPY (Napiwotzki et al., 2001), which took advantage of the then-new ESO 8 m Very Large Telescope (VLT) to acquire unprecedented high-resolution, high-signal-to-noise ratio (S/N), multi-epoch spectroscopy of about 800 bright WDs. A number of newly discovered double WDs from the SPY survey were then followed-up and published (e.g., Napiwotzki et al., 2002; Karl et al., 2003a,b; Nelemans et al., 2005). To date, there are about 100 close double-WD systems with known orbital parameters, including seven eclipsing double WDs (see Brown et al., 2017, and references therein). However, when we exclude systems containing extremely low-mass (ELM) WDs with $M < 0.3\ M_\odot$ (e.g., Brown et al., 2016), this number is reduced to $\sim$30.

To estimate the WD binary fraction and its merger rate using a well-defined statistical approach, Maoz et al. (2012) devised a method that characterises the local WD population by measuring the distribution of ΔRV_{max}, the maximal radial-velocity (RV) shift between different epochs of the same WD. They showed that even with noisy and sparsely sampled data (with as little as two epochs per system), the ΔRV_{max} distribution can provide meaningful constraints on the WD binary fraction and on the binary separation distribution, as well as on the double-WD merger rate. The 'core' of the ΔRV_{max} distribution, at low RV differences where most of the systems reside, is the result of the measurement uncertainty, and WDs that are found in this area of the distribution are most likely single. However, systems that end up in the 'tail' of the distribution are most probably real detections of double WDs. The observed ΔRV_{max} distribution is then compared to simulated distributions with varying binary fraction and separation distribution parameters: populations with a high binary fraction or a separation distribution favouring systems at closer separations introduce a higher fraction of 'tail' systems. Thus, this method allows probing of the double-WD population characteristics without the need of follow-up observations and full orbital solutions of the binary candidates.

The ΔRV_{max} method requires the use of a large spectroscopic sample, with at least two epochs per WD. Exploiting the multiple subexposures per object taken by the Sloan Digital Sky Survey (SDSS) to allow cosmic ray rejection, Badenes and Maoz (2012) applied this method to a large sample of ~4,000 WDs that have SDSS spectroscopic data. They found that the double-WD merger rate is of the order of magnitude of the SN Ia rate in the MW (based on the rate measured for SBc galaxies and the approximate Galactic mass). However, because of the large measurement errors of the SDSS spectra ($\sim 80\,\mathrm{km\,s^{-1}}$), only a few, high-orbital-velocity, candidate systems could be detected with small-number statistics, and the measured merger rate was highly uncertain. The low RV precision was mostly the result of the low resolution of the SDSS spectra, combined with the fact that most WDs show only very broad hydrogen Balmer absorption lines in their optical spectra. Thus, only double WD systems with ΔRV_{max} higher than $\sim 250\,\mathrm{km\,s^{-1}}$ (or with orbital separations smaller than ~0.05 au), constituting 15 out of the ~4000 WDs, were detectable in the SDSS sample.

In order to improve the results and remove the degeneracy between the model parameters, a more precise multi-epoch spectroscopic sample was needed – a requirement fulfilled by the above-mentioned ESO-SPY sample (Napiwotzki et al., 2001). Owing to the high resolution of the SPY sample, finer details in the WD spectra could be used to extract the RV with a precision of $\sim 1-2\,\mathrm{km\,s^{-1}}$ per epoch. Applying the ΔRV_{max} method to a sample of 439 WDs from SPY, Maoz and Hallakoun (2017) were able to further constrain the WD binary fraction and separation distribution, this time to separations out to ~4 au, with 43 candidate double-WD systems with $\Delta RV_{max} > 10\,\mathrm{km\,s^{-1}}$. Since the sample was smaller by an order of magnitude compared to the SDSS sample of Badenes and Maoz (2012), it was not sensitive to rare systems with very small separations (<0.05 au), and thus the two samples were in fact complementary. In a recent study, Maoz et al. (2018) combine the SDSS and SPY results to obtain improved constraints on the double-WD population and its merger rate. They find that the WD binary fraction out to 4 au is about 10%, and that the Galactic specific WD merger rate is $\sim 10^{-11}\,\mathrm{yr^{-1}}$ per WD. This rate is 4.5 to 7 times higher than the MW's specific SN Ia rate, implying that there are indeed more than enough double-WDs to account for the observed SN Ia rate.

These results have further implications for the WD mass function. As Maoz et al. (2018) note, when the double-WD merger rate is integrated over the age of the Galaxy, it implies that 8.5 to 11% of all WDs ever formed have already merged with another WD. Assuming that most of the mergers result in a more-massive WD rather than an SN Ia explosion, we expect that about 10% of the WDs are double-WD merger products. This is consistent with the 15 to 20% "high-mass bump" at $\sim 1\,M_\odot$, observed in the WD mass function when including faint WDs (see Maoz et al., 2018, and references therein).

These numbers suggest that ~12 to 18% of the double-WD mergers result in an SN Ia explosion. However, if a core density of at least $\sim 2\times 10^7\,\mathrm{g\,cm^{-3}}$ is needed for the production of $\sim 0.5\,M_\odot$ of ^{56}Ni observed in SNe Ia, then the merging binary should consist of at least one WD more massive than $\sim 0.9\,M_\odot$ (B. Katz, private communication; Moll et al., 2014). This hints at a sequence of two mergers: one merger to create the massive

WD, which later on merges with a third WD and explodes as an SN Ia. As Maoz et al. (2018) note, this requires that nearly all of the merged double WDs have originated in triple systems. This is in contrast with the observed fraction (10 to 20 %) of triple systems among main-sequence A-type stars (Duchêne and Kraus, 2013; Leigh and Geller, 2013), which are the progenitors of most present-day WDs. This discrepancy might be resolved if the Galactic SN Ia rate is overestimated, or if the WD number density is underestimated. A better estimate of the WD number density is expected in the near future, with the results of the Gaia mission. Nevertheless, the progenitor-population situation looks brighter for the DD scenario, with only a factor of a few discrepancies, compared to the more severe situation for the SD progenitor population.

12.4 Single Degenerates: Is This It?

In the last 20 years the paradigm has changed quite substantially, both from the observational (with the advent of massive SN searches) and the theoretical (with the deployment of 3D explosion simulations) point of view. In spite of the initial success (and popularity) of SDs, the DD scenario has become much more physically viable.

All these new acquisitions have modified our perception, so much that the preference has moved from SD to DD. At the 2013 Leiden workshop on Type Ia progenitors, we ran a poll, asking the participants to give percentages (adding up to 100%) for the various progenitor channels. A total of 47 valid ballots were received, and these are the results: SD: 22%, DD: 60%, CD: 7%, Other: 11%. Considering that CDs are a flavour of DDs, the popularity of double degenerates emerges very clearly. Of course, not all problems are solved. For instance, CO + CO mergers are likely too rare; this therefore requires sub-Chandrasekhar or CO + He WD mergers. It remains to be seen whether this can explain the bulk of normal Type Ia in terms of observed properties (and not just their rates).

A few days before the start of the workshop on 'The Impact of Binaries on Stellar Evolution', we asked for a number of expert opinions. We report them here, as they provide a nice complement to what we have tried to put together in this certainly incomplete summary.

1. 'A review of all the proposed models reveals that each one of them still encounters a few significant difficulties. Consequently, the inescapable conclusion may be that Type Ia supernovae can be produced by a number of progenitor systems' (Mario Livio).
2. 'The thermonuclear explosion of a CO white dwarf has successfully explained the basic observed features of Type Ia supernovae. Both the Chandrasekhar and the sub-Chandrasekhar mass models have been examined. However, no clear observational indication rejects how the white dwarf mass grows until C ignition, i.e., whether the white dwarf accretes H/He-rich matter from its binary companion [SD] or whether two CO white dwarfs merge [DD]' (Kenichi Nomoto).
3. 'All men are created equal – but not Type Ia supernovae. Nature does it right!' (Wolfgang Hillebrandt).

4. 'An SN Ia is the outcome of detonating 1 solar mass of C and O with $\rho_{max} \approx 0.5–2 \times 10^8$ g cm^{-3}' (Stan Woosley).
5. 'I recognise the need to consider a variety of models that might apply to the broad category of "Type Ia" and the limitations of the SD model, but still have reservations about DD models and sub-Chandra models (low central densities). I still think a delayed-detonation model, for its flaws, is the standard to beat in terms of reproducing the spectral evolution. Igniting carbon near the Chandrasekhar mass seems to work best for typical Type Ia, but there are variations that might work for both SD and DD. I'm intrigued by spin-up/spin-down models' (Craig Wheeler).
6. 'Just as all roads lead to Rome, and one size does not fit all, I think it's becoming increasingly clear that Type Ia supernovae are produced by several evolutionary paths leading to different progenitor systems and white dwarf masses' (Alex Filippenko).
7. 'Do not be a slave to fashion! Scientific ideas ebb and flow. Though evidence for interaction is sparse and strong evidence for partners is missing, keep an open mind. The single degenerate model is wounded, but not dead. Perhaps nature finds more than one way to explode a white dwarf' (Robert Kirshner).
8. 'Probably more than one way to make an SN Ia, but seems the primary track involves a white dwarf + another compact star' (Brian Schmidt).
9. 'At least two types of progenitor systems can produce SNe-Ia: SD and/or DD. When the WDs reach the Chandrasekhar mass, they can explode in different ways: detonation, deflagration, delayed detonation ... There are some occurrences in which WDs explode as sub-Chandra or super-Chandra. Therefore, from a phenomenological point of view, we potentially have a large variety of outcomes. To some extent, this fact seems supported by observation. In conclusion, there are different progenitors, different lifetimes, different sizes, different ages, different chemical compositions, and probably different spins. Nevertheless, this variety is characterised by a common ending: after the explosion, nothing is left: "Much ado about nothing" or shall we think again?' (Massimo Della Valle).
10. 'Binaries in all cases, white dwarf in any case, leftover companion in no case (seen so far). Case to be made for binary evolution to produce the massive white dwarfs' (Bruno Leibundgut).
11. 'As we are hesitant to choose between SD and DD, so may have been Nature, perhaps making prompt and late Ia's, respectively. Few, but not so few, intermediate mass binaries end up with a spectacular Type Ia display, as few of the more massive ones make black hole mergers and solar masses in gravitational waves. For us, understanding these paths is understanding how binaries evolve through two common-envelope phases, which makes our job quite difficult, as uncertainties multiply uncertainties. But don't forget, haemoglobin comes from Ia's as chlorophyll from core-collapse supernovae. We have a special link to Ia's, hence an obligation to understand how they come about' (Alvio Renzini).
12. 'In one sentence: Beware assumptions! Adding another: Pre-and post explosion signatures of DD and SD models can be very much alike' (Rosanne Di Stefano).

13. 'For typical SN Ia that eject about a Chandra mass, I like the canonical SD model (perhaps a DD model in which the merger product spins down long enough to become rather like the SD model, although with a total mass at least a bit above Chandra rather than a bit beneath it, might be OK), but if some SN Ia eject substantially less than a Chandra mass yet are basically symmetric, then the DD model with a Shen-Moore very low-mass helium shell may be best for them' (David Branch).
14. 'This matter can be addressed by looking at what is the major challenge for each of the two scenarios, SD and DD, based solely on the MW SN Ia rate and on what we know about the putative progenitor populations' (Dani Maoz).

References

Badenes, C., and Maoz, D. 2012. The Merger Rate of Binary White Dwarfs in the Galactic Disk. *ApJ L*, **749**, L11.

Badenes, C., Hughes, J. P., Bravo, E., and Langer, N. 2007. Are the Models for Type Ia Supernova Progenitors Consistent with the Properties of Supernova Remnants? *ApJ*, **662**, 472–486.

Badenes, C., Mullally, F., Thompson, S. E., and Lupton, R. H. 2009. First Results from the SWARMS Survey. SDSS 1257+5428: A Nearby, Massive White Dwarf Binary with a Likely Neutron Star or Black Hole Companion. *ApJ*, **707**, 971–978.

Bedin, L. R., Ruiz-Lapuente, P., González Hernández, J. I., et al. 2014. Improved Hubble Space Telescope Proper Motions for Tycho-G and Other Stars in the Remnant of Tycho's Supernova 1572. *MNRAS*, **439**, 354–371.

Bloom, J. S., Kasen, D., Shen, K. J., et al. 2012. A Compact Degenerate Primary-Star Progenitor of SN 2011fe. *ApJ L*, **744**, L17.

Bragaglia, A., Greggio, L., Renzini, A., and D'Odorico, S. 1990. Double Degenerates among DA White Dwarfs. *ApJ L*, **365**, L13–L17.

Brown, W. R., Gianninas, A., Kilic, M., Kenyon, S. J., and Allende Prieto, C. 2016. The ELM Survey. VII. Orbital Properties of Low-Mass White Dwarf Binaries. *ApJ*, **818**, 155.

Brown, W. R., Kilic, M., Kosakowski, A., and Gianninas, A. 2017. Discovery of a Detached, Eclipsing 40 Minute Period Double White Dwarf Binary and a Friend: Implications for He+CO White Dwarf Mergers. *ApJ*, **847**, 10.

Bulla, M., Sim, S. A., Pakmor, R., et al. 2016. Type Ia Supernovae from Violent Mergers of Carbon-Oxygen White Dwarfs: Polarization Signatures. *MNRAS*, **455**, 1060–1070.

Cikota, A., Patat, F., Cikota, S., Spyromilio, J., and Rau, G. 2017. Common Continuum Polarization Properties: A Possible Link between Proto-Planetary Nebulae and Type Ia Supernova Progenitors. *MNRAS*, **471**, 2111–2116.

Di Stefano, R. 2010. The Progenitors of Type Ia Supernovae. II. Are They Double-Degenerate Binaries? The Symbiotic Channel. *ApJ*, **719**, 474–482.

Duchêne, G., and Kraus, A. 2013. Stellar Multiplicity. *ARA&A*, **51**, 269–310.

Foss, D., Wade, R. A., and Green, R. F. 1991. Limits on the Space Density of Double Degenerates as Type IA Supernova Progenitors. *ApJ*, **374**, 281–287.

Hachisu, I., Kato, M., and Nomoto, K. 1996. A New Model for Progenitor Systems of Type IA Supernovae. *ApJ L*, **470**, L97.

Hoyle, F., and Fowler, W. A. 1960. Nucleosynthesis in Supernovae. *ApJ*, **132**, 565–590.

Karl, C., Napiwotzki, R., Heber, U., et al. 2003a. Double Degenerates from the Supernova Ia Progenitor Survey (SPY). de Martino, D., Silvotti, R., Solheim, J.-E., and Kalytis, R. (eds), *NATO ASIB Proc. 105: White Dwarfs*, **105**, 43. Dordrecht: Kluwer Academic.

Karl, C. A., Napiwotzki, R., Nelemans, G., et al. 2003b. Binaries Discovered by the SPY Project. III. HE 2209-1444: A Massive, Short Period Double Degenerate. *A&A*, **410**, 663–669.

Kasen, D. 2010. Seeing the Collision of a Supernova with Its Companion Star. *ApJ*, **708**, 1025–1031.

Kashi, A., and Soker, N. 2011. A Circumbinary Disc in the Final Stages of Common Envelope and the Core-Degenerate Scenario for Type Ia Supernovae. *MNRAS*, **417**, 1466–1479.

Kelly, P. L., Fox, O. D., Filippenko, A. V., et al. 2014. Constraints on the Progenitor System of the Type Ia Supernova 2014J from Pre-Explosion Hubble Space Telescope Imaging. *ApJ*, **790**, 3.

Kerzendorf, W. E., Schmidt, B. P., Asplund, M., et al. 2009. Subaru High-Resolution Spectroscopy of Star G in the Tycho Supernova Remnant. *ApJ*, **701**, 1665–1672.

Kerzendorf, W. E., Schmidt, B. P., Laird, J. B., Podsiadlowski, P., and Bessell, M. S. 2012. Hunting for the Progenitor of SN 1006: High-Resolution Spectroscopic Search with the FLAMES Instrument. *ApJ*, **759**, 7.

Kerzendorf, W. E., Yong, D., Schmidt, B. P., et al. 2013. A High-Resolution Spectroscopic Search for the Remaining Donor for Tycho's Supernova. *ApJ*, **774**, 99.

Kerzendorf, W. E., Childress, M., Scharwächter, J., Do, T., and Schmidt, B. P. 2014. A Reconnaissance of the Possible Donor Stars to the Kepler Supernova. *ApJ*, **782**, 27.

Leigh, N. W. C., and Geller, A. M. 2013. The Dynamical Significance of Triple Star Systems in Star Clusters. *MNRAS*, **432**, 2474–2479.

Li, W., Bloom, J. S., Podsiadlowski, P., et al. 2011. Exclusion of a Luminous Red Giant as a Companion Star to the Progenitor of Supernova SN 2011fe. *Nature*, **480**, 348–350.

Livio, M. 2001. Type Ia Supernovae and Their Implications for Cosmology. Livio, M., Panagia, N., and Sahu, K. (eds), *Supernovae and Gamma-Ray Bursts: The Greatest Explosions since the Big Bang*, **13**, 334–355. Cambridge: Cambridge University Press.

Maguire, K., Sullivan, M., Patat, F., et al. 2013. A Statistical Analysis of Circumstellar Material in Type Ia Supernovae. *MNRAS*, **436**, 222–240.

Maoz, D., and Hallakoun, N. 2017. The Binary Fraction, Separation Distribution, and Merger Rate of White Dwarfs from SPY. *MNRAS*, **467**, 1414–1425.

Maoz, D., Badenes, C., and Bickerton, S. J. 2012. Characterizing the Galactic White Dwarf Binary Population with Sparsely Sampled Radial Velocity Data. *ApJ*, **751**, 143.

Maoz, D., Mannucci, F., and Nelemans, G. 2014. Observational Clues to the Progenitors of Type Ia Supernovae. *ARA&A*, **52**, 107–170.

Maoz, D., Hallakoun, N., and Badenes, C. 2018. The Separation Distribution and Merger Rate of Double White Dwarfs: Improved Constraints. *MNRAS*, **476**, 2584

Marsh, T. R., Dhillon, V. S., and Duck, S. R. 1995. Low-Mass White Dwarfs Need Friends: Five New Double-Degenerate Close Binary Stars. *MNRAS*, **275**, 828.

Maxted, P. F. L., and Marsh, T. R. 1999. The Fraction of Double Degenerates among DA White Dwarfs. *MNRAS*, **307**, 122–132.

Moll, R., Raskin, C., Kasen, D., and Woosley, S. E. 2014. Type Ia Supernovae from Merging White Dwarfs. I. Prompt Detonations. *ApJ*, **785**, 105.

Napiwotzki, R., Christlieb, N., Drechsel, H., et al. 2001. Search for Progenitors of Supernovae Type Ia with SPY. *Astron. Nachr.*, **322**, 411–418.

Napiwotzki, R., Koester, D., Nelemans, G., et al. 2002. Binaries Discovered by the SPY Project. II. HE 1414-0848: A Double Degenerate with a Mass Close to the Chandrasekhar Limit. *A&A*, **386**, 957–963.

Napiwotzki, R., Yungelson, L., Nelemans, G., et al. 2004. Double Degenerates and Progenitors of Supernovae Type Ia. Hilditch, R. W., Hensberge, H., and Pavlovski, K. (eds), *Spectroscopically and Spatially Resolving the Components of the Close Binary Stars*. *ASPC*, **318**, 402–410. San Francisco: Astronomical Society of the Pacific.

Nelemans, G., Napiwotzki, R., Karl, C., et al. 2005. Binaries Discovered by the SPY Project. IV. Five Single-Lined DA Double White Dwarfs. *A&A*, **440**, 1087–1095.

Nomoto, K., and Iben, I. Jr. 1985. Carbon Ignition in a Rapidly Accreting Degenerate Dwarf: A Clue to the Nature of the Merging Process in Close Binaries. *ApJ*, **297**, 531–537.

Nugent, P. E., Sullivan, M., Cenko, S. B., et al. 2011. Supernova SN 2011fe from an Exploding Carbon-Oxygen White Dwarf Star. *Nature*, **480**, 344–347.

Pagnotta, A., and Schaefer, B. E. 2014. Identifying and Quantifying Recurrent Novae Masquerading as Classical Novae. *ApJ*, **788**, 164.

Patat, F., Baade, D., Höflich, P., et al. 2009. VLT Spectropolarimetry of the Fast Expanding Type Ia SN 2006X. *A&A*, **508**, 229–246.

Patat, F., Chandra, P., Chevalier, R., et al. 2007. Detection of Circumstellar Material in a Normal Type Ia Supernova. *Science*, **317**, 924.

Patat, F. 2009. Seeking for the Progenitors of Type Ia Supernovae. Giobbi, G., Tornambe, A., Raimondo, G., Limongi, M., Antonelli, L. A., Menci, N., and Brocato, E. (eds), *American Institute of Physics Conference Series*. Am. Inst. of Physics Conf. Ser., **1111**, 299–306. College Park: AIP Publishing.

Patat, F., Cox, N. L. J., Parrent, J., and Branch, D. 2010. Studying the Small Scale ISM Structure with Supernovae. *A&A*, **514**, A78.

Patat, F., Taubenberger, S., Cox, N. L. J., et al. 2015. Properties of Extragalactic Dust Inferred from Linear Polarimetry of Type Ia Supernovae. *A&A*, **577**, A53.

Robinson, E. L., and Shafter, A. W. 1987. An Upper Limit to the Space Density of Short-Period Noninteracting Binary White Dwarfs. *ApJ*, **322**, 296.

Roelofs, G., Bassa, C., Voss, R., and Nelemans, G. 2008. On the Detection of the Progenitor of the Type Ia Supernova 2007. *MNRAS*, **391**, 290–296.

Ruiz-Lapuente, P., Comeron, F., Méndez, J., et al. 2004. The Binary Progenitor of Tycho Brahe's 1572 Supernova. *Nature*, **431**, 1069–1072.

Schaefer, B. E., and Pagnotta, A. 2012. An Absence of Ex-Companion Stars in the Type Ia Supernova Remnant SNR 0509-67.5. *Nature*, **481**, 164–166.

Sternberg, A., Gal-Yam, A., Simon, J. D., et al. 2011. Circumstellar Material in Type Ia Supernovae via Sodium Absorption Features. *Science*, **333**, 856.

Tornambé, A., and Piersanti, L. 2013. Pre-Explosive Observational Properties of Type Ia Supernovae. *MNRAS*, **431**, 1812–1822.

Tutukov, A. V., and Yungelson, L. R. 1981. Evolutionary Scenario for Close Binary Systems of Low and Moderate Masses. *Nauchnye Informatsii*, **49**, 3.

Voss, R., and Nelemans, G. 2008. Discovery of the Progenitor of the Type Ia Supernova 2007. *Nature*, **451**, 802–804.

Whelan, J., and Iben, I. Jr. 1973. Binaries and Supernovae of Type I. *ApJ*, **186**, 1007–1014.

Yoon, S.-C., and Langer, N. 2004. Helium Accreting CO White Dwarfs with Rotation: Helium Novae Instead of Double Detonation. *A&A*, **419**(May), 645–652.

13

Binary Interactions and Gamma-Ray Bursts

NIAL R. TANVIR

13.1 Introduction

The first known gamma-ray burst (GRB) was detected just over 50 years ago, at the time of writing, in July 1967 (Strong and Klebesadel, 1976). Subsequently, we came to understand that GRBs fall into at least two classes (Mazets et al., 1981; Kouveliotou et al., 1993) distinguished primarily by the duration of their prompt γ-ray flash. These are the so-called long-soft and short-hard varieties (hereafter long-GRB and short-GRB respectively). The dividing line between the two classes is roughly $\sim$2 s, although ascribing individual bursts to one class or the other can be difficult in some cases (e.g., Levan et al., 2007).

The discovery of X-ray (Costa et al., 1997) and optical (van Paradijs et al., 1997) afterglows led to the identification of host galaxies and then to the determination of redshifts (Metzger et al., 1997), confirming that long-GRBs are at cosmological distances. Early observations of well-localised long-GRBs revealed that they always occur in star-forming galaxies, and specifically in regions of intense star formation (Fruchter et al., 2006; Svensson et al., 2010). This suggested an origin in the core collapse of massive stars, which was confirmed by direct observation of supernovae accompanying them in many low-redshift cases (Galama et al., 1998; Hjorth et al., 2003; Xu et al., 2013). Spectroscopically, these are broad-lined type Ic supernovae, indicating envelope-stripped stars. This is consistent with the requirement for jet launching since to evacuate a passage through an extended envelope would require the central engine to be active for longer than the typical GRB duration (e.g., Zhang et al., 2004).

Similar breakthroughs for short-GRBs were slower in coming due to their lower detected numbers and typically fainter afterglows, but ultimately it was confirmed that they too are associated with galaxies at cosmological distances (Gehrels et al., 2005; Bloom et al., 2006). Here evidence has steadily accumulated supporting of the idea that short-GRBs result from the merger of neutron stars, but otherwise require the launching of an ultrarelativistic jet similar to long-GRBs.

Thus the short-GRBs have an intimate connection to binary stars, whereas it remains an open question whether the progenitors of long-GRBs also require binary evolution. The question is of interest not only in terms of our understanding of GRBs themselves, but also in their use as probes of the Universe.

In this review, I discuss the potential role of binarity in creating gamma-ray burst progenitors, and explore the importance of understanding long-GRB progenitors for their role as probes of high-redshift star formation, in particular in the era of reionisation. I also consider the importance of the breakthrough electromagnetic detection of a binary neutron star merger identified by via its gravitational wave emission.

13.2 Long-Duration Gamma-Ray Bursts from Binary Stars

A key question is whether the progenitors of long-GRBs are single or binary stars, or possibly channels involving both. High specific angular momentum seems to be an essential requirement for powering a GRB, to ensure production of a massive accretion disk (Fryer et al., 2005, 2007) and/or a rapidly rotating magnetar remnant (Wheeler et al., 2000; Metzger et al., 2011) that can act as the central engine. Furthermore, the progenitor must also have lost any extended envelope in order for the relativistic jet to break out during the timescale for which the central engine is active (tens of seconds). Retaining a high mass rapidly rotating core, with no hydrogen envelope, is problematic as substantial mass loss (which does occur for many massive stars, particularly at higher metallicities) naturally also removes angular momentum.

The leading single-star hypothesis invokes chemically homogeneous evolution (Yoon et al., 2005, 2006). Here sufficiently high rotation rates lead to a high degree of internal mixing within the star, and hence full hydrogen burning. Such stars never go through a red supergiant phase and experience little mass loss. Thus they would naturally produce a very massive, rapidly rotating core at the end of their lives. This model, though, works well at metallicities below about 0.1 solar, which does not sit comfortably with the observation that long-GRBs occur in environments up to and possibly in excess of solar metallicity (e.g., Savaglio et al., 2012).

Alternative progenitors involving binary evolution have been considered. This is appealing since most massive stars are in binary systems, many of which will interact at some point during their evolution (Sana et al., 2012), and such interactions can lead to removal of hydrogen-rich envelopes (e.g., Podsiadlowski et al., 1992). Furthermore, it provides a potential reservoir of additional angular momentum, namely that of the binary orbit (Woosley et al., 2006), whether it leads to spin-up through tidal locking, mass transfer or complete merging (e.g., Fryer et al., 2005).

One possibility studied in some detail is that proposed by Podsiadlowski et al. (2010). They suggested that during a common-envelope stage in the late evolution of binaries with a high-mass primary and moderate-mass secondary, hydrogen from the secondary may accrete onto the helium-burning shell of the primary leading to the explosive ejection of the envelope (specifically, they analysed the case of an $18\,M_\odot$ primary and secondaries in the range $1–5\,M_\odot$). This would leave a massive CO core (and low-mass companion) that then has lost little angular momentum. Additional spin-up may occur through tidal interaction with the secondary, creating the conditions to produce a long-GRB. This route is also expected to be enhanced at low metallicity, but should occur more rarely at metallicities up to around solar.

13.2.1 GRBs as Probes of Reionisation

An important question in cosmology is when and how the intergalactic medium (IGM) was reionised. Spectroscopy of high-redshift quasars, for example, shows that the IGM has been predominantly ionised since a redshift of at least $z \sim 6$ (McGreer et al., 2015). Observations of the E-mode polarisation signature of the cosmic microwave background provides a measure of the electron-scattering optical depth, and the most recent results from *Planck* have pointed to reionisation occurring predominantly around $z \sim 8$ to 10 (Planck Collaboration et al., 2016). At these times, the abundance of bright quasars was very low, and so it has long been supposed that reionisation must have been brought about and sustained by extreme ultraviolet (EUV) radiation from massive stars.

Proving this hypothesis depends on several poorly understood factors. Most important of these are the star formation rate density, which determines the production rate of ionising radiation, and the fraction of that radiation that escapes its birth galaxy to the intergalactic medium. Both of these are hard to measure and have been sources of controversy.

The High-Redshift Galaxy Luminosity Function

Detecting very high-redshift galaxies requires deep observations, and samples are subject to contamination by misidentified lower-redshift sources (nearly all cases rely on photometric redshift estimates). Furthermore, samples are also necessarily incomplete, in the sense that any given detection limit will only probe some way down the luminosity function. This issue is highlighted by the recent work of Bouwens et al. (2017), who pointed out that the apparently small (but difficult to measure) sizes of faint galaxies over a wide range of redshift has a significant impact on the completeness corrections, particularly at z > 6.

Several GRBs have now been detected at redshifts, either spectroscopic (e.g., Tanvir et al., 2009, 2017a) or photometric (e.g., Cucchiara et al., 2011; Bolmer et al., 2018), placing them within the era of reionisation. Although the rate of GRBs detected by *Swift* at $z > 6$ is low (analysis of a high completeness subsample by Perley et al., 2016, indicates a rate of 1–2%), the extreme luminosity of GRBs and their afterglows means, for example, that they can provide detailed abundance information even for very faint high-z galaxies (e.g., Hartoog et al., 2015). Thus, despite being few in number, GRBs offer a powerful alternative route to probing high-redshift star formation and star-forming environments.

Since GRBs are detected independently of their host luminosity, they therefore allow us to deduce the presence of star-forming galaxies even below the detection threshold of current technology (Tanvir et al., 2012; Trenti et al., 2012). Such studies, although still based on relatively small numbers, confirm that the bulk of star formation at redshifts greater than $z \sim 6$ is occurring in galaxies below the effective *HST* deep field detection limit (Tanvir et al., 2012; McGuire et al., 2016). This is consistent with a steep (power law index $\alpha \sim 2$) faint-end slope of the galaxy luminosity function during the era of reionisation.

The Escape Fraction of Ionising Radiation

Nonetheless, despite a high star formation rate density, stars can have only driven reionization if a sufficiently high fraction of the ionising photons that they create escape their host galaxies into the IGM. This is typically not the case, since extreme UV is very easily absorbed by neutral gas or dust, whereas escape fractions of at least 10 to 20% at $z \sim 8$ are required to bring about reionisation (e.g., Faisst, 2016). Directly measuring radiation below the Lyman limit at high redshift is rendered unfeasible by the near-complete IGM absorption. Spectroscopy of GRB afterglows frequently allow us to measure the neutral hydrogen column density directly from modelling the Lyman-α absorption line. This sets a lower limit to the opacity to ionising radiation along that line of sight, and by averaging many GRBs we can infer the average opacity to massive stars that are the progenitors of long-GRBs.

Early studies of samples of ~30 GRBs, mainly in the range $z = 2$–4, concluded an upper limit of $f_{\rm esc} < 7.5\%$ (Chen et al., 2007; Fynbo et al., 2009). A new and more than four

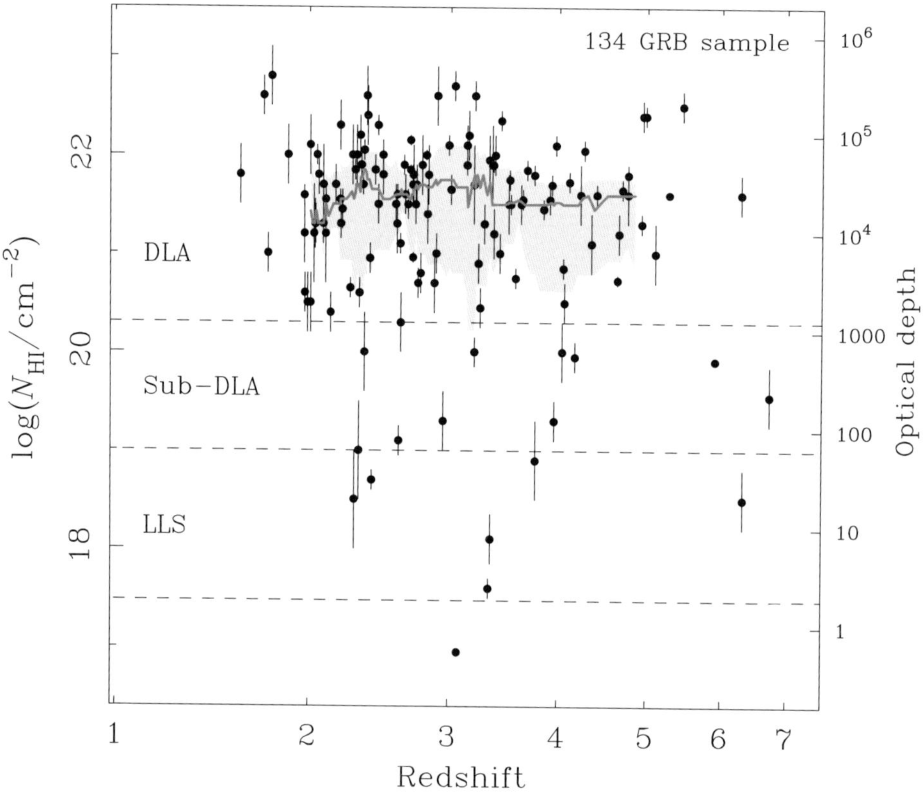

Figure 13.1 Neutral hydrogen column density as a function of redshift for a sample of long-GRBs derived from spectroscopy of their afterglows. The corresponding optical depth at the Lyman limit is shown on the right-hand axis. The large majority are damped Lyman-α systems, with a small proportion being classified as sub–damped Lyman-α (DLA) systems, Lyman Limit systems (LLS) and below. All sight lines apart from the two with the lowest column densities are essentially opaque ($\tau > 8$) to extreme ultraviolet ionising radiation. The running median (solid line) and interquartile range (grey shading) of 20 points show little evidence for significant variation with redshift to at least $z \sim 5$.

times larger sample of GRBs in the range $z = 1.6$–6.7 has recently been analysed by Tanvir et al. (2018), which finds all but two sight lines are essentially opaque (see Figure 13.1), indicating a 98% upper limit to the escape fraction of $f_{esc} < 1.5\%$. Unless there is significant evolution above $z \sim 5$ towards lower column-density sight lines, this does indicate a serious difficulty for the favoured explanation of reionisation.

This estimate is subject to potential systematic biases, although many of these, such as the selection against identifying and obtaining spectroscopy for highly extincted afterglows, work in the direction of making the limit more stringent, and it is hard to see how a significant number of very low-column systems could have been missed. A fundamental assumption, though, which also applies to many other uses of GRBs as probes, is that they are representative of the massive stars largely responsible for the far- and extreme-UV production. From an empirical point of view, the strong correlation of long-GRBs with star-forming galaxies, and young star-forming regions within those galaxies (e.g., Fruchter et al., 2006), does support this assumption, at least for subsolar metallicity star formation, but a fuller understanding of the progenitor systems is urgently required to justify applying these tests.

13.3 Short-Duration Gamma-Ray Bursts from Binary Stars

The mergers of compact binaries, involving two neutron stars or a neutron star and stellar-mass black hole, have long been considered as possible progenitors of cosmological GRBs (Eichler et al., 1989). Following the association of long-GRBs with massive star core collapse, such compact binary mergers became the favoured hypothesis for the progenitors of short-GRBs. From an observational point of view, the association of short-GRBs with host galaxies exhibiting a wide range of star formation histories (Fong et al., 2013a) supported a progenitor with a long delay time following a star formation episode.

The wide spread of locations, both within and also outside of putative host galaxies, is also consistent with the progenitors having received substantial natal kicks in some cases, in the supernovae that formed the compact objects (Fong and Berger, 2013; Tunnicliffe et al., 2014), or equally with some proportion being formed dynamically in remote globular clusters (e.g., Church et al., 2011).

This hypothesis received significant support from the discovery in 2013 of evidence of a rebrightening in the near-infrared following a short-GRB (GRB 130603B), which was consistent with expectations of so-called 'kilonova'[1] emission (Tanvir et al., 2013; Berger et al., 2013). Kilonovae are expected due to the radioactive decay of heavy elements, in particular those created by rapid neutron capture (the r-process) in the highly neutron-rich material ejected in the merger (Li et al., 1998). Subsequently, further possible examples of kilonova bumps were noted in the light curves of other short-GRBs (Yang et al., 2015; Jin et al., 2016). If indeed substantial r-process nucleosynthesis is occurring in neutron star mergers, then they may be the dominant production site of many heavy elements in

[1] Also often referred to as a 'macronova'.

the universe (Freiburghaus et al., 1999). The formation and evolution of stellar systems leading to such compact binaries are therefore of significant interest (Tauris et al., 2017), determining the lifetime distribution and population statistics.

13.3.1 GW 170817

In a dramatic recent development, the first high-confidence binary neutron star merger to be detected in gravitational waves was identified by LIGO/Virgo (Abbott et al., 2017a). Remarkably, this event, GW 170817, also produced a short-GRB, detected by both *Fermi*/GBM and *INTEGRAL*/SPI-ACS (Abbott et al., 2017b), and a bright kilonova (Abbott et al., 2017c; Arcavi et al., 2017; Coulter et al., 2017; Cowperthwaite et al., 2017; Evans et al., 2017; Kasliwal et al., 2017; Nicholl et al., 2017; Smartt et al., 2017; Tanaka et al., 2017; Tanvir et al., 2017b; Valenti et al., 2017).

GW 170817 has certainly established the association of binary neutron star mergers with kilonovae, which increases the confidence that the signal seen following GRB 130603B was indeed also a kilonova. However, the accompanying short-GRB was undoubtedly peculiar in being intrinsically by far the faintest seen to date (at only 40 Mpc, it is more than 10$\times$ closer than any previous short-GRB with a well-established distance), with any afterglow component being initially undetected, although a rising X-ray and radio afterglow appeared after a couple of weeks (Haggard et al., 2017; Hallinan et al., 2017; Troja et al., 2017). In fact, there has already been considerable discussion as to the source of the gamma rays and the nature of any relativistic component. Emission of γ-rays, beginning $\sim$2 s after the merger itself, does seem to require a relativistic jet, but the absence of an early afterglow suggests that our line of sight was outside the ultrarelativistic core of the jet. In this scenario, the γ-rays are associated with the mildly relativistic cocoon surrounding the jet that breaks out of the debris surrounding the merger (e.g., Gottlieb et al., 2017). The rising afterglow, which has recently been detected in the optical at $\Delta t > 100$ days with *HST*, is inconsistent with predictions of off-axis emission from a sharp-edged jet, but may then be largely due to emission from a shock created by the cocoon, or that created by a structured jet with lower velocity material at larger angles from the central jet axis (e.g., Lazzati et al., 2017; Mooley et al., 2017; Lyman et al., 2018; Margutti et al., 2018). Distinguishing these possibilities may be possible with very late time observations.

13.4 Conclusions

Binary stellar evolution is of importance for both short and long classes of GRBs. The former determine the populations of neutron-star compact binaries that create short-GRBs when they merge, and will be of intense interest in coming years with the opening of multi-messenger electromagnetic plus gravitational wave astronomy. Long-GRBs may very well also be produced by one or more binary evolution pathways, and in any case, understanding how representative GRB progenitors are of massive star populations more generally, including binaries, is crucial for their use as cosmological probes and in particular for understanding the role of massive stars in bringing about and sustaining the reionised IGM.

References

Abbott, B. P., Abbott, R., Abbott, T. D., et al. 2017a. GW170817: Observation of Gravitational Waves from a Binary Neutron Star Inspiral. *Physical Review Letters*, **119**, 161101.

Abbott, B. P., Abbott, R., Abbott, T. D., et al. 2017b. Gravitational Waves and Gamma-Rays from a Binary Neutron Star Merger: GW170817 and GRB 170817A. *ApJ*, **848**, L13.

Abbott, B. P., Abbott, R., Abbott, T. D., et al. 2017c. Multi-Messenger Observations of a Binary Neutron Star Merger. *ApJ*, **848**, L12.

Arcavi, I., Hosseinzadeh, G., Howell, D. A., et al. 2017. Optical Emission from a Kilonova Following a Gravitational-Wave-Detected Neutron-Star Merger. *Nature*, **551**, 64.

Berger, E., Fong, W., Chornock, R. 2013. An R-Process Kilonova Associated with the Short-Hard GRB 130603B. *ApJ*, **774**, L23.

Bloom, J. S., Prochaska, J. X., Pooley, D., et al. 2006. Closing in on a Short-Hard Burst Progenitor: Constraints from Early-Time Optical Imaging and Spectroscopy of a Possible Host Galaxy of GRB 050509b. *ApJ*, **638**, 354.

Bolmer, J., Greiner, J., Krühler, T., et al. 2018. Dust Reddening and Extinction Curves towards Gamma-Ray Bursts at $z > 4$. *A&A*, **609**, A62.

Bouwens, R. J., Illingworth, G. D., Oesch, P. A., et al. 2017. Extremely Small Sizes for Faint $z \sim 2$-8 Galaxies in the Hubble Frontier Fields: A Key Input for Establishing Their Volume Density and UV Emissivity. *ApJ*, **843**, 41.

Chen, H.-W., Prochaska, J. X., Gnedin, N. Y. 2007. A New Constraint on the Escape Fraction in Distant Galaxies Using γ-Ray Burst Afterglow Spectroscopy. *ApJ*, **667**, L125.

Church, R. P., Levan, A. J., Davies, M. B., et al. 2011. Implications for the Origin of Short Gamma-Ray Bursts from Their Observed Positions around Their Host Galaxies. *MNRAS*, **413**, 2004.

Costa, E., Frontera, F., Heise, J., et al. 1997. Discovery of an X-Ray Afterglow Associated with the Gamma-Ray Burst of 28 February 1997. *Nature*, **387**, 783.

Coulter, D. A., Foley, R. J., Kilpatrick, C. D., et al. 2017. Swope Supernova Survey 2017a (SSS17a), the Optical Counterpart to a Gravitational Wave Source. *Science*, **358**, 1556.

Cowperthwaite, P. S., Berger, E., Villar, V. A., et al. 2017. The Electromagnetic Counterpart of the Binary Neutron Star Merger LIGO/Virgo GW170817. II. UV, Optical, and Near-Infrared Light Curves and Comparison to Kilonova Models. *ApJ*, **848**, L17.

Cucchiara, A., Levan, A. J., Fox, D. B., et al. 2011. A Photometric Redshift of $z \sim 9.4$ for GRB 090429B. *ApJ*, **736**, 7.

Eichler, D., Livio, M., Piran, T., et al. 1989. Nucleosynthesis, Neutrino Bursts and Gamma-Rays from Coalescing Neutron Stars. *Nature*, **340**, 126.

Evans, P. A., Cenko, S. B., Kennea, J. A., et al. 2017. Swift and NuSTAR Observations of GW170817: Detection of a Blue Kilonova. *Science*, **358**, 1565.

Faisst, A. L. 2016. Revisiting the Lyman Continuum Escape Crisis: Predictions for $z > 6$ from Local Galaxies. *ApJ*, **829**, 99.

Fong, W., Berger, E., Chornock, R., et al. 2013. Demographics of the Galaxies Hosting Short-Duration Gamma-Ray Bursts. *ApJ*, **769**, 56.

Fong, W., Berger, E. 2013. The Locations of Short Gamma-Ray Bursts as Evidence for Compact Object Binary Progenitors. *ApJ*, **776**, 18.

Freiburghaus, C., Rosswog, S., Thielemann, F.-K. 1999. R-Process in Neutron Star Mergers. *ApJ*, **525**, L121.

Tauris, T. M., Kramer, M., Freire, P. C. C., et al. 2017. Formation of Double Neutron Star Systems. *ApJ*, **846**, 170.

Trenti, M., Perna, R., Levesque, E. M., et al. 2012. Gamma-Ray Burst Host Galaxy Surveys at Redshift $z > 4$: Probes of Star Formation Rate and Cosmic Reionization. *ApJ*, **749**, L38.

Troja, E., Piro, L., van Eerten, H., et al. 2017. The X-Ray Counterpart to the Gravitational-Wave Event GW170817. *Nature*, **551**, 71.

Tunnicliffe, R. L., Levan, A. J., Tanvir, N. R., et al. 2014. On the Nature of the 'Hostless' Short GRBs. *MNRAS*, **437**, 1495.

Valenti, S., David, Sand, J., et al. 2017. The Discovery of the Electromagnetic Counterpart of GW170817: Kilonova AT 2017gfo/DLT17ck. *ApJ*, **848**, L24.

van Paradijs, J., Groot, P. J., Galama, T., et al. 1997. Transient Optical Emission from the Error Box of the Gamma-Ray Burst of 28 February 1997. *Nature*, **386**, 686.

Wheeler, J. C., Yi, I., Höflich, P., et al. 2000. Asymmetric Supernovae, Pulsars, Magnetars, and Gamma-Ray Bursts. *ApJ*, **537**, 810.

Woosley, S. E., Heger, A. 2006. The Progenitor Stars of Gamma-Ray Bursts. *ApJ*, **637**, 914.

Xu, D., de Ugarte Postigo, A., Leloudas, G., et al. 2013. Discovery of the Broad-Lined Type Ic SN 2013cq Associated with the Very Energetic GRB 130427A. *ApJ*, **776**, 98.

Yang, B., Jin, Z.-P., Li, X., et al. 2015. A Possible Macronova in the Late Afterglow of the Long-Short Burst GRB 060614. *Nature Comm.*, **6**, 7323.

Yoon, S.-C., Langer, N. 2005. Evolution of Rapidly Rotating Metal-Poor Massive Stars towards Gamma-Ray Bursts. *A&A*, **443**, 643.

Yoon, S.-C., Langer, N., Norman, C. 2006. Single Star Progenitors of Long Gamma-Ray Bursts. I. Model Grids and Redshift Dependent GRB Rate. *A&A*, **460**, 199.

Zhang, W., Woosley, S. E., Heger, A. 2004. The Propagation and Eruption of Relativistic Jets from the Stellar Progenitors of Gamma-Ray Bursts. *ApJ*, **608**, 365.

14

Binaries as Sources of Gravitational Waves

GIJS NELEMANS

14.1 Gravitational Waves and Their Detection

With the discovery of first the binary black hole merger GW150914 (Abbott et al., 2016e) and recently the binary neutron star merger GW170817 (Abbott et al., 2017e,f), the field of gravitational wave (GW) astrophysics has been opened in a spectacular fashion.

Since the original prediction of GW (Einstein, 1916, 1918), the field has made a long journey. Since the 1960s, the importance of GW for binary stars and the importance of binaries as GW sources has been recognised (e.g., Paczyński, 1967; Press and Thorne, 1972). Since the 1970s, binary stars, in particular those with neutron star (NS), white dwarf (WD) and (hypothetical) black hole (BH) components, have been considered the most promising sources for detectable GW (e.g., Clark et al., 1979; Tutukov and Yungelson, 1979; Nelemans et al., 2001; Belczynski et al., 2002). With the development of laser interferometer detectors such as LIGO and Virgo, the field gradually evolved to the stage where real detections of GW on Earth could be expected. At the same time, a long history of proposed GW detectors in space has culminated in the selection of the LISA mission as the first space detector (Amaro-Seoane et al., 2017).

For excellent reviews about GW and GW sources, I refer, e.g., to Schutz (1989) and Sathyaprakash and Schutz (2009). For binaries, the effect of GW emission is a decrease in orbital energy and angular momentum (e.g., Peters, 1964) leading, in general, to a shortening of the orbital period and circularisation of the orbit. This will stop once the stars start to interact, which, in most cases, leads to merger of the two stars. Far from the sources, the effect of GW is a minuscule effect on the metric, which can in principle be measured.

14.1.1 Detectors: Status and Future

The way the minute ripples in space time can be measured is essentially by determining the time difference that a light ray spends traversing a 'fixed' distance using laser interferometry (Weiss, 1972). This clever idea, and the skills and endurance to turn the idea into a working detector, was awarded the 2017 Nobel Prize in Physics for Weiss, Thorne and Barish. GW come with different frequencies/wavelengths (depending on the size of the source), and detectors need to be properly sized to be sensitive to certain waves (very roughly source size $\approx$ wavelength $\approx$ detector size).

The currently operational *ground-based laser interferometers advanced LIGO* (Aasi et al., 2015; Abbott et al., 2016c) and *advanced Virgo* (Acernese et al., 2015) are 4 and 3 km long vacuum instruments in which laser beams are first accurately stabilised (both in frequency and in power) and then split over the long arms. In the arms, the lasers are reflected hundreds of times in Fabry–Perot cavities, increasing the effective size of the detector, before being combined again and tuned in such a way that almost perfect destructive interference is achieved. Since this means most of the light is reflected back towards the laser, a power-recycling mirror is placed in the path of the laser, reflecting the light back into the interferometer, vastly increasing the power of the light stored in the interferometer and thus increasing the sensitivity of the instrument. This is needed because the length changes are so small that they are an incredibly small fraction ($\sim 10^{-10}$) of the 1μ wavelength of the laser and thus lead to only tiny changes in the interferometer power output.

In addition to achieving the necessary accuracy of the measurement, the detectors need to be isolated from environmental disturbances and carefully monitored to determine any external noise components. Seismic noise is the most important noise source at low frequencies and is suppressed by many orders of magnitude via seismic isolation mechanisms. After all these measures, there are still large numbers of transient noise fluctuations that can mimic real GW signals, collectively known as triggers. By searching for coincident and identical triggers in multiple detectors, the trigger background can be very strongly reduced. However, even then the number of triggers increases very sharply when going to weaker triggers, so there is a fairly clean separation between noise and real signals (see, e.g., figure 4 of Abbott et al., 2016e).

The two *advanced LIGO* and the *advanced Virgo* detectors are currently offline to undergo further improvement to increase the sensitivity. The plan is for the detectors to come online again for the year-long third observing run (O3) about a year after the end of O2 (Abbott et al., 2016f), i.e., in the fall of 2018 and the beginning of 2019. After that, further improvements are expected before the detectors reach their design sensitivity. Additional detectors, KAGRA in Japan (Aso et al., 2013) and LIGO India (Iyer et al., 2011) are expected to join the GW detector network in the early 2020s. Further improvements on the sites of the current detectors are envisioned in the later 2020s (termed A+ and LIGO Voyager for LIGO sites).

On a longer time scale, a third generation of detectors is planned, such as the European Einstein Telescope or the US Cosmic Explorer, each with significantly larger arms (10–40 km), possibly underground to reduce the environmental noise. The sensitivities of these detectors would be at least a factor of 10 better than the advanced detectors (e.g., Sathyaprakash et al., 2012). Boosted by the success of the second-generation detectors, the preparations for the third-generation detectors is taking off.[1]

[1] https://gwic.ligo.org/3Gsubcomm/.

For larger GW wavelengths, the *ESA LISA mission* is planned for a launch in 2034 (Amaro-Seoane et al., 2017). It consists of three identical spacecraft that exchange laser beams in an interferometer with arm's lengths of 2.5 million km. Although the basic measurement idea is the same as for the ground-based detectors, it is not possible to actually reflect the laser beam back along the arms, since the divergence of the laser beam would lead to a signal that is much too weak. Instead, the technique of time delay interferometry (Armstrong et al., 1999) is used, in which at each time and in each of the space craft the phase information of the incoming laser beam is measured by comparing it to a local reference. All this information is later combined into a virtual interferometer by adding up signals with the appropriate time shifts to determine the interference that *would have* resulted if the light had actually followed that path and would have been physically combined. The advantage of this technique is that the laser noise largely cancels and that different types of virtual interferometers can be constructed (referred to as different TDI configurations/variables). The LISA mission is currently studied by ESA with a significant contribution from NASA, and the final adoption of the mission is expected in the early 2020s.

14.1.2 Order of Magnitude Estimates: Detection Horizons

Gravitational waves are produced by the sources with a changing quadrupole moment and thus binary stars are about the most optimal sources for GW generation (e.g., Thorne, 1997). There is a nice paradox that GW are energetically at the same time the weakest and the strongest phenomena in the Universe. The GW luminosity of a binary GW source is given by Peters and Matthews (1963),

$$L_{\rm GW} = \frac{32G^4}{5c^5}\frac{M_1^2M_2^2(M_1+M_2)}{a^5}$$

which due to the extremely small prefactor $G^4/c^5 \sim 10^{-81}$ vanishes almost completely for any human-made experimental setup and even for normal binary stars is very small (e.g., the GW luminosity of Algol is $\sim 10^{-6}$ times the solar luminosity). On the other hand (forgetting factors of a few difference between M_1, M_2, M_1+M_2), the same equation can be written as

$$L_{\rm GW} \approx \frac{32c^5}{5G}\left(\frac{GM}{c^2a}\right)^5$$

which can approach $c^5/G \sim 10^{59}$ erg/s for systems with $v \approx c$, i.e., $GM/a \approx c^2$ and thus can reach above 10^{26} solar luminosities! This condition ($v \sim c$) is of course quite extreme, but for binary BHs that are about to merge can be approached.

The frequency of the emitted GW is simply related to the orbital frequency $f_{\rm GW} = 2f_{\rm orb} \sim GM/a^3$. For merging black holes, the separation at merger scales with the horizon size and thus mass ($a \sim R \sim M$), i.e., $f_{\rm merge} \sim 1/M^2$: more massive BHs merge at lower frequencies.

The (instantaneous) amplitude h of the GW scales with $\sqrt{L_{\rm GW})/(f_{\rm GW}d)}$ and thus $h \propto \mathcal{M}^{5/3} f_{\rm GW}^{2/3} d^{-1}$, with 'chirp mass' $\mathcal{M} = (M_1M_2)^{2/5}/(M_1+M_2)^{1/5}$ and d the distance.

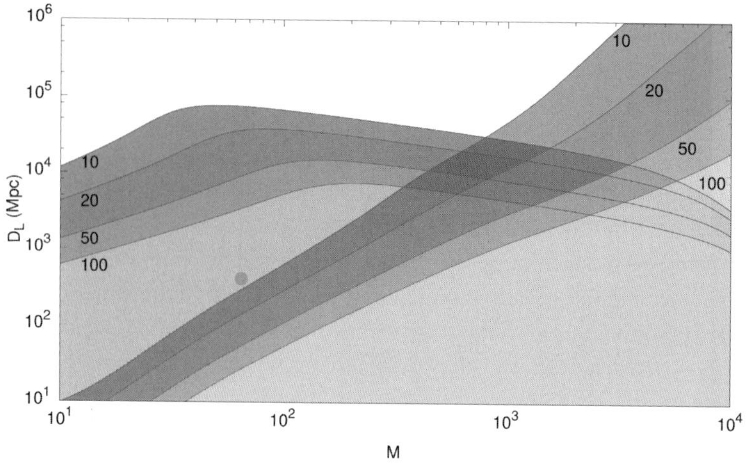

Figure 14.1 Detection horizons as a function of the total mass for equal mass binaries for future detectors. The expected S/N is given by the contours. GW150914 is indicated by the dot. Figure courtesy Neil Cornish.

We thus can immediately see that for a binary that is not changing its frequency during the observation, the signal scales with the (chirp) mass to the power 5/3, while for such a source the signal becomes stronger for higher frequency. For systems that do evolve, such as the observed LIGO/Virgo sources, this picture is not correct, since there are a limited number of GW cycles observed, which is inversely proportional to $\mathcal{M}^{5/3}$. The signal strength increases with the square root of the number of cycles, so scales with $\mathcal{M}^{5/6}$.

In Figure 14.1, the distance to which binaries can be observed with ET and LISA is shown as a function of total mass (assuming equal mass binaries). The plot roughly agrees with the scalings derived previously, except for masses above several tens of solar masses in ET, where the signals slowly move out of the ET sensitivity band (and the maximum distance thus starts to go down again with mass). Because GW get redshifted, the maximum sensitivity corresponds to increasingly lower mass systems at larger distances. From this figure, the enormous promise of GW astrophysics is immediately clear, since BH binaries can be observed essentially throughout the whole Universe with these instruments!

14.2 What Can We Learn about Stars and Binaries from GW Measurements?

For many people, GW detections are primarily interesting from the point of view of testing general relativity (Abbott et al., 2016g), having independent cosmological measurements (Abbott et al., 2017a) or detecting signals from the Early Universe or from unknown sources. For us, GW measurements are a new way to study binary systems with compact objects, for which there are many open questions. After listing the most important open questions, I will discuss what we know and can learn about these questions from GW observations.

14.2.1 Open Questions about Compact Objects

Compact objects form at the end of the lives of stars. In this way, stellar evolution and compact object formation are intimately linked. If we want to understand the formation of NSs and BHs, we have to know how the massive stars evolve from which they form (e.g., Sukhbold et al., 2016). The most pressing issues are stellar winds and the influence of rotation on the internal evolution of the stars. But even if these were known, it is not clear how the final state of the star translates into the formation of the compact object. The way the core collapses and the possible ensuing supernova (SN) explosion are still very unclear as are the resulting masses and spins of the compact objects (e.g., Ugliano et al., 2012). From NS observations, it is clear that the collapse leads to an asymmetric kick (e.g., Lyne and Lorimer, 1994). The open question is if this also holds for BHs (e.g., Repetto et al., 2017). All of this is further complicated by binary interactions that lead to mass transfer and influence the evolution of the stars. In addition, the binary evolution processes (mass transfer and its resulting loss of mass and angular momentum from the system, including common-envelope evolution) influence the further evolution of the system and are typically rather poorly understood. Furthermore, if we want to understand compact objects and binaries in the global context of the evolution of galaxies, including the Milky Way, we also have to understand all these processes in the changing environment, in particular metallicity. With the new tool of studying compact objects in binaries with GW, we can hope to shed a new light on the combined effect of all the uncertainties and thus learn about them.

14.2.2 Black Holes

BHs are the most enigmatic objects, not only by their nature, but also in their formation. We simply do not know which stars form BHs, if this happens via initial formation of a neutron star that later collapses or not and how often (if at all) BH formation is accompanied by an SN (e.g., Ugliano et al., 2012).

A particularly important question is the existence of asymmetric kicks at the formation of BHs. Several individual black hole X-ray binaries have been studied in detail in order to investigate if proof can be given that BH kicks exist (e.g., Nelemans et al., 1999; Willems et al., 2005; Fragos et al., 2009). Although kicks are possible or even plausible, there is not unambiguous evidence. An alternative approach is to look at the positions of the observed BH X-ray binaries in the Milky Way and compare them to those of neutron-star X-ray binaries (White and van Paradijs, 1996; Jonker and Nelemans, 2004; Repetto et al., 2012). The more recent of these studies show that the BH systems are found at similar height above the plane as NS systems or at least surprisingly high up if BHs do not receive significant ($\sim$100km/s) kicks. However, NS systems may have formed in a different way. Also, the observed BH systems are differently biased because the companion stars have to be studied in detail in order to show that the compact object is so massive that it has to be a BH. In a systematic study, all these effects are taken into account and the local height of BH X-ray binaries is compared to a simulated population for different kicks and rescaled to the local Galactic potential such that all systems can be compared (Repetto and Nelemans, 2015;

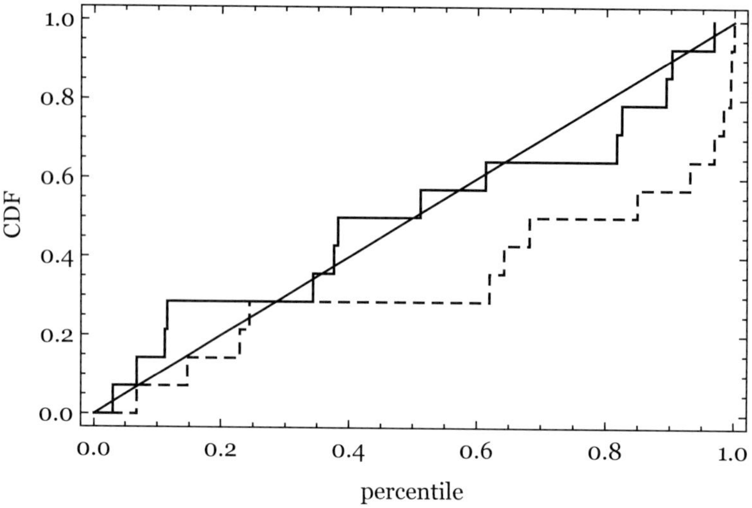

Figure 14.2 Cumulative distribution of the local height above the plane of black hole X-ray binaries compared to a simulated population that originates in the disk with low (dashed) and high (solid) kicks. From Repetto et al. (2017). Reproduced by permission of the AAS.

Repetto et al., 2017). The result is shown in Figure 14.2 and suggests that indeed at least some BH get relatively high kicks.

At the same time, simulations of the formation of binary BH as GW sources suggest that (too) large kicks disrupt too many systems, leading to a very low merger rate (e.g., Belczynski et al., 2016a). The LIGO/Virgo detections have spectacularly opened a new way to (statistically) study the BH population. Including the most recent announced detection, GW170608, five high-confidence detections and one candidate (LVT151012) detection have been reported (Abbott et al., 2016b, 2017c,b,d). They span a wide range of masses (Figure 14.3) ranging from 'heavy' BH with masses over 30 $M_{\odot}$ to ones in the range of masses known from X-ray binaries (Casares and Jonker, 2014). The rapid increase in sources makes for an exciting expectation that soon we will be able to put real constraints on the mass distribution of binary BH. Looking at Figure 14.3, it will be the question whether there is a bimodal distribution with low- and high-mass systems, or a continuum, especially with the uncertain character of the intermediate system LVT151012.

Apart from the masses, the observations have made it possible to estimate the binary BH merger rate. All detections, including the newer ones, are consistent with the latest rate determination (Abbott et al., 2017b) of 12 to 213 /yr/Gpc^3. The last piece of information that is available from the data is an estimate of the 'effective spin', i.e., the mass weighted component of the spin parallel to the orbital angular momentum; see Figure 14.4. As can be seen, most of the systems have effective spins close to zero.

In our current understanding, there are several ways in which binary BH that merge can form (apart from 'exotic' scenarios such as primordial BH, pop III stars, etc.); see Abbott et al. (2016a). The first is the same path that has been put forward for the double NS

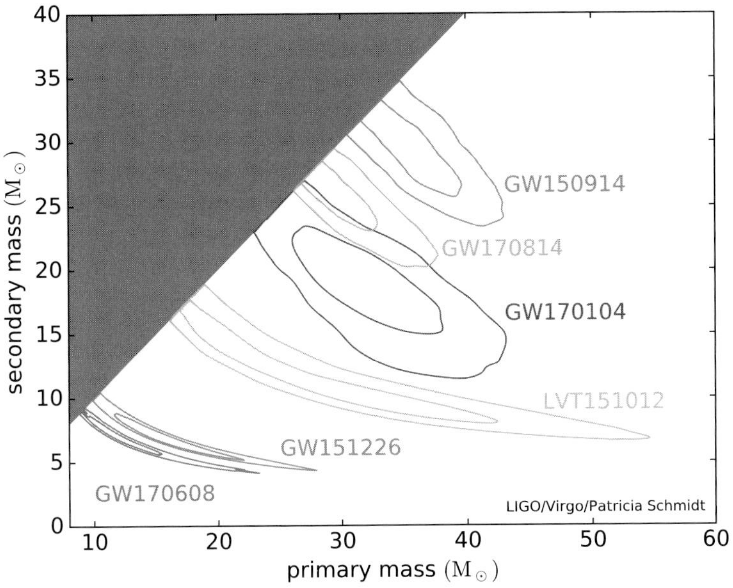

Figure 14.3 Mass measurements of the current binary BH GW detections. Courtesy LIGO/Virgo collaboration/Patricia Schmidt.

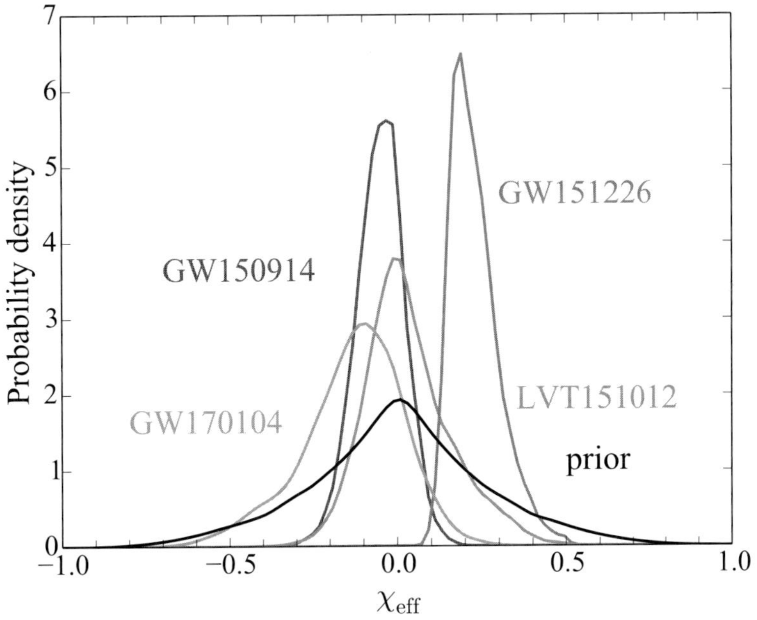

Figure 14.4 Effective spin measurements of the current binary BH GW detections. From Abbott et al. (2017b). Reproduced by permission of the AAS.

systems found in the 1970s (e.g., Flannery and van den Heuvel, 1975) but then for more massive stars (e.g., Belczynski et al., 2016a,b). There are several variations on this scenario, in particular avoiding the common-envelope stage, either by invoking stable mass transfer, possible due to the large masses of the BH (van den Heuvel et al., 2017) or due to the rapid rotation of the massive stars, leading to so-called homogeneous evolution, in which expansion to giant dimensions is prevented by enhanced mixing (de Mink and Mandel, 2016). A fundamentally different formation scenario is via dynamical interactions in dense stellar environments (Sigurdsson and Hernquist, 1993; Portegies Zwart and McMillan, 2000) that are found to be a viable mechanism to produce mergers such as GW150914 (e.g., Mapelli, 2016; Rodriguez et al., 2016; Park et al., 2017). There is a lot of excitement about the idea that the effective spins will be the key to distinguishing different formation channels or inferring BH kicks. The idea is that for dynamically formed systems, the angles between the BH spins would be random, while for binaries they would be more or less aligned, unless the BH get very large kicks. However, as long as there is no clear indication that the BH spin is directly related to the spin of the progenitor star, these considerations are not very convincing. The reason to doubt the relation between progenitor spin and BH spin is the observed very large misalignment between the spins in the double pulsar (Breton et al., 2008).

Another extremely nice and promising result is the realisation that GW150914-like sources are also promising LISA sources (Sesana, 2016). Apart from a multitude of nice measurements when combining LISA and ground-based measurements, there is some chance to detect any remaining eccentricity in the orbits in the LISA band, which would point towards dynamical formation (e.g., Breivik et al., 2016).

It is clear that in the coming years we will have a steady increase in the number of detected binary BHs. With improved detector sensitivity, first with the current second generation of detectors and of course more spectacularly with the third-generation detectors, the numbers could start to add up very significantly. This will allow fantastic studies. My personal expectation is that with several tens of detections, we will get a more solid view of the mass and spin distributions, maybe seeing the first evidence for different populations if they exist. With several hundreds of detections, we can start to bin up the measurements, for instance in redshift bins, and study the evolution of the system properties. At the same time, the overall mass and spin distributions, including, e.g., mass gaps, upper and lower limits to BH masses, etc., will be determined quite accurately. Finally, for the third-generation detectors, with thousands and thousands of detections, we can really determine all these properties as a function of the evolution of the universe!

14.2.3 Neutron Stars and Electromagnetic Follow-Up

Since the discovery of PSR J1913+16 (Hulse and Taylor, 1975), and maybe even before, double NS have been the prime example GW sources (see, e.g., Phinney, 1991). The discovery of the orbital change due to GW (Taylor and Weisberg, 1982) started a process that recently reached a spectacular highlight with the discovery of the first double NS GW source GW170817, which also has been detected in gamma-rays, optical/infrared, X-rays and radio emission (see Abbott et al., 2017f,e).

The current knowledge about the 15 Galactic double NS is nicely summarised in Tauris et al. (2017). From these systems, it is clear that there is a sizeable population of double NS that will merge in the Galaxy. The source properties, in particular the fact that at short orbital periods there are systems with low and with high eccentricities, suggest that there may be more than one way to form them. It has been suggested that the low eccentricities arise from NS with small kicks, formed in electron-capture SN (Podsiadlowski et al., 2004), that occur much more frequently in close binary systems than in single stars or wide binaries. However, the recent finding that even a significant fraction of normal radio pulsars has low velocities (Verbunt et al., 2017) raises some questions about the one-to-one link between low kicks and electron capture supernovae (SNe).

The formation of double NS has been outlined already in the 1970s (De Loore et al., 1975; Flannery and van den Heuvel, 1975; Massevitch et al., 1976). It starts with two massive stars of which, after a phase of mass transfer, the primary explodes in an SN. If the orbit remains bound, the system temporarily takes the form of a high-mass X-ray binary in which the NS accretes from the stellar wind of the companion after which it goes through a second phase of mass transfer that leads to strong inward spiral (van den Heuvel and De Loore, 1973). The resulting close binary of a helium-burning star and an NS may become an X-ray binary again (like Cyg X-3) after which the helium star explodes to become the second NS. GW emission subsequently brings the system to merger.

The expectation, based on simulations, that after the merger of two NS, some of the material is ejected and the possibility that in the merger (temporarily) a massive, rapidly spinning NS is formed have led to the suggestion that the merger of two NS should be accompanied by an electromagnetic signal. In particular, the radioactive decay of the ejected low-density NS material, expected to have a brightness in between novae and SN, is often called kilonova or macronova (see Metzger, 2017, for a review). The basic physics is very elegantly described in Li and Paczyński (1998) and Kulkarni (2005) and shows that the electromagnetic signal of the expanding ejecta essentially depends on the injected energy and the opacity of the material. Barnes and Kasen (2013) showed that the opacity of the material may be much higher than Thomson scattering. For still relatively low opacities, the emission comes out early and fast and is blue; for high opacities, the emission comes out later and slower and is red.

In order to fully exploit the promise of multimessenger GW astronomy, the GW signals need to be quickly identified and shared with electromagnetic (EM) partners. The experience in the first observing run (Abbott et al., 2016d) has been used to improve the procedures, and after the discovery of GW170817 the alert went out within one hour. In the first day, several tens of observatories have reacted and observed the counterpart in all bands (see Abbott et al., 2017f; Figure 14.5). The optical/infrared counterpart is consistent with the predictions of the kilonova models (Cowperthwaite et al., 2017; Drout et al., 2017; Kasliwal et al., 2017; Levan et al., 2017; Pian et al., 2017; Smartt et al., 2017) although the

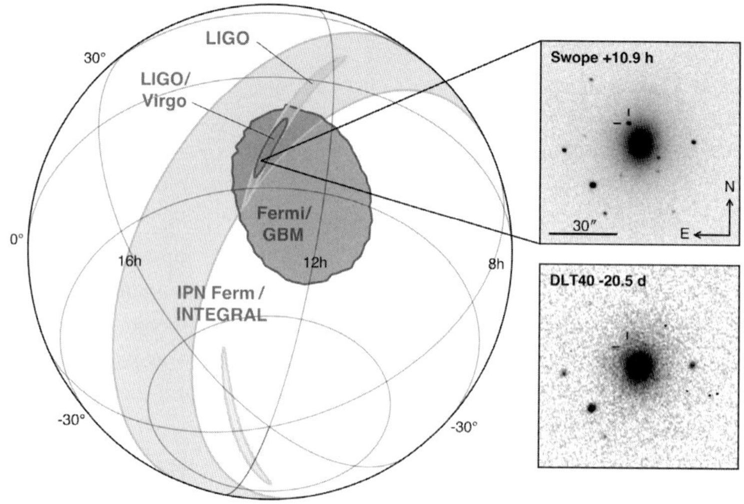

Figure 14.5 Fermi, Integral and LIGO/Virgo localisation of the GW170817 and the subsequent detected optical counterpart. From Abbott et al. (2017f). Reproduced by permission of the AAS.

bright transient implies a quite high ejecta mass ($\sim$0.03 $M_{\odot}$). The masses inferred from the GW detection are similar to the ones of the known double NS (Abbott et al., 2017e).

Even though it is only one detection, it can serve to make an estimate of the merger rate of double NS: 1540^{+3200}_{-1220}/Gpc^3/yr. Although within the range of expectations (e.g., Abadie et al., 2010), it is quite high compared to the recent theoretical models and not easy to reconcile with the models without overestimating the binary BH merger rate (Chruslinska et al., 2017). This means that the next observing run will very quickly show if this single detection was a statistical outlier, or if the merger rate really is so high.

With this in mind, and the question whether the high ejecta mass and thus bright kilonova are typical or not, it is important to systematically investigate the GW error boxes in the future. For that, the BlackGEM telescope array, for instance, was developed. It consists of three 65 cm telescopes, each with a 2.7 square degree field of view that can reach mag 23 in 5 min integration (Bloemen et al., 2016).

For double NS, the expectation is that *LISA* will observe these systems only in the Milky Way, so expected numbers are low (e.g., Nelemans et al., 2001). However, a very nice feature is that the GW strength of the double NS with periods shorter than about 30 minutes is such that the measurement of the population in that period range will be *complete* (eLISA Consortium et al., 2013).

Overall, the promise of GW astrophysics for the study of NS binaries lies mainly in the determination of the rates and properties of the systems that can be related to the preceding binary evolution. In particular, if EM signals are also detected, the study can be extended to included detailed studies of the parent population and the ages of the mergers, probing the delay time between the formation of the double NS and the merger.

14.2.4 White Dwarfs and Electromagnetic Data

For WD, the LIGO and Virgo detectors are too small. Indeed, any ground-based detector will have difficulty detecting any WD, so we have to turn to LISA. It was realised early on (Evans et al., 1987) that double WD would form a large population of detectable sources, and even so large that they could form a confusion noise, or foreground noise (e.g., Hils et al., 1990). These systems are now found regularly optically (e.g., Marsh, 2011; Brown et al., 2017). With the current design of the mission (Amaro-Seoane et al., 2017), the sensitivity of LISA is such that with a recent model for the Galactic population of double WD (Toonen et al., 2012), the expectation is that it will detect this foreground, in addition to maybe 25,000 individually resolved systems (Cornish and Robson, 2017; Korol et al., 2017). A small but not negligible fraction will also be detectable with EM instruments (e.g., Littenberg et al., 2013).

The emission of GW will bring close double WD closer and eventually leads to Roche-lobe overflow of the lowest-mass WD. Contrary to the case of two NS, there is a possibility for stable mass transfer (see Marsh et al., 2003). The expected properties of such interacting double WD correspond closely to the observed AM canum venaticorum (CVn) systems (see Solheim, 2010). These are characterised by very short orbital periods (5–60 min), helium accretion discs around WD and very low-mass, unseen companions that need to be degenerate in order to fit in the orbits.

Some of the observed systems are expected to be individually detected by LISA and are referred to as verification binaries (e.g., Ströer and Vecchio, 2006; Shah et al., 2015; Brown et al., 2017). Between now and the launch of LISA, there will be different instruments that

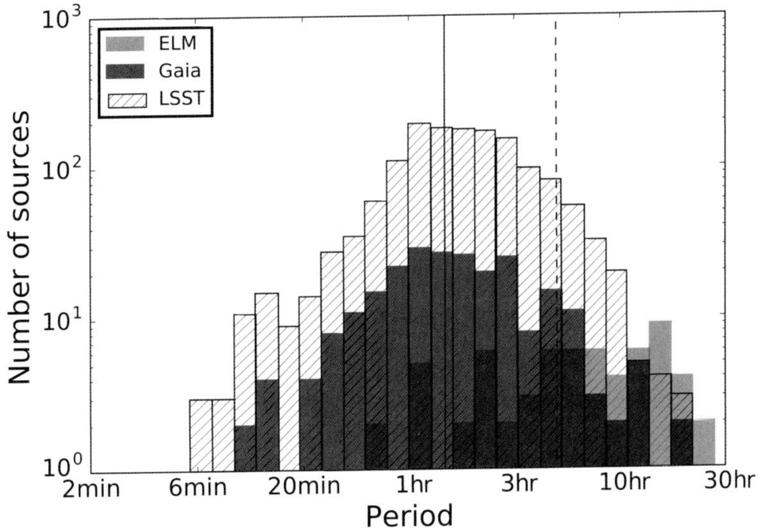

Figure 14.6 Expected LISA double WD sources for which an optical counterpart can be detected by LSST or Gaia. From Korol et al. (2017).

have the potential to increase the number of verification binaries significantly, in particular Gaia and LSST (Breivik et al., 2017; Korol et al., 2017; and see Figure 14.6). The combined data from GW and EM observations, in particular period evolution measurements, e.g., through eclipse timings, have the potential to greatly improve our understanding of the evolution into double WD and the small fraction that turns into AM CVn systems (Shah and Nelemans, 2014a,b; Breivik et al., 2017)

We can expect to get information about the binary evolution leading to the formation of double WD and the initial binary populations for low-mass stars from the numbers and properties of the large sample of detected systems. In particular, when the (chirp) masses can be measured (only in a subset of the individual systems), constraints on the common envelope and the stability of mass transfer between two WD can be inferred. Finally, the systems with measured distances (even fewer or the systems with complementary EM data) can be used to study the Galactic distribution of systems and, since some of these will be located throughout the Galaxy, the structure of the Milky Way (e.g., Korol et al., 2018).

14.3 Conclusions

We now really can say that GW astrophysics has started and is not just something for the future. At least for the next decade, binaries will be the key GW sources. With the plans for increased sensitivity, the prospects are excellent, not only for more detections, but also for sufficiently large samples that allow statistical studies. For binary BH, we can expect many detections. At the moment, their formation is unclear, and hopefully the statistical studies of the population will start to give directions on how to tackle the question of their origin. For double NS, the first question is how many detections we will see in the next few years. In addition, the first detection was such a rich event, with detections in all EM bands, that we will have to see how typical such an event is. In any case, the study of EM counterparts will provide valuable extra information that will make it easier to use them to study binary evolution. Indeed, if the merger rate is as high as suggested by this first detection, we may already have to go back to the drawing board for their evolution. In the long run, detectors such as LISA and ET will start the phase of mass detections, with thousands of sources expected. The WD binaries detected by LISA will provide new information on low-mass binary formation and the delicate process of mass transfer between WD. Before the LISA launch, many new systems will be detected, and long-term monitoring of (eclipsing) sources will provide crucial information to complement the LISA measurements. Overall, it is clear that our understanding of binary evolution will get a boost from all these new data!

References

Aasi, J., Abbott, B. P., Abbott, R., et al. 2015. Advanced LIGO. *Classical and Quantum Gravity*, **32**(7), 074001.

Abadie, J., Abbott, B. P., Abbott, R., et al. 2010. TOPICAL REVIEW: Predictions for the Rates of Compact Binary Coalescences Observable by Ground-Based Gravitational-Wave Detectors. *Classical and Quantum Gravity*, **27**(17), 173001.

Abbott, B. P., Abbott, R., Abbott, T. D., et al. 2016a. Astrophysical Implications of the Binary Black-Hole Merger GW150914. *ApJ L*, **818**, L22.

Abbott, B. P., Abbott, R., Abbott, T. D., et al. 2016b. Binary Black Hole Mergers in the First Advanced LIGO Observing Run. *Physical Review X*, **6**, 041015.

Abbott, B. P., Abbott, R., Abbott, T. D., et al. 2016c. GW150914: The Advanced LIGO Detectors in the Era of First Discoveries. *Physical Review Letters*, **116**, 131103.

Abbott, B. P., Abbott, R., Abbott, T. D., et al. 2016d. Localization and Broadband Follow-Up of the Gravitational-Wave Transient GW150914. *ApJ L*, **826**, L13.

Abbott, B. P., Abbott, R., Abbott, T. D., et al. 2016e. Observation of Gravitational Waves from a Binary Black Hole Merger. *Physical Review Letters*, **116**, 061102.

Abbott, B. P., Abbott, R., Abbott, T. D., et al. 2016f. Prospects for Observing and Localizing Gravitational-Wave Transients with Advanced LIGO and Advanced Virgo. *Living Reviews in Relativity*, **19**, 1. *arXiv e-prints,* 1304.0670.

Abbott, B. P., Abbott, R., Abbott, T. D., et al. 2016g. Tests of General Relativity with GW150914. *Physical Review Letters*, **116**, 221101.

Abbott, B. P., Abbott, R., Abbott, T. D., et al. 2017a. A Gravitational-Wave Standard Siren Measurement of the Hubble Constant. *Nature*, **551**, 85–88.

Abbott, B. P., Abbott, R., Abbott, T. D., et al. 2017b. GW170104: Observation of a 50-Solar-Mass Binary Black Hole Coalescence at Redshift 0.2. *Physical Review Letters*, **118**, 221101.

Abbott, B. P., Abbott, R., Abbott, T. D., et al. 2017c. GW170608: Observation of a 19-Solar-Mass Binary Black Hole Coalescence. *ApJ L*, **851**, L35.

Abbott, B. P., Abbott, R., Abbott, T. D., et al. 2017d. GW170814: A Three-Detector Observation of Gravitational Waves from a Binary Black Hole Coalescence. *Physical Review Letters*, **119**, 141101.

Abbott, B. P., Abbott, R., Abbott, T. D., et al. 2017e. GW170817: Observation of Gravitational Waves from a Binary Neutron Star Inspiral. *Physical Review Letters*, **119**, 161101.

Abbott, B. P., Abbott, R., Abbott, T. D., et al. 2017f. Multi-Messenger Observations of a Binary Neutron Star Merger. *ApJ L*, **848**, L12.

Acernese, F., Agathos, M., Agatsuma, K., et al. 2015. Advanced Virgo: A Second-Generation Interferometric Gravitational Wave Detector. *Classical and Quantum Gravity*, **32**, 024001.

Amaro-Seoane, P., Audley, H., Babak, S., et al. 2017. Laser Interferometer Space Antenna. *ArXiv e-prints*, 1702.00786.

Armstrong, J. W., Estabrook, F. B., and Tinto, M. 1999. Time-Delay Interferometry for Space-Based Gravitational Wave Searches. *ApJ*, **527**, 814–826.

Aso, Y., Michimura, Y., Somiya, K., et al. 2013. Interferometer Design of the KAGRA Gravitational Wave Detector. *PRD*, **88**(4), 043007.

Barnes, J., and Kasen, D. 2013. Effect of a High Opacity on the Light Curves of Radioactively Powered Transients from Compact Object Mergers. *ApJ*, **775**, 18.

Belczynski, K., Holz, D. E., Bulik, T., and O'Shaughnessy R. 2016b. The First Gravitational-Wave Source from the Isolated Evolution of Two Stars in the 40–100 Solar Mass Range. *Nature*, **534**, 512–515.

Belczynski, K., Kalogera, V., and Bulik, T. 2002. A Comprehensive Study of Binary Compact Objects as Gravitational Wave Sources: Evolutionary Channels, Rates, and Physical Properties. *ApJ*, **572**, 407–431.

Belczynski, K., Repetto, S., Holz, D. E., et al. 2016a. Compact Binary Merger Rates: Comparison with LIGO/Virgo Upper Limits. *ApJ*, **819**, 108.

Bloemen, S., Groot, P., Woudt, P., et al. 2016. MeerLICHT and BlackGEM: Custom-Built Telescopes to Detect Faint Optical Transients. *Ground-based and Airborne Telescopes VI. Proc. SPIE*, **9906**, 990664.

Breivik, K., Kremer, K., Bueno, M., et al. 2017. Characterizing Accreting Double White Dwarf Binaries with LISA and Gaia. *ArXiv e-prints*, 1710.08370.

Breivik, K., Rodriguez, C. L., Larson, S. L., Kalogera, V., and Rasio, F. A. 2016. Distinguishing between Formation Channels for Binary Black Holes with LISA. *ApJ L*, **830**, L18.

Breton, R. P., Kaspi, V. M., Kramer, M., et al. 2008. Relativistic Spin Precession in the Double Pulsar. *Science*, **321**, 104.

Brown, W. R., Kilic, M., Kosakowski, A., and Gianninas, A. 2017. Discovery of a Detached, Eclipsing 40 Minute Period Double White Dwarf Binary and a Friend: Implications for He+CO White Dwarf Mergers. *ApJ*, **847**, 10.

Casares, J., and Jonker, P. G. 2014. Mass Measurements of Stellar and Intermediate-Mass Black Holes. *Space Sci. Rev.*, **183**, 223–252.

Chruslinska, M., Belczynski, K., Klencki, J., and Benacquista M. 2017. Double Neutron Stars: Merger Rates Revisited. *ArXiv e-prints*, 1708.07885.

Clark, J. P. A., van den Heuvel, E. P. J., and Sutantyo, W. 1979. Formation of Neutron Star Binaries and Their Importance for Gravitational Radiation. *A&A*, **72**, 120–128.

Cornish, N., and Robson, T. 2017. Galactic Binary Science with the New LISA Design. *Journal of Physics Conf. Ser.*, **840**, 012024.

Cowperthwaite, P. S., Berger, E., Villar, V. A., et al. 2017. The Electromagnetic Counterpart of the Binary Neutron Star Merger LIGO/Virgo GW170817. II. UV, Optical, and Near-Infrared Light Curves and Comparison to Kilonova Models. *ApJ L*, **848**, L17.

De Loore, C., De Greve, J. P., and de Cuyper, J. P. 1975. Evolution of Massive Close Binaries. II. The POST X-Ray Binary Stage: Origin of Run-Away and Binary Pulsars. *Astrophys.&Sp.Sci.*, **36**, 219–225.

de Mink, S. E., and Mandel, I. 2016. The Chemically Homogeneous Evolutionary Channel for Binary Black Hole Mergers: Rates and Properties of Gravitational-Wave Events Detectable by Advanced LIGO. *MNRAS*, **460**, 3545–3553.

Drout, M. R., Piro, A. L., Shappee, B. J., et al. 2017. Light Curves of the Neutron Star Merger GW170817/SSS17a: Implications for R-Process Nucleosynthesis. *Science*, **358**, 1570–1574.

Einstein, A. 1916. Näherungsweise Integration der Feldgleichungen der Gravitation. *Sitzungsberichte der Königlich Preußischen Akademie der Wissenschaften (Berlin), 688–696.*

Einstein, A. 1918. Über Gravitationswellen. *Sitzungsberichte der Königlich Preußischen Akademie der Wissenschaften (Berlin), 154–167.*

eLISA Consortium, Amaro Seoane, P., Aoudia, S., et al. 2013. The Gravitational Universe. *ArXiv e-prints*, 1305.5720.

Evans, C. R., Iben, I., and Smarr, L. 1987. Degenerate Dwarf Binaries as Promising Sources of Gravitational Radiation. *ApJ*, **323**, 129–139.

Flannery, B. P., and van den Heuvel, E. P. J. 1975. On the Origin of the Binary Pulsar PSR 1913 + 16. *A&A*, **39**, 61–67.

Fragos, T., Willems, B., Kalogera, V., et al. 2009. Understanding Compact Object Formation and Natal Kicks. II. The Case of XTE J1118 + 480. *ApJ*, **697**, 1057–1070.

Hils, D., Bender, P. L., and Webbink, R. F. 1990. Gravitational Radiation from the Galaxy. *ApJ*, **360**, 75–94.

Hulse, R. A., and Taylor, J. H. 1975. Discovery of a Pulsar in a Binary System. *ApJ L*, **195**, L51–L53.

Iyer, B., et al. 2011. *LIGO India*. Tech. Rept. LIGO-M1100296. https://dcc.ligo.org/LIGO-M1100296/public.

Jonker, P. G., and Nelemans, G. 2004. The Distances to Galactic Low-Mass X-Ray Binaries: Consequences for Black Hole Luminosities and Kicks. *MNRAS*, **354**, 355.

Kasliwal, M. M., Nakar, E., Singer, L. P., et al. 2017. Illuminating Gravitational Waves: A Concordant Picture of Photons from a Neutron Star Merger. *Science*, **358**, 1559–1565.

Korol, V., Rossi, E. M., and Barausse, E. 2018. A Multi-Messenger Study of the Milky Way's Stellar Disc and Bulge with LISA, Gaia and LSST. *ArXiv e-prints*, 1806.03306.

Korol, V., Rossi, E. M., Groot, P. J., et al. 2017. Prospects for Detection of Detached Double White Dwarf Binaries with Gaia, LSST and LISA. *MNRAS*, **470**, 1894–1910.

Kulkarni, S. R. 2005. Modeling Supernova-Like Explosions Associated with Gamma-Ray Bursts with Short Durations. *ArXiv e-prints*, 0510256.

Levan, A. J., Lyman, J. D., Tanvir, N. R., et al. 2017. The Environment of the Binary Neutron Star Merger GW170817. *ApJ L*, **848**, L28.

Li, L.-X., and Paczyński, B. 1998. Transient Events from Neutron Star Mergers. *ApJ L*, **507**, L59–L62.

Littenberg, T. B., Larson, S. L., Nelemans, G., and Cornish, N. J. 2013. Prospects for Observing Ultracompact Binaries with Space-Based Gravitational Wave Interferometers and Optical Telescopes. *MNRAS*, **429**, 2361–2365.

Lyne, A. G., and Lorimer, D. R. 1994. High Birth Velocities of Radio Pulsars. *Nature*, **369**, 127–129.

Mapelli M. 2016. Massive Black Hole Binaries from Runaway Collisions: The Impact of Metallicity. *MNRAS*, **459**, 3432–3446.

Marsh T. R. 2011. Double White Dwarfs and LISA. *Classical and Quantum Gravity*, **28**, 094019.

Marsh, T. R., Nelemans, G., and Steeghs, D. 2003. Direct Impact Accretors: Evolutionary Links between Detached and Semi-Detached White Dwarfs. de Martino D., Kalytis R., Silvotti R., Solheim J., (eds), *White Dwarfs, Proc. XIII Workshop on White Dwarfs*, Kluwer, p. 275.

Massevitch, A. G., Tutukov, A. V., and Yungelson, L. R. 1976. Evolution of Massive Close Binaries and Formation of Neutron Stars and Black Holes. *Astrophys.&Sp.Sci.*, **40**, 115–133.

Metzger, B. D. 2017. Kilonovae. *Living Reviews in Relativity*, **20**, 3.

Nelemans, G., Tauris, T. M., and van den Heuvel, E. P. J. 1999. Constraints on Mass Ejection in Black Hole Formation Derived from Black Hole X-Ray Binaries. *A&A*, **352**, L87–L90.

Nelemans, G., Yungelson, L. R., and Portegies Zwart, S. F. 2001. The Gravitational Wave Signal from the Galactic Disk Population of Binaries Containing Two Compact Objects. *A&A*, **375**, 890–898.

Paczyński, B. 1967. Gravitational Waves and the Evolution of Close Binaries. *Acta Astron.*, **17**, 287.

Park, D., Kim, C., Lee, H. M., Bae, Y.-B., and Belczynski, K. 2017. Black Hole Binaries Dynamically Formed in Globular Clusters. *MNRAS*, **469**, 4665–4674.

Peters, P. C. 1964. Gravitational Radiation and the Motion of Two Point Masses. *Physical Review*, **136**, 1224–1232.

Peters, P. C., and Matthews, J. 1963. Gravitational Radiation from Point Masses in a Keplerian Orbit. *Phys. Rev.*, **131**, 435.

Phinney, E. S. 1991. The Rate of Neutron Star Binary Mergers in the Universe: Minimal Predictions for Gravity Wave Detectors. *ApJ L*, **380**, L17–L21.

Pian, E., D'Avanzo, P., Benetti, S., et al. 2017. Spectroscopic Identification of R-Process Nucleosynthesis in a Double Neutron-Star Merger. *Nature*, **551**, 67–70.

Podsiadlowski, P., Langer, N., Poelarends, A. J. T., et al. 2004. The Effects of Binary Evolution on the Dynamics of Core Collapse and Neutron Star Kicks. *ApJ*, **612**, 1044–1051.

Portegies Zwart, S. F., and McMillan, S. L. W. 2000. Black Hole Mergers in the Universe. *ApJ L*, **528**, L17–L20.

Press, W., and Thorne, K. S. 1972. Gravitational-Wave Astronomy. *ARA&A*, **10**, 335–374.

Repetto, S., and Nelemans, G. 2015. Constraining the Formation of Black Holes in Short-Period Black Hole Low-Mass X-Ray Binaries. *MNRAS*, **453**, 3341–3355.

Repetto, S., Davies, M. B., and Sigurdsson, S. 2012. Investigating Stellar-Mass Black Hole Kicks. *MNRAS*, **425**, 2799–2809.

Repetto, S., Igoshev, A. P., and Nelemans, G. 2017. The Galactic Distribution of X-Ray Binaries and Its Implications for Compact Object Formation and Natal Kicks. *MNRAS*, **467**, 298–310.

Rodriguez, C. L., Chatterjee, S., and Rasio, F. A. 2016. Binary Black Hole Mergers from Globular Clusters: Masses, Merger Rates, and the Impact of Stellar Evolution. *PRD*, **93**(8), 084029.

Sathyaprakash, B., Abernathy, M., Acernese, F., et al. 2012. Scientific Objectives of Einstein Telescope. *Classical and Quantum Gravity*, **29**, 124013.

Sathyaprakash, B. S., and Schutz, B. F. 2009. Physics, Astrophysics and Cosmology with Gravitational Waves. *Living Reviews in Relativity*, **12**, 2.

Schutz, B. F. 1989. Gravitational Wave Sources and Their Detectability. *Classical and Quantum Gravity*, **6**, 1761–1780.

Sesana, A. 2016. Prospects for Multiband Gravitational-Wave Astronomy after GW150914. *Physical Review Letters*, **116**, 231102.

Shah, S., and Nelemans, G. 2014a. Constraining Parameters of White-Dwarf Binaries Using Gravitational-Wave and Electromagnetic Observations. *ApJ*, **790**, 161.

Shah, S., and Nelemans, G. 2014b. Measuring Tides and Binary Parameters from Gravitational Wave Data and Eclipsing Timings of Detached White Dwarf Binaries. *ApJ*, **791**, 76.

Shah, S., Larson, S. L., and Brown, W. 2015. Ultra-Compact Binaries as Gravitational Wave Sources. *Journal of Physics Conference Series*, **610**, 012003.

Sigurdsson, S., and Hernquist, L. 1993. Primordial Black Holes in Globular Clusters. *Nature*, **364**, 423–425.

Smartt, S. J., Chen, T.-W., Jerkstrand, A., et al. 2017. A Kilonova as the Electromagnetic Counterpart to a Gravitational-Wave Source. *Nature*, **551**, 75–79.

Solheim, J.-E. 2010. AM CVn Stars: Status and Challenges. *PASP*, **122**, 1133–1163.

Ströer, A., and Vecchio A. 2006. The LISA Verification Binaries. *Classical and Quantum Gravity*, **23**, 809.

Sukhbold, T., Ertl, T., Woosley, S. E., Brown, J. M., and Janka, H.-T. 2016. Core-Collapse Supernovae from 9 to 120 Solar Masses Based on Neutrino-Powered Explosions. *ApJ*, **821**, 38.

Tauris, T. M., Kramer, M., Freire, P. C. C., et al. 2017. Formation of Double Neutron Star Systems. *ApJ*, **846**, 170.

Taylor, J. H., and Weisberg, J. M. 1982. A New Test of General Relativity: Gravitational Radiation and the Binary Pulsar PSR 1913+16. *ApJ*, **253**, 908–920.

Thorne, K. S. 1997. Gravitational Radiation: A New Window onto the Universe. *Rev. Mex. Astron. Astrofis.*, **10**, 1–28.

Toonen, S., Nelemans, G., and Portegies Zwart, S. 2012. Supernova Type Ia Progenitors from Merging Double White Dwarfs: Using a New Population Synthesis Model. *A&A*, **546**, A70.

Tutukov, A. V., and Yungelson, L. R. 1979. On the Influence of Emission of Gravitational Waves on the Evolution of Low-Mass Close Binary Stars. *Acta Astron.*, **29**, 665–680.

Ugliano, M., Janka, H.-T., Marek, A., and Arcones, A. 2012. Progenitor-Explosion Connection and Remnant Birth Masses for Neutrino-Driven Supernovae of Iron-Core Progenitors. *ApJ*, **757**, 69.

van den Heuvel, E. P. J., and De Loore C. 1973. The Nature of X-Ray Binaries III: Evolution of Massive Close Binaries with One Collapsed Component – with a Possible Application to Cygnus X-3. *A&A*, **25**, 387–395.

van den Heuvel, E. P. J., Portegies Zwart, S. F., and de Mink, S. E. 2017. Forming Short-Period Wolf–Rayet X-Ray Binaries and Double Black Holes through Stable Mass Transfer. *MNRAS*, **471**, 4256–4264.

Verbunt, F., Igoshev, A., and Cator, E. 2017. The Observed Velocity Distribution of Young Pulsars. *ArXiv e-prints*, 1708.08281.

Weiss, R. 1972. Electromagnetically Coupled Broadband Gravitational Wave Antenna. *Quarterly Progress Report, Research Laboratory of Electronics of MIT*, **105**, 54.

White, N. E., and van Paradijs, J. 1996. The Galactic Distribution of Black Hole Candidates in Low-Mass X-Ray Binary Systems. *ApJ*, **473**, L25–L29.

Willems, B., Henninger, M., Levin, T., et al. 2005. Understanding Compact Object Formation and Natal Kicks I: Calculation Methods and the Case of GRO J1655-40. *ApJ*, **625**, 324.

15

The Impact of Binaries on the Stellar Initial Mass Function

PAVEL KROUPA AND TEREZA JERABKOVA

15.1 Introduction

If the number of stars in the initial stellar mass interval m, $m + \mathrm{d}m$ is $\mathrm{d}N$, then

$$\mathrm{d}N = \xi(m)\,\mathrm{d}m, \tag{15.1}$$

where $\xi(m)$ is the stellar initial mass function (IMF), which is, formally, the complete ensemble of stars with their initial masses that form together. Because stars do not form at the same instant in time, the stellar IMF does not exist as such. Indeed, even if we were able to measure the mass of each star, it is not possible to empirically measure the IMF: at any given time, a stellar population is observed, some massive stars have either already evolved or been ejected from the system, the low-mass stars may not have formed yet in entirety and/or some have been lost through stellar dynamical processes and stellar mergers may have occurred.

The IMF is a mathematical shortcut, a *hilfskonstrukt*, allowing the modelling of stellar populations, the stars of which form in spatially and temporarily correlated star formation events (CSFEs), i.e., in density maxima in molecular clouds that are fed by convergent filamentary gas flows (Kirk et al., 2013; Burkert, 2017; Hacar et al., 2017b). Theory has not been very successful in quantifying how the gas in a galaxy transforms into stars, due to the immense complexities and multitude of physical processes involved, but observations inform that stars form in CFSEs, which for practical purposes can be called embedded clusters (Lada and Lada, 2003; Lada, 2010; Gieles et al., 2012; Megeath et al., 2016; Kroupa et al., 2018). These are typically of subparsec scale and assemble over a timescale of a million years (Marks and Kroupa, 2012), having spatial dimensions comparable to the molecular cloud filaments and the intersection thereof (André et al., 2016; Lu et al., 2018). These embedded clusters appear to form mass segregated as the Atacama Large Millimeter Array (ALMA) and other observations have shown (Kirk and Myers, 2011; Alfaro and Román-Zúñiga, 2018; Plunkett et al., 2018) and may contain from a dozen binaries to many millions of stars, as a result of the large range of gravity-assisted density fluctuations, filaments and inflows in molecular clouds, which themselves are merely momentary condensations of the complex interstellar medium of a galaxy. Data on globular clusters also indicate these to have formed mass segregated (Haghi et al., 2015; Zonoozi et al.,

2017). The flat birth radius–mass relation (Marks and Kroupa, 2012) and primordial mass segregation may be directly related to the existence of the most-massive-star vs. embedded-cluster-stellar-mass (the $m_{\rm max} - M_{\rm ecl}$) relation (Weidner and Kroupa, 2006).

Star formation in the embedded clusters is feedback regulated, most likely as a result of the sum of outflows and stellar radiation compensating the depth of the potential of the embedded cluster and individual protostars. This is evident in the majority of the gas being expelled, as shown by observations (Megeath et al., 2016) and magnetohydrodynamical simulations (Machida and Matsumoto, 2012; Bate et al., 2014; Federrath et al., 2014; Federrath, 2015, 2016). Noteworthy is that observed CSFEs spanning the range of a few 10 to a few $10^5 M_\odot$ in stars (the Taurus-Auriga embedded clusters, the Orion Nebula Cluster, NGC3603 and R136, all of which have been observed in much detail) are dynamically and physically well reproduced if the star-formation efficiency (SFE) is about 33%, the gas expulsion occurs with about 10 km/s (Whitmore et al., 1999; Zhang et al., 2001; Kroupa, 2005; Smith et al., 2005), and the embedded phase lasts about 0.6 Myr (Kroupa et al., 2001; Kroupa and Bouvier, 2003; Banerjee and Kroupa, 2013, 2014, 2018, 2017). It is an interesting problem to understand which mechanism disperses the gas from these systems. Observations suggest that even T Tauri associations lose their residual gas on a timescale of about a million years (Neuhäuser et al., 1998), and Hansen et al. (2012) perform gravo-radiation-hydrodynamics simulations of low-mass star formation in turbulent clouds according to which about two-thirds of the gas is blown out of the low-mass embedded clusters. The individual embedded clusters therefore should expand significantly within a few million years (Banerjee and Kroupa, 2017). Direct evidence for this expansion has been documented. Getman et al. (2018) note that the core radii of a large ensemble of very young clusters expand by a factor of about 10 within 1 to 3.5 Myr, and and previously Brandner (2008) provided a useful collation of the size changes for very young clusters as a function of their age. Banerjee and Kroupa (2017) show that the expansion can be achieved only via the expulsion of residual gas if SFE $\approx$1/3 because binary-star heating and mass loss through evolving stars do not suffice, even in unison, to expand the clusters sufficiently within the short time ($\approx$10 Myr). The gas expulsion unbinds a large fraction of modest-mass embedded clusters (Brinkmann et al., 2017) for which observational evidence has been found (Weidner et al., 2007) with potentially important implications for the structure of Galactic disks (Kroupa, 2002b). This leads to an understanding of OB associations as forming from many expanding post–gas-expulsion embedded clusters, all of which formed over a few million years within a molecular cloud that may itself be expanding or contracting such that the kinematical field of the young stars is complex and may not show an overall expansion of the OB association (e.g., Wright and Mamajek 2018).

Since the timescales of stellar and galaxy evolution, which much of astrophysics is concerned with, are longer than a few million years, the stellar populations may, to a good degree of approximation, be described by the IMF, if, for example, the evolution of a star cluster or its spectral energy distribution is to be quantified. Because the appearance and evolution of astrophysical systems sensitively depends on the relative fraction of the short-lived massive versus the long-lived low-mass stars, much emphasis has been placed on constraining the detailed shape of the IMF and of the possible variation of this shape with physical conditions of the star-forming gas. Reviews of this topic can be found in Chabrier (2003); Bastian et al. (2010); and Kroupa et al. (2013).

15.2 The Nature of the IMF of Stars and in Galaxies

Here an overview of the IMF is given in view of recent developments. Before continuing, the important point needs to be stressed that the IMF (normalised to, e.g., unit stellar mass) is mathematically strictly identical to the galaxywide IMF (gwIMF, the IMF of all stars forming in a galaxy and normalised in the same manner) if the process of star formation is mathematically equivalent to random sampling from an invariant IMF without constraints. Only in this case will many low-mass CSFEs together have the same IMF as one massive embedded cluster with the same number of stars. That is,

$$\mathrm{gwIMF} = \mathrm{IMF}, \tag{15.2}$$

if the invariant IMF is a probability density distribution function (PDF), where invariant means

$$\mathrm{IMF}(x_1, \rho_1, t_1, T_1, M_{\mathrm{tot},1}, \ldots) = \mathrm{IMF}(x_2, \rho_2, t_2, T_2, M_{\mathrm{tot},2}, \ldots), \tag{15.3}$$

for any $\mathrm{parameter}_{1,2}$. That is, the IMF, which is invariant in any local physical parameter (in one CSFE, i.e., in one embedded cluster), will result in the same gwIMF. Note that this implicitly assumes that massive stars can form in very low-density regions.

Thus, by constraining observationally gwIMF, this one fundamentally important concept is tested (Kroupa et al., 2013). Indeed, the Milky Way field IMF, extracted from the Galactic field stellar population, has been found to be top-light relative to the canonical IMF (Fig. 15.3), as shown by Scalo (1986) and Rybizki and Just (2015). For dwarf late-type galaxies with small star-formation rates (SFRs), it appears to be top-light (lacking massive stars, Lee et al., 2009; Pflamm-Altenburg et al., 2009; Watts et al., 2018), and for late-type galaxies with high SFRs and elliptical galaxies that formed as major early starbursts, the galaxywide IMF is top-heavy (Matteucci, 1994; Gibson and Matteucci, 1997; van Dokkum, 2008; Gunawardhana et al., 2011; Romano et al., 2017). The survey by Hsu et al. (2012, 2013) of very young stars in the low-density star-forming region of the Orion A cloud south of the dense Orion Nebula Cluster, where only low-mass embedded clusters are found, has revealed significant evidence for a top-light IMF that differs from the IMF found in the Orion Nebula Cluster.

These results suggest that the IMF may not be a probability density distribution function. Progress on this issue has been achieved recently by interpreting the IMF as an optimal sampling distribution function (Kroupa et al., 2013; Schulz et al., 2015; Yan et al., 2017). *A variable x is distributed according to an optimally sampled distribution function if, upon binning and independent of bin size, the Poisson scatter is zero in each bin.*

The physical interpretation of optimal sampling is perfect feedback self-regulation of the star-formation process, possibly with a high regularity of stellar masses along their hosting filaments and fibres. While optimal sampling is an extreme mathematical assumption, and nature most likely has some randomisation process when embedded clusters form, the lack of variation of the shape of the stellar IMF and the lack of scatter in the $m_{\rm max} - M_{\rm ecl}$ relation indicate that star formation may be closer to optimal than random sampling (Kroupa et al., 2013).

This is intimately linked to the existence of the $m_{\rm max} - M_{\rm ecl}$ relation, according to which an embedded cluster of stellar mass $M_{\rm ecl}$ has a most massive star of mass $m_{\rm max}$ (Weidner et al., 2013; Megeath et al., 2016; Ramírez Alegría et al., 2016; Stephens et al., 2017) that can be potentially physically related to the density of fibre structures in molecular clouds (Hacar et al., 2017a, 2018). Because of this relation, Equation 15.2 does not hold, because many low-mass CSFEs add only low-mass stars while the few if any massive CFEs do not add enough massive stars to the galaxy if the CFSEs are distributed as a Salpeter-like power-law embedded cluster mass function (Yan et al., 2017). The existence of the $m_{\rm max} - M_{\rm ecl}$ relation may also pose important implications for the existence of multiple populations/starbursts in star clusters, as discussed by Bekki et al. (2017) and Kroupa et al. (2018).

15.3 The Stellar Mass–Luminosity Relation

Stellar masses cannot be measured directly in most instances, and so the IMF can only be inferred by transforming the luminosity of a star to its mass using the mass–luminosity relation (other techniques such as measuring the gravity at the relevant stellar photosphere are unrealistic for large ensembles of stars). Given the uncertainties involved, this only works reliably for main-sequence stars. Immediately the problem becomes apparent that low-mass stars have not yet reached the main sequence when the massive stars have already evolved away from it, which is why $\xi(m)$ is but a *hilfskonstrukt*. For an ensemble of single main-sequence stars, which can be created for an observed population by artificially correcting all stars to their zero-age main sequence, we have

$$\xi(m) = \frac{\mathrm{d}N}{\mathrm{d}m} = -\frac{\mathrm{d}N}{\mathrm{d}M_x}\frac{\mathrm{d}M_x}{\mathrm{d}m}, \tag{15.4}$$

where M_x is the absolute magnitude of a star of mass m in the photometric pass- (e.g., V-) band. The $M_x(m)$ function is theoretically uncertain, in particular near $m = 0.3\ M_\odot$, below which the stars become fully convective and molecular hydrogen becomes an important opacity source and contributor to the mean molecular weight. The theory of the internal constitution of stars remains uncertain due to the complex physics of convection, radiation transport, rotation and magnetic fields. For $m < 0.1\ M_\odot$ the cores of the stars become electron degenerate, causing the $M_x(m)$ function to become very steep, because a small change in mass does not lead to a corresponding change in the central density. Thus, small changes in m cause large changes in M_x, with major uncertainties. Empirical constraints on the $M_x(m)$ function, in connection with theoretical models, have yielded good constraints on the IMF (Kroupa et al., 1993; Kroupa, 2002a; Chabrier, 2003). A noteworthy outcome of this work was the realisation that the function $-dm/dM_x$ has a pronounced maximum near $M_V \approx 11.5(m \approx 0.3M_\odot)$ (Kroupa et al., 1990, the Kroupa–Tout–Gilmore or KTG peak). It is evident in the stellar luminosity function, $\Psi_V = dN/dM_V$, of all resolved stellar populations (Kroupa and Tout, 1997; Kroupa, 2002a). The implication is that the IMF does not have a turnover at this mass. It is noteworthy that just by simply counting stars on the sky to construct Ψ_X, their internal constitution is probed.

15.4 Binary Stars

Given that the observed maximum in the stellar luminosity function, $\Psi_X = \Psi(M_X)$, defined as $dN = -\Psi_X \, dM_x$, of all populations has the KTG peak, the remaining bias affecting the interpretation of Ψ_X into ξ via Equation 15.4 is through unresolved multiple stars. The observer can only count the number of main-sequence stars as a function of their luminosity, i.e., construct $\Psi(M_x)$, but typically 50% of all main-sequence stars with about $m <$few $M_\odot$ are binaries with some being also triple and quadruple systems (Goodwin et al., 2007; Duchêne and Kraus, 2013). Note that most stars with mass $<$few$M_\odot$ should form as binaries (Goodwin and Kroupa, 2005). Figure 15.1 illustrates that, when including very young stars of all masses, it transpires that the binary fraction is close to 1 and independent of primary star mass. Only the old, Galactic field main-sequence stars show a trend of decreasing binary fraction with decreasing primary mass. This is very well accounted for if all stars form as binaries in embedded clusters (Thies et al., 2015). The birth population of binaries with primaries $<$few$M_\odot$ has well-defined distribution functions (periods, mass ratios and eccentricities), and is well described by randomly sampling the IMF for the component masses, leading to a flat mass-ratio distribution function (Kroupa, 1995a; Belloni et al., 2017). As emphasised in the latter reference, when modelling a stellar population it is essential to first draw the desired number of stars from the IMF and then to randomly pair the stars from this ensemble to create the binaries, because a different procedure would affect the IMF.

Detailed calculations and modelling of the Milky Way field population revealed that the deep star counts (reaching stars to distances beyond about 20 pc, which rely on the method of photometric parallax to estimate the stellar distances) do not resolve most of the multiple systems, or they miss the companions due to the flux limit or due to glare from the primary. Taking this bias into account, and also the Malmquist and Lutz–Kelker biases, leads to a good reproduction of the various empirical determinations of Ψ_V (Kroupa et al., 1993). Thus the novel approach used by Kroupa et al. (1993) to arrive at the canonical IMF (for $m < 1\ M_\odot$) was to employ the observational constraints on the $M_V(m)$ relation, the deep and the nearby (i.e., parallax-based) star counts simultaneously, to solve for one IMF and for the thickness of the Galactic disk. The approach was to assume a binary population in the Milky Way disk and to match the deep star counts, which do not resolve multiple systems, for the same IMF that also matches the nearby star counts in which multiple systems are resolved. More recent constraints on the IMF from MW disk star count data have not led to a revision of these results (Bochanski et al., 2010).

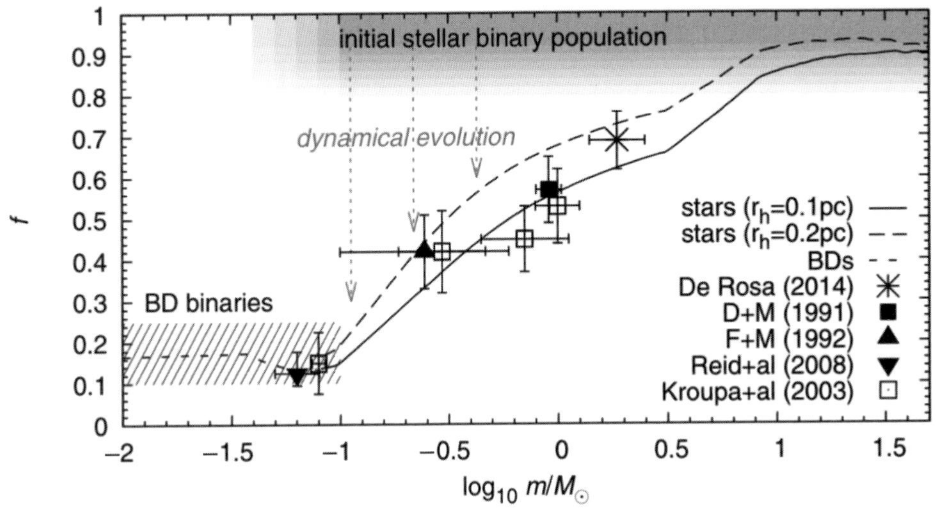

Figure 15.1 The binary fraction, f, as a function of primary star mass, m. The data (shown by various symbols) demonstrate that for old, main-sequence Galactic field stars the binary fraction decreases with decreasing primary mass. Young (i.e., TTauri) stars, however, have a binary fraction near unity independently of mass. This is shown as the shaded region termed 'initial stellar binary population'. It extends to equal-age massive stars, which have a very high multiplicity fraction as well (Sana et al., 2012; Moe and Di Stefano, 2017). The difference for young stars with $f \approx 1$ and for brown dwarfs (BDs) with $f \approx 0.15$ implies that BDs form with stars but follow their own distribution functions (Marks et al., 2015). It is definitely incorrect to argue that the trend of decreasing f with decreasing m implies that BDs form as an extension of the stellar population. An observational approach to test this statement is suggested by Marks et al. (2017). The old main-sequence field stellar binary properties can be well understood if all stars form as binaries in embedded clusters with half-mass radii $r_{\rm h}$ (Marks and Kroupa, 2012; Belloni et al., 2018) in which the binary systems are dissolved over time as the clusters evolve ('dynamical evolution' arrows). The data are, from top to bottom: De Rosa et al. (2014); Duquennoy and Mayor (1991); Fischer and Marcy (1992); Reid et al. (2008); Kroupa et al. (2003). Adapted from Thies et al. (2015). Reproduced by permission of the AAS.

Thus, the IMF as derived from the MW disk stellar population can be written as a two-part power-law form

$$\xi(m) = k_i m^{-\alpha_i}, \tag{15.5}$$

where k_i ensure the appropriate normalisation and continuity, and $\alpha_1 = 1.3, 0.08 < m/M_\odot \leq 0.5$ and $\alpha_2 = 2.3, 0.5 < m/M_\odot \leq 1$ being the Salpeter power-law index. A log-normal description is also possible and is indistinguishable from the preceding simpler form (Kroupa et al., 2013).

Taking care of modelling binary systems is an essential part of arriving at the preceding result, as neglect of unresolved binary systems leads to different biases in populations with different dynamical histories; see Figure 15.2. Thus, an observer may readily derive different 'IMFs' in two populations that in fact have the same IMF but have different binary populations.

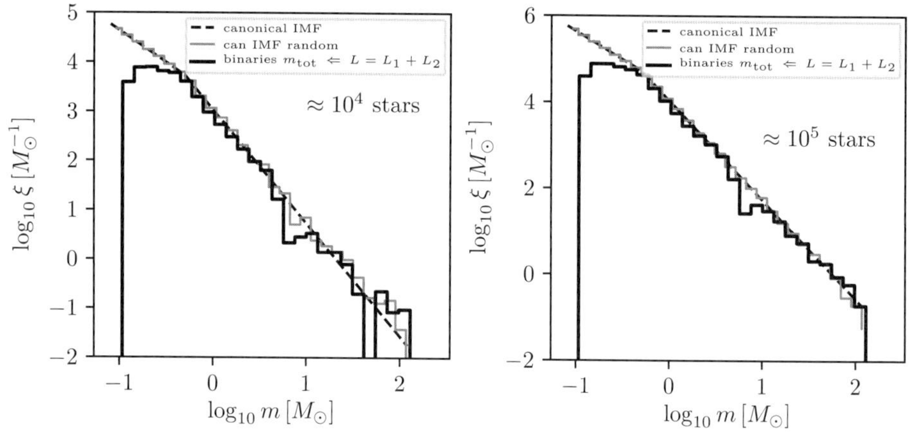

Figure 15.2 To demonstrate the effect of unresolved binaries on IMF estimates from observations, the system IMF is plotted as the thick, solid histogram assuming all stars are in binaries. The distribution of stars drawn randomly from the stellar IMF is shown as the solid grey histogram, and the canonical IMF is the dashed line. Stars are paired into binaries as follows: stars more massive than 5 $M_\odot$ are assumed to have a mass-ratio distribution, which is uniform in the range between 0.1 and 1, as in Oh et al. (2015), while primaries less massive than 5 $M_\odot$ have companions drawn randomly from the IMF leading to a flat mass-ratio distribution. Important here is to follow the correct procedure as explained in Oh et al. (2015) and Oh and Kroupa (2016). Note that here the zero-age main-sequence luminosities of the stars in a binary are added to provide the luminosity of the unresolved binary system, which is then converted into a mass using the main-sequence mass–luminosity relation from Nebojsa (2004) and Salaris and Cassisi (2006) to define the luminosities of stars of mass m. The feature around 5 $M_\odot$ is caused by the approximate treatment of the bimodal binary distribution. The unresolved binaries can significantly affect the low-mass end of the IMF if not corrected for appropriately, but have a negligible effect on the massive-star IMF (see also Weidner et al., 2009).

To achieve insights into this problem, the distribution function of the initial binary population needs to be derived, in equivalence to the IMF, in order to then be able to calculate the present-day binary populations for different systems, allowing for the dissolution of the binaries and the ejection of stars through stellar-dynamical encounters. This means mathematically formulating the birth distributions of orbital periods, mass ratios and eccentricities again as *hilfskonstrukts* since they are not observable and do not exist at any instant in time, because, for example, wide binaries may be dissolved in an embedded cluster before other stars have formed. This needs to be tackled to arrive at a full understanding of the Galactic field population and of individual star clusters. For late-type stars, much progress has been achieved but requires the inclusion of highly accurate Nbody modelling currently only possible with the Aarseth direct Nbody codes and the Giersz–MOCCA Monte Carlo code for massive clusters (Kroupa, 1995a; Belloni et al., 2017). The consistency check of using these birth binary distribution functions, evolving typical birth CSFEs until they have dissolved to the Galactic field, yields stellar luminosity functions in very good agreement with the observed trigonometrical parallax–based and photometric parallax–based luminosity functions and also with the observed field binary population (Kroupa, 1995b).

Figure 15.2 visualises the effect of binaries on the IMF by first drawing a list of stars from the IMF. Each star is given a luminosity, and stars weighing $m < 5\ M_\odot$ are paired randomly from this list, while more massive primaries obtain a secondary if the mass-ratio lies in the range 0.1 to 1, both the primary and secondary being taken from this same list, the stars being then removed from the list. This procedure, of only using stars in the list, is important in order to not change the IMF; for further details, see Oh et al. (2015) and Oh and Kroupa (2016). The stellar system 'masses' are calculated from the combined luminosity of the binaries, and this yields the 'system IMF'. It is virtually identical above about 1 $M_\odot$to the stellar IMF but has a very significant deficit of less massive stars. Figure 15.2 demonstrates the maximal error that can be done when assuming a stellar population has no binaries. Typical stellar populations have a binary fraction near 50%, but the bias at the low-mass end is still very large, as is shown in Kroupa et al. (2013), for example. Because only a system IMF can be deduced from the observations, unless every binary is resolved, the arrived-at IMF would have a turnover in the range 0.3 to 0.5 $M_\odot$, which however is not present in reality.

A similar degree of work constraining the birth distributions of binaries, as is available for $m <$ few $M_\odot$ stars (Kroupa, 1995a), is not yet available for massive binaries. The observational constraints on the present-day binary properties of massive stars have been quantified well recently (Sana et al., 2012; Moe and Di Stefano, 2017; see also Chapter 10), but these need to be evolved backwards through the dynamical environment of their birth clusters to distill the likely birth distribution functions. While this is is a mature procedure for late-type stars (Kroupa, 1995a; Belloni et al., 2017), only the first steps into this direction have been made by Oh et al. (2015) and Oh and Kroupa (2016).

Concerning the massive stars, the seminal work of Massey (2003) showed the observationally constrained IMF to be the canonical invariant Salpeter power-law with $\alpha_3 = 2.3, m > 1\ M_\odot$, independent of metallicity and density for very young stellar populations available in the Local Group of galaxies (see Equation 15.5 for the low-mass stars). Impressive here is that many different teams using different telescopes and observing different clusters and OB associations have shown the IMF to be essentially invariant with the canonical Salpeter–Massey index $\alpha_3 = 2.3, m > 1\ M_\odot$ (figure 5 in Kroupa, 2002a). That is, the shape of the IMF not showing the expected Poisson scatter and the $m_{\rm max} - M_{\rm ecl}$ relation having little or no intrinsic scatter (Weidner et al., 2013) together confirm the IMF to be closer to an optimal density distribution function (Section 15.2). Binary and higher-order multiple systems do not affect α_3 significantly (Weidner et al., 2009; Figure 15.2), although changes in component masses through binary-stellar evolution affect the power-law index to some degree (Schneider et al., 2015), and the merging of binary components induced in the stellar-dynamically violent environment of massive embedded clusters lead to the appearance of supercanonical ($m > 150\ M_\odot$) stars (Banerjee et al., 2012).

However, given the large internal binding energy of binaries amongst massive stars, massive stars are ejected preferentially from their birth clusters. Noteworthy here is that the canonical IMF is perfectly consistent with the observationally derived IMF in modest

M31 star clusters by Weisz et al. (2015), given that massive stars are ejected from these (figure 8 in Oh and Kroupa 2016). When observed, these massive stars are missing, leading to a deficit of massive stars in the young clusters. This has been noted to be the case for the Orion Nebula Cluster (Pflamm-Altenburg and Kroupa, 2006). This bias has been corrected for in the starburst cluster R136 in the Dor 30 star-forming region in the Large Magellanic Cloud, showing that this $\approx 10^5\ M_\odot$ very young low-metallicity cluster is likely to have been born with a top-heavy IMF (Banerjee and Kroupa, 2012). This has been confirmed recently by independent observations (Schneider et al., 2018; see also Kalari et al. 2018 for another similar case).

15.5 The IMF Is a Systematically Varying Function

From a theoretical point of view (Adams and Fatuzzo, 1996; Adams and Laughlin, 1996; Larson, 1998; Dib et al., 2007), it has been relatively straightforward to understand that the IMF ought to become top-heavy with decreasing metallicity (less capability for the gas to cool) and increasing density (more coagulations of gas clumps, stronger converging flows). In regions of high star-formation rate, cosmic rays produced by the high rate of supernovae penetrate molecular clouds, leading to higher ambient temperatures and thus most probably also to top-heavy IMFs (Papadopoulos et al., 2011; Papadopoulos and Thi, 2013). The IMF may become somewhat bottom-heavy in high-metallicity clouds and bottom-light at low metallicities for the same reasons. Essentially, the preceding theoretical framework largely rests on star formation depending on the local Jeans mass and/or being self-regulated through the various feedback processes that act once protostars appear, whereby photons have a larger interaction cross section with higher-metallicity gas, while it is modulated by the gravitational potential, which, if deep enough, may cause protostellar cores to merge before individual stars can form in them. On the other hand, the theory based on gravo-turbulent molecular clouds (Padoan et al., 2007; Hennebelle and Chabrier, 2008; Hopkins, 2013) appears to be challenged by recent and rather surprising observations according to which stars form in fine, phase-space coherent fibres and filaments (Hacar et al., 2017a, 2018; Bresnahan et al., 2018), which can only appear when the cloud has stopped being turbulent on the relevant scales. A strongly turbulent cloud can also not lead to a bottom-heavy IMF as suggested by Chabrier et al. (2014), because the shocks destroy low-mass protostellar cores before they can collapse (Bertelli Motta et al., 2016; Liptai et al., 2017). Given the immensely complex physical processes and boundary conditions acting during star formation, it is not surprising that it has until now not been possible to arrive at a convincing theoretical description of the properties of the IMF such that it is consistent with observed stellar and brown dwarf populations.

Rather than relying on theoretical arguments, the line of thought followed here is that the properties of the stellar IMF are constrained from observed simple stellar populations (containing stars of the same metallicity and age) in order to extract rules according to which it changes with conditions. Such an approach allows star-formation theory to be tested and, independently of how successful star-formation theory may be, it ensures that the IMF is consistent with resolved stellar populations. It allows larger systems such as galaxies to be described as being composed of many simple populations. An embedded, open or globular cluster is typically composed of a simple population.

The top-heaviness referred to earlier has been confirmed by direct star counts in the 30 Dor region (Schneider et al., 2018). Other direct evidence for a top-heavy IMF in a massive low-metallicity star cluster has also recently been published (Kalari et al., 2018). An elaborate analysis of the evolution of Milky Way globular clusters (GCs) and of ultracompact dwarf galaxies (UCDs) has previously (Dabringhausen et al., 2009, 2012; Marks et al., 2012) allowed a quantification of the dependency of the shape of the IMF on metallicity and density of the embedded-cluster forming cloud core,

$$\alpha_{1,2,3} = \mathrm{fn}([\mathrm{Fe/H}], \rho_{\mathrm{ecl}}), \tag{15.6}$$

where [Fe/H] is the iron abundance relative to the Solar value and ρ_{ecl} is the mass-volume-density (stellar plus gas mass) of the embedded cluster prior to gas expulsion (eq. 12, 14 and 15 with 0.87 replaced by -0.87 in Marks et al. 2012). According to these results the IMF becomes increasingly top-heavy above a star-formation-rate density of about $0.1\,M_\odot/(\mathrm{yr}\ \mathrm{pc}^3)$ within the embedded-cluster-forming cloud core on a pc scale (Marks et al., 2012; Kroupa et al., 2013). Further evidence for this systematic variation of the stellar IMF with metallicity and density comes from the mass-to-light ratia of star clusters in the Andromeda galaxy (Hasani Zonoozi et al., 2016; Haghi et al., 2017). The high rate of core-collapse supernovae in the central UCD-type objects in the star-bursting galaxy Arp 220 (Dabringhausen et al., 2012), the larger rates of type Ib and IIb relative to 'normal' type II supernovae in Arp 299 (Anderson et al., 2011) as well as the different ratios of supernova types in star-forming (metal-poor) dwarf relative to (metal-rich) giant galaxies found in the Palomar Transient Factory (Arcavi et al., 2010) all suggest the aforementioned systematic MF variation.

This variation of the IMF is visualised in Figure 15.3, noting that the bottom-lightness at low metallicity may be related to the bottom-light trend with decreasing metallicity found by Gennaro et al. (2018) for ultrafaint dwarf galaxies (Jeřábková et al., 2018). The top-heaviness at low metallicity and high density may well be relevant for the emergence of multiple populations of stars in globular clusters (Prantzos and Charbonnel, 2006; Bekki et al., 2017).

With this we now have, for the first time, a mathematical formulation of a systematically varying IMF that is consistent with resolved nearby very young and open star clusters, GCs and UCDs and, it appears, also whole galaxies (Kroupa and Weidner, 2003; Kroupa et al., 2013; Recchi and Kroupa, 2015; Yan et al., 2017; Jeřábková et al., 2018; Watts et al., 2018). Elliptical galaxies, which formed with star-formation rates larger than a few thousand $M_\odot/\mathrm{yr}$, will have enriched with metals rapidly, as already inferred by Matteucci (1994) and Gibson and Matteucci (1997), leading, in their innermost supersolar abundance regions, to stellar IMFs that are bottom-heavy (Conroy and van Dokkum, 2012), since the high-metallicity individual starburst embedded clusters forming in the innermost region of

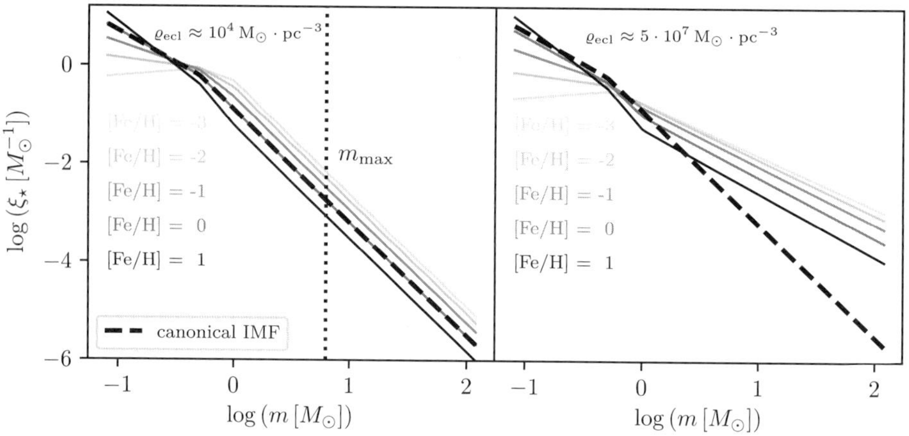

Figure 15.3 The variable IMF in comparison to the canonical IMF ($\alpha_1 = 1.3, \alpha_2 = \alpha_3 = 2.3$) shown as the dashed line. There is significant observational evidence that the IMF is not universal, that is, that it depends on environment; see Equation 15.6. Based on Marks et al. (2012), it is shown here how the IMF can vary with density and metallicity. In each panel, the metallicity and density-dependent IMF is shown as a solid line of thickness as given by the key in the panel. Left panel: low-density case, corresponding to an embedded star cluster with a stellar mass of $M_{ecl} = 100\, M_\odot$. The vertical line demonstrates the expected high-mass star cutoff based on the $m_{max} - M_{ecl}$ relation (Weidner et al., 2013), that is, such an embedded cluster would not contain stars more massive than m_{max} unless a binary merges. Right panel: high-density case, corresponding a $10^8\, M_\odot$ star cluster in which $m_{max} \approx 150\, M_\odot$. The density, ρ_{ecl}, is given in terms of the mass in gas and stars within the embedded cluster, as defined in equation 7 in Yan et al. (2017). All IMFs are normalised to the same stellar mass.

such systems will have had bottom-heavy IMFs (equation 12 in Marks et al., 2012). The application of Equation 15.6 to galaxies, by adding all IMFs in all forming embedded clusters, resolves the Hα radial cutoff versus the UV-extended Galactic-disk problem (Pflamm-Altenburg and Kroupa, 2008); implies all galaxies, independent of mass, to consume their gas on an about 3 Gyr timescale, solving the stellar-mass buildup time problem of dwarf galaxies (Pflamm-Altenburg and Kroupa, 2009); and leads to the observed mass–metallicity relation of galaxies (Köppen et al., 2007; Fontanot et al., 2017).

An application of Equation 15.6 to the early Universe suggests that the very young and very massive GC and UCD progenitors must have appeared quasarlike with immense luminosities and supernova II rates, opening tests of IMF variation to high-redshift studies (Jeřábková et al., 2017).

References

Adams, F. C., and Fatuzzo, M. 1996. A Theory of the Initial Mass Function for Star Formation in Molecular Clouds. *ApJ*, **464**, 256.

Adams, F. C., and Laughlin, G. 1996. Implications of White Dwarf Galactic Halos. *ApJ*, **468**, 586.

Alfaro, E. J., and Román-Zúñiga, C. 2018. Primordial Mass and Density Segregation in a Young Molecular Cloud. *ArXiv e-prints*, 1804.06961.

Anderson, J. P., Habergham, S. M., and James, P. A. 2011. On the Multiple Supernova Population of Arp 299: Constraints on Progenitor Properties and Host Galaxy Star Formation Characteristics. *MNRAS*, **416**, 567–579.

André, P., Revéret, V., Könyves, V., et al. 2016. Characterizing Filaments in Regions of High-Mass Star Formation: High-Resolution Submilimeter Imaging of the Massive Star-Forming Complex NGC 6334 with ArTéMiS. *A&A*, **592**, A54.

Arcavi, I., Gal-Yam, A., Kasliwal, M. M., et al. 2010. Core-Collapse Supernovae from the Palomar Transient Factory: Indications for a Different Population in Dwarf Galaxies. *ApJ*, **721**, 777–784.

Banerjee, S., and Kroupa, P. 2012. On the True Shape of the Upper End of the Stellar Initial Mass Function: The Case of R136. *A&A*, **547**, A23.

Banerjee, S., and Kroupa, P. 2013. Did the Infant R136 and NGC 3603 Clusters Undergo Residual Gas Expulsion? *ApJ*, **764**, 29.

Banerjee, S., and Kroupa, P. 2014. A Perfect Starburst Cluster Made in One Go: The NGC 3603 Young Cluster. *ApJ*, **787**, 158.

Banerjee, S., and Kroupa, P. 2017. How Can Young Massive Clusters Reach Their Present-Day Sizes? *A&A*, **597**, A28.

Banerjee, S., and Kroupa, P. 2018. Formation of Very Young Massive Clusters and Implications for Globular Clusters. *Astrophysics and Space Science Library*, **424**, 143.

Banerjee, S., Kroupa, P., and Oh, S. 2012. The Emergence of Super-Canonical Stars in R136-Type Starburst Clusters. *MNRAS*, **426**, 1416–1426.

Bastian, N., Covey, K. R., and Meyer, M. R. 2010. A Universal Stellar Initial Mass Function? A Critical Look at Variations. *ARA&A*, **48**, 339–389.

Bate, M. R., Tricco, T. S., and Price, D. J. 2014. Collapse of a Molecular Cloud Core to Stellar Densities: Stellar-Core and Outflow Formation in Radiation Magnetohydrodynamic Simulations. *MNRAS*, **437**, 77–95.

Bekki, K., Jeřábková, T., and Kroupa, P. 2017. The Origin of Discrete Multiple Stellar Populations in Globular Clusters. *MNRAS*, **471**, 2242–2253.

Belloni, D., Askar, A., Giersz, M., et al. 2017. On the Initial Binary Population for Star Cluster Simulations. *MNRAS*, **471**, 2812–2828.

Belloni, D., Kroupa, P., Rocha-Pinto, H. J., et al. 2018. Dynamical Equivalence, the Origin of the Galactic Field Stellar and Binary Population, and the Initial Radius-Mass Relation of Embedded Clusters. *MNRAS*, **474**, 3740–3745.

Bertelli Motta, C., Clark, P. C., Glover, S. C. O., et al. 2016. The IMF as a Function of Supersonic Turbulence. *MNRAS*, **462**, 4171–4182.

Bochanski, J. J., Hawley, S. L., Covey, K. R., et al. 2010. The Luminosity and Mass Functions of Low-Mass Stars in the Galactic Disk. II. The Field. *AJ*, **139**, 2679–2699.

Brandner, W. 2008. Westerlund 1 and Its Galactic Siblings: Observation Confronts Theory. *ArXiv e-prints*, 0803.1974.

Bresnahan, D., Ward-Thompson, D., Kirk, J. M., et al. 2018. The Dense Cores and Filamentary Structure of the Molecular Cloud in Corona Australis. Herschel SPIRE and PACS Observations from the Herschel Gould Belt Survey. *ArXiv e-prints*, 1801.07805.

Brinkmann, N., Banerjee, S., Motwani, B., et al. 2017. The Bound Fraction of Young Star Clusters. *A&A*, **600**, A49.

Burkert, A. 2017. A Bathtub Model for the Star-Forming Interstellar Medium. *Mem. Soc. Astr. Ital.*, **88**, 533.

Chabrier, G. 2003. Galactic Stellar and Substellar Initial Mass Function. *PASP*, **115**, 763–795.

Chabrier, G., Hennebelle, P., and Charlot, S. 2014. Variations of the Stellar Initial Mass Function in the Progenitors of Massive Early-Type Galaxies and in Extreme Starburst Environments. *ApJ*, **796**, 75.

Conroy, C., and van Dokkum, P. G. 2012. The Stellar Initial Mass Function in Early-Type Galaxies from Absorption Line Spectroscopy. II. Results. *ApJ*, **760**, 71.

Dabringhausen, J., Kroupa, P., and Baumgardt, H. 2009. A Top-Heavy Stellar Initial Mass Function in Starbursts as an Explanation for the High Mass-to-Light Ratios of Ultra-Compact Dwarf Galaxies. *MNRAS*, **394**, 1529–1543.

Dabringhausen, J., Kroupa, P., Pflamm-Altenburg, J., and Mieske, S. 2012. Low-Mass X-Ray Binaries Indicate a Top-heavy Stellar Initial Mass Function in Ultracompact Dwarf Galaxies. *ApJ*, **747**, 72.

De Rosa, R. J., Patience, J., Wilson, P. A., et al. 2014. The VAST Survey. III. The Multiplicity of A-Type Stars within 75 Pc. *MNRAS*, **437**, 1216–1240.

Dib, S., Kim, J., and Shadmehri, M. 2007. The Origin of the Arches Stellar Cluster Mass Function. *MNRAS*, **381**, L40–L44.

Duchêne, G., and Kraus, A. 2013. Stellar Multiplicity. *ARA&A*, **51**, 269–310.

Duquennoy, A., and Mayor, M. 1991. Multiplicity among Solar-Type Stars in the Solar Neighbourhood. II. Distribution of the Orbital Elements in an Unbiased Sample. *A&A*, **248**, 485–524.

Federrath, C. 2015. Inefficient Star Formation through Turbulence, Magnetic Fields and Feedback. *MNRAS*, **450**, 4035–4042.

Federrath, C. 2016. On the Universality of Interstellar Filaments: Theory Meets Simulations and Observations. *MNRAS*, **457**, 375–388.

Federrath, C., Schrön, M., Banerjee, R., et al. 2014. Modeling Jet and Outflow Feedback during Star Cluster Formation. *ApJ*, **790**, 128.

Fischer, D. A., and Marcy, G. W. 1992. Multiplicity among M Dwarfs. *ApJ*, **396**, 178–194.

Fontanot, F., De Lucia, G., Hirschmann, M., et al. 2017. Variations of the Stellar Initial Mass Function in Semi-Analytical Models: Implications for the Mass Assembly and the Chemical Enrichment of Galaxies in the GAEA Model. *MNRAS*, **464**, 3812–3824.

Gennaro, M., Tchernyshyov, K., Brown, T. M., et al. 2018. Evidence of a Non-Universal Stellar Initial Mass Function: Insights from HST Optical Imaging of Six Ultra-Faint Dwarf Milky Way Satellites. *ApJ*, **855**, 20.

Getman, K. V., Kuhn, M. A., Feigelson, E. D., et al. 2018. Young Star Clusters in Nearby Molecular Clouds. *MNRAS*, **477**, 298–324.

Gibson, B. K., and Matteucci, F. 1997. Infall Models of Elliptical Galaxies: Further Evidence for a Top-Heavy Initial Mass Function. *MNRAS*, **291**, L8–L12.

Gieles, M., Moeckel, N., and Clarke, C. J. 2012. Do All Stars in the Solar Neighbourhood Form in Clusters? A Cautionary Note on the Use of the Distribution of Surface Densities. *MNRAS*, **426**, L11–L15.

Goodwin, S. P., and Kroupa, P. 2005. Limits on the Primordial Stellar Multiplicity. *A&A*, **439**, 565–569.

Goodwin, S. P., Kroupa, P., Goodman, A., and Burkert, A. 2007. The Fragmentation of Cores and the Initial Binary Population. *Protostars and Planets V*, 133–147.

Gunawardhana, M. L. P., Hopkins, A. M., Sharp, R. G., et al., 2011. Galaxy and Mass Assembly (GAMA): The Star Formation Rate Dependence of the Stellar Initial Mass Function. *MNRAS*, **415**, 1647–1662.

Hacar, A., Alves, J., Tafalla, M., et al. 2017b. Gravitational Collapse of the OMC-1 Region. *A&A*, **602**, L2.

Hacar, A., Tafalla, M., and Alves, J. 2017a. Fibers in the NGC 1333 Proto-Cluster. *A&A*, **606**, A123.

Hacar, A., Tafalla, M., Forbrich, J., et al. 2018. An ALMA Study of the Orion Integral Filament. I. Evidence for Narrow Fibers in a Massive Cloud. *A&A*, **610**, A77.

Haghi, H., Khalaj, P., Hasani Zonoozi, A., et al. 2017. A Possible Solution for the M/L-[Fe/H] Relation of Globular Clusters in M31. II. The Age–Metallicity Relation. *ApJ*, **839**, 60.

Haghi, H., Zonoozi, A. H., Kroupa, P., et al. 2015. Possible Smoking-Gun Evidence for Initial Mass Segregation in Re-Virialized Post-Gas Expulsion Globular Clusters. *MNRAS*, **454**, 3872–3885.

Hansen, C. E., Klein, R. I., McKee, C. F., et al. 2012. Feedback Effects on Low-Mass Star Formation. *ApJ*, **747**, 22.

Hasani Zonoozi, A., Haghi, H., and Kroupa, P. 2016. A Possible Solution for the M/L–[Fe/H] Relation of Globular Clusters in M31: A Metallicity and Density Dependent Top-Heavy IMF. *ApJ*, **826**, 89.

Hennebelle, P., and Chabrier, G. 2008. Analytical Theory for the Initial Mass Function: CO Clumps and Prestellar Cores. *ApJ*, **684**, 395–410.

Hopkins, P. F. 2013. Variations in the Stellar CMF and IMF: From Bottom to Top. *MNRAS*, **433**, 170–177.

Hsu, W.-H., Hartmann, L., Allen, L., et al. 2012. The Low-Mass Stellar Population in L1641: Evidence for Environmental Dependence of the Stellar Initial Mass Function. *ApJ*, **752**, 59.

Hsu, W.-H., Hartmann, L., Allen, L. et al. 2013. Evidence for Environmental Dependence of the Upper Stellar Initial Mass Function in Orion A. *ApJ*, **764**, 114.

Jeřábková, T., Kroupa, P., Dabringhausen, J., et al. 2017. The Formation of Ultra Compact Dwarf Galaxies and Massive Globular Clusters: Quasar-Like Objects to Test for a Variable Stellar Initial Mass Function. *A&A*, **608**, A53.

Jeřábková, T., Zonoozi, A. H., Kroupa, P., et al. 2018. The Impact of Metallicity and Star Formation Rate on the Galaxy-Wide Initial Mass Function. *A&A*, submitted.

Kalari, V. M., Carraro, G., Evans, C. J., et al. 2018. The Magellanic Bridge Cluster NGC 796: Deep Optical AO Imaging Reveals the Stellar Content and Initial Mass Function of a Massive Open Cluster. *ApJ*, **857**, 132.

Kirk, H., and Myers, P. C. 2011. Young Stellar Groups and Their Most Massive Stars. *ApJ*, **727**, 64.

Kirk, H., Myers, P. C., Bourke, T. L., et al. 2013. Filamentary Accretion Flows in the Embedded Serpens South Protocluster. *ApJ*, **766**, 115.

Köppen, J., Weidner, C., and Kroupa, P. 2007. A Possible Origin of the Mass–Metallicity Relation of Galaxies. *MNRAS*, **375**, 673–684.

Kroupa, P. 1995a. The Dynamical Properties of Stellar Systems in the Galactic Disc. *MNRAS*, **277**, 1507.

Kroupa, P. 1995b. Unification of the Nearby and Photometric Stellar Luminosity Functions. *ApJ*, **453**, 358–368.

Kroupa, P. 2002a. The Initial Mass Function of Stars: Evidence for Uniformity in Variable Systems. *Science*, **295**, 82–91.

Kroupa, P. 2002b. Thickening of Galactic Discs through Clustered Star Formation. *MNRAS*, **330**, 707–718.

Kroupa, P. 2005. The Fundamental Building Blocks of Galaxies. Page 629 of: Turon, C., O'Flaherty, K. S., and Perryman, M. A. C. (eds), *The Three-Dimensional Universe with Gaia*. ESA Special Publication, vol. 576. Paris: European Space Agency.

Kroupa, P., Aarseth, S., and Hurley, J. 2001. The Formation of a Bound Star Cluster: From the Orion Nebula Cluster to the Pleiades. *MNRAS*, **321**, 699–712.
Kroupa, P., and Bouvier, J. 2003. The Dynamical Evolution of Taurus-Auriga–type Aggregates. *MNRAS*, **346**, 343–353.
Kroupa, P., Bouvier, J., Duchêne, G., et al. 2003. On the Universal Outcome of Star Formation: Is There a Link between Stars and Brown Dwarfs? *MNRAS*, **346**, 354–368.
Kroupa, P., Jerabkova, T., Dinnbier, F., et al. 2018. Evidence for Feedback and Stellar-Dynamically Regulated Bursty Star Cluster Formation: The Case of the Orion Nebula Cluster. *A&A*, **612**, A74.
Kroupa, P., and Tout, C. A. 1997. The Theoretical Mass-Magnitude Relation of Low Mass Stars and Its Metallicity Dependence. *MNRAS*, **287**, 402–414.
Kroupa, P., and Weidner, C. 2003. Galactic-Field Initial Mass Functions of Massive Stars. *ApJ*, **598**, 1076–1078.
Kroupa, P., Tout, C. A., and Gilmore, G. 1990. The Low-Luminosity Stellar Mass Function. *MNRAS*, **244**, 76–85.
Kroupa, P., Tout, C. A., and Gilmore, G. 1993. The Distribution of Low-Mass Stars in the Galactic Disc. *MNRAS*, **262**, 545–587.
Kroupa, P., Weidner, C., Pflamm-Altenburg, J., et al. 2013. The Stellar and Sub-Stellar Initial Mass Function of Simple and Composite Populations. Page 115 of Oswalt, T. D., and Gilmore, G. (eds), *Planets, Stars and Stellar Systems. Volume 5: Galactic Structure and Stellar Populations*. New York: Springer.
Lada, C. J. 2010. The Physics and Modes of Star Cluster Formation: Observations. *Philosophical Transactions of the Royal Society of London Series A*, **368**, 713–731.
Lada, C. J., and Lada, E. A. 2003. Embedded Clusters in Molecular Clouds. *ARA&A*, **41**, 57–115.
Larson, R. B. 1998. Early Star Formation and the Evolution of the Stellar Initial Mass Function in Galaxies. *MNRAS*, **301**, 569–581.
Lee, J. C., Gil de Paz, A., Tremonti, C., et al. 2009. Comparison of Hα and UV Star Formation Rates in the Local Volume: Systematic Discrepancies for Dwarf Galaxies. *ApJ*, **706**, 599–613.
Liptai, D., Price, D. J., Wurster, J., et al. 2017. Does Turbulence Determine the Initial Mass Function? *MNRAS*, **465**, 105–110.
Lu, X., Zhang, Q., Liu, H. B., et al. 2018. Filamentary Fragmentation and Accretion in High-Mass Star-Forming Molecular Clouds. *ApJ*, **855**, 9.
Machida, M. N., and Matsumoto, T. 2012. Impact of Protostellar Outflow on Star Formation: Effects of the Initial Cloud Mass. *MNRAS*, **421**, 588–607.
Marks, M., Janson, M., Kroupa, P., et al. 2015. M-Dwarf Binaries as Tracers of Star and Brown Dwarf Formation. *MNRAS*, **452**, 1014–1025.
Marks, M., and Kroupa, P. 2012. Inverse Dynamical Population Synthesis: Constraining the Initial Conditions of Young Stellar Clusters by Studying Their Binary Populations. *A&A*, **543**, A8.
Marks, M., Kroupa, P., Dabringhausen, J. et al. 2012. Evidence for Top-Heavy Stellar Initial Mass Functions with Increasing Density and Decreasing Metallicity. *MNRAS*, **422**, 2246–2254.
Marks, M., Martín, E. L., Béjar, V. J. S., et al. 2017. Using Binary Statistics in Taurus-Auriga to Distinguish between brown dwarf formation processes. *A&A*, **605**, A11.
Massey, P. 2003. Massive Stars in the Local Group: Implications for Stellar Evolution and Star Formation. *ARA&A*, **41**, 15–56.

Matteucci, F. 1994. Abundance Ratios in Ellipticals and Galaxy Formation. *A&A*, **288**, 57–64.

Megeath, S. T., Gutermuth, R., Muzerolle, J., et al. 2016. The Spitzer Space Telescope Survey of the Orion A and B Molecular Clouds. II. The Spatial Distribution and Demographics of Dusty Young Stellar Objects. *AJ*, **151**, 5.

Moe, M., and Di Stefano, R. 2017. Mind Your Ps and Qs: The Interrelation between Period (P) and Mass-Ratio (Q) Distributions of Binary Stars. *ApJ S*, **230**, 15.

Nebojsa, Duric. 2004. Book Review: Advanced Astrophysics. Cambridge, Cambridge University Press 2004, XIV+296 pp., ISBN 0-521-52571-3, ISBN 0-521-81967-9. *Orion: Zeitschrift für Amateur-Astronomie*, **62**, 36.

Neuhäuser, R., Frink, S., Röser, S. R., et al. 1998. Spatial Distribution and Kinematics of T Tauri Stars Discovered with ROSAT. *Acta Historica Astronomiae*, **3**, 206.

Oh, S., and Kroupa, P. 2016. Dynamical Ejections of Massive Stars from Young Star Clusters under Diverse Initial Conditions. *A&A*, **590**, A107.

Oh, S., Kroupa, P., and Pflamm-Altenburg, J. 2015. Dependency of Dynamical Ejections of O Stars on the Masses of Very Young Star Clusters. *ApJ*, **805**, 92.

Padoan, P., Nordlund, Å., Kritsuk, A. G., et al. 2007. Two Regimes of Turbulent Fragmentation and the Stellar Initial Mass Function from Primordial to Present-Day Star Formation. *ApJ*, **661**, 972–981.

Papadopoulos, P. P., and Thi, W.-F. 2013. The Initial Conditions of Star Formation: Cosmic Rays as the Fundamental Regulators. Page 41 of: Torres, D. F., and Reimer, O. (eds), *Cosmic Rays in Star-Forming Environments*. Advances in Solid State Physics, vol. 34. Berlin, Heidelberg: Springer-Verlag.

Papadopoulos, P. P., Thi, W.-F., Miniai, F., et al. 2011. Extreme Cosmic Ray Dominated Regions: A New Paradigm for High Star Formation Density Events in the Universe. *MNRAS*, **414**, 1705–1714.

Pflamm-Altenburg, J., and Kroupa, P. 2006. A Highly Abnormal Massive Star Mass Function in the Orion Nebula Cluster and the Dynamical Decay of Trapezium Systems. *MNRAS*, **373**, 295–304.

Pflamm-Altenburg, J., and Kroupa, P. 2008. Clustered Star Formation as a Natural Explanation for the Hα Cut-off in Disk Galaxies. *Nature*, **455**, 641–643.

Pflamm-Altenburg, J., and Kroupa, P. 2009. The Fundamental Gas Depletion and Stellar-Mass Buildup Times of Star-Forming Galaxies. *ApJ*, **706**, 516–524.

Pflamm-Altenburg, J., Weidner, C., and Kroupa, P. 2009. Diverging UV and Hα Fluxes of Star-Forming Galaxies Predicted by the IGIMF Theory. *MNRAS*, **395**, 394–400.

Plunkett, A. L., Fernández-López, M., Arce, H. G., et al. 2018. Distribution of Serpens South Protostars Revealed with ALMA. *ArXiv e-prints*, 1804.02405.

Prantzos, N., and Charbonnel, C. 2006. On the Self-Enrichment Scenario of Galactic globular Clusters: Constraints on the IMF. *A&A*, **458**, 135–149.

Ramírez Alegría, S., Borissova, J., Chené, A.-N., et al. 2016. Massive Open Star Clusters Using the VVV Survey V. Young Clusters with an OB Stellar Population. *A&A*, **588**, A40.

Recchi, S., and Kroupa, P. 2015. The Chemical Evolution of Galaxies with a Variable Integrated Galactic Initial Mass Function. *MNRAS*, **446**, 4168–4175.

Reid, I. N., Cruz, K. L., Burgasser, A. J., et al. 2008. L-Dwarf Binaries in the 20-Parsec Sample. *AJ*, **135**, 580–587.

Romano, D., Matteucci, F., Zhang, Z.-Y., et al. 2017. The Evolution of CNO Isotopes: A New Window on Cosmic Star Formation History and the Stellar IMF in the Age of ALMA. *MNRAS*, **470**, 401–415.

Rybizki, J., and Just, A. 2015. Towards a fully consistent Milky Way Disc Model. III. Constraining the Initial Mass Function. *MNRAS*, **447**, 3880–3891.
Salaris, Maurizio, and Cassisi, Santi. 2006. Book Review: Evolution of Stars and Stellar Populations/John Wiley & Sons, 2005. *The Observatory*, **126**, 306.
Sana, H., de Mink, S. E., de Koter, A., et al. 2012. Binary Interaction Dominates the Evolution of Massive Stars. *Science*, **337**, 444.
Scalo, J. M. 1986. The Stellar Initial Mass Function. *Fundamentals of Cosmic Physics*, **11**, 1–278.
Schneider, F. R. N., Izzard, R. G., Langer, N., et al. 2015. Evolution of Mass Functions of Coeval Stars through Wind Mass Loss and Binary Interactions. *ApJ*, **805**, 20.
Schneider, F. R. N., Sana, H., Evans, C. J., et al. 2018. An Excess of Massive Stars in the Local 30 Doradus Starburst. *Science*, **359**, 69–71.
Schulz, C., Pflamm-Altenburg, J., and Kroupa, P. 2015. Mass Distributions of Star Clusters for Different Star Formation Histories in a Galaxy Cluster Environment. *A&A*, **582**, A93.
Smith, N., Stassun, K. G., and Bally, J. 2005. Opening the Treasure Chest: A Newborn Star Cluster Emerges from Its Dust Pillar in Carina. *AJ*, **129**, 888–899.
Stephens, I. W., Gouliermis, D., Looney, L. W., et al. 2017. Stellar Clusterings around 'Isolated' Massive YSOs in the LMC. *ApJ*, **834**, 94.
Thies, I., Pflamm-Altenburg, J., Kroupa, P., et al. 2015. Characterizing the Brown Dwarf Formation Channels from the Initial Mass Function and Binary-star Dynamics. *ApJ*, **800**, 72.
van Dokkum, P. G. 2008. Evidence of Cosmic Evolution of the Stellar Initial Mass Function. *ApJ*, **674**, 29–50.
Watts, A. B., Meurer, G. R., Lagos, C. D. P., et al. 2018. Star Formation in the Outskirts of DDO 154: A Top-Light IMF in a Nearly Dormant Disc. *MNRAS*, **477**, 5554–5567.
Weidner, C., and Kroupa, P. 2006. The Maximum Stellar Mass, Star-Cluster Formation and Composite Stellar Populations. *MNRAS*, **365**, 1333–1347.
Weidner, C., Kroupa, P., Nürnberger, D. E. A., et al. 2007. On the Infant Weight Loss of Low- to Intermediate-Mass Star Clusters. *MNRAS*, **376**, 1879–1885.
Weidner, C., Kroupa, P., and Maschberger, T. 2009. The Influence of Multiple Stars on the High-Mass Stellar Initial Mass Function and Age Dating of Young Massive Star Clusters. *MNRAS*, **393**, 663–680.
Weidner, C., Kroupa, P., and Pflamm-Altenburg, J. 2013. The m_{max}-M_{ecl} Relation, the IMF and IGIMF: Probabilistically Sampled Functions. *MNRAS*, **434**, 84–101.
Weisz, D. R., Johnson, L. C., Foreman-Mackey, D., et al. 2015. The High-Mass Stellar Initial Mass Function in M31 Clusters. *ApJ*, **806**, 198.
Whitmore, B. C., Zhang, Q., Leitherer, C., et al. 1999. The Luminosity Function of Young Star Clusters in 'the Antennae' Galaxies (NGC 4038-4039). *AJ*, **118**, 1551–1576.
Wright, N. J., and Mamajek, E. E. 2018. The Kinematics of the Scorpius-Centaurus OB Association from Gaia DR1. *MNRAS*, **476**, 381–398.
Yan, Z., Jerabkova, T., and Kroupa, P. 2017. The Optimally Sampled Galaxy-Wide Stellar Initial Mass Function: Observational Tests and the Publicly Available GalIMF Code. *A&A*, **607**, A126.
Zhang, Q., Fall, S. M., and Whitmore, B. C. 2001. A Multiwavelength Study of the Young Star Clusters and Interstellar Medium in the Antennae Galaxies. *ApJ*, **561**, 727–750.
Zonoozi, A. H., Haghi, H., Kroupa, P., et al. 2017. Direct N-body Simulations of Globular Clusters. III. Palomar 4 on an Eccentric Orbit. *MNRAS*, **467**, 758–767.

16

The Formation of Binary Stars: Insights from Theory and Observation

CATHIE J. CLARKE

16.1 Introduction

The high-binary and higher-order multiplicity fraction amongst stellar populations of all ages demonstrates that binary formation is an integral part of star formation. Thus it has often been suggested that whereas the reproduction of the stellar initial mass function (IMF) is the most obvious observational diagnostic of the star formation process, in fact multiplicity statistics provide a much more exacting dataset that theory should strive to replicate. Much of the interest in collating binary statistics has thus revolved around the constraints that such data can place on star formation theories.

In the context of the present volume, however, the main interest is perhaps instead more empirical: it is necessary to *know* the observational statistics (see Chapters 2 and 10) because – particularly in respect of the closest binaries and their higher-order multiplicity – these provide crucial inputs into stellar evolution calculations. From this perspective, whether these statistics are understood theoretically would be a secondary matter were it not for the fact that good binary statistics are only available for volumes centred on the solar neighbourhood (e.g., Raghavan et al., 2010; Tokovinin, 2014). Star formation theory becomes relevant here *if* we have a theory that is robust enough to allow extrapolation from a local situation that is observationally calibrated into regimes of, e.g., metallicity or stellar density where observational data are sparser. This chapter reviews the successes and failures of binary star formation theory in explaining observations. This requires simulations that produce statistically significant samples of binaries, and it is for this reason that we restrict ourselves to (abundant) low-mass stars. For a discussion of binary formation in massive stars, see Zinnecker and Yorke (2007) and Meyer et al. (2018).

At the beginning of the twenty-first century, it would have seemed that we were close to achieving the goal of being able to reproduce binary statistics from numerical simulations (e.g., Bate et al., 2002a; Bate, Bonnell and Bromm, 2002, 2003): insights from large-scale binary simulations are reviewed in Section 16.3. At that time, it was evident that the simulations omitted important physics (particularly magnetic fields and a proper treatment of thermodynamics) and yet nevertheless reproduced local statistics admirably (Bate, 2009a). While a number of groups worked towards the goal of including a more comprehensive set of physical processes, it was to be hoped that such developments would not materially affect the earlier success in reproducing local binary statistics. Unfortunately, this has proved *not* to be the case. In the phrase of Hennebelle and Teyssier (2008), inclusion of a realistic level of magnetic support in the star-forming gas gives rise to a 'fragmentation crisis': the magnetically mediated removal of angular momentum tends to suppress disc formation and associated fragmentation (see also Machida et al., 2008). This remains an issue after nearly a decade of further experimentation using a variety of codes. While it has been possible to gain an understanding of factors (such as initial magnetic field topology) that can exacerbate or mitigate this problem, the computational expense of these experiments means that it is presently unfeasible to generate statistically significant samples of simulated binaries for comparison with observations. In this sense, the field stands roughly where it was around 1990 in regard to purely hydrodynamical (nonmagnetised) simulations (see Larson, 1978; Boss and Bodenheimer, 1979; Boss, 1986, 1991; Bonnell et al., 1991, 1992; Chapman at al., 1992, for classic early binary formation simulations).

Any review of binary star formation at this stage needs to acknowledge this central theoretical difficulty, and this will be discussed in detail in Section 16.4. On the other hand – while theory struggles with achieving the fragmentation that provides the fundamental building block of binary formation – there are additional clues about the nature of multiple formation that are being revealed from recent observations. Apart from the datasets discussed in detail in the Chapter 2, the last decade has seen two highly relevant developments. First of all, our knowledge of higher-order multiplicity has increased dramatically (Tokovinin, 2014), and the high incidence of triples and quadruples is of potential interest to the binary stellar evolution community inasmuch as the Kozai–Lidov mechanism (Kozai, 1962; Lidov, 1962) then provides a way to drive inner pairs into close proximity. However, it also reinforces the notion that – whatever the theoretical difficulties involved in achieving fragmentation – star formation must involve fragmentation into small N groups rather than isolated single or binary systems. The insights provided by the statistics of higher-order multiple systems are discussed in Section 16.5. The second observational window on the process of multiple star formation is provided by the Atacama Large Millimetre Array (ALMA) and the Karl Jansky Very Large Array (VLA), which have the sensitivity and resolution to provide images of multiple *proto*binary fragments and their associated distributed emission. These facilities are currently yielding dramatic images of *wide* binary/multiple systems in the making; as multi-object spectrographs are now yielding suitable samples of premain–sequence *spectroscopic* binaries, it will also become possible to characterise the environments of the close pairs that are of particular interest to the stellar evolution community. Some recent highlights are summarised in Section 16.6. Before turning to these issues, however, a brief thumbnail sketch of possible binary/multiple formation scenarios is provided to motivate the following discussion (Section 16.2).

16.2 Overview of Binary Formation Mechanisms

Figure 16.1 is a schematic depiction of binary formation scenarios from the proceedings of a workshop on 'Binaries as Tracers of Stellar Formation' held in Bettmeralp, Switzerland,

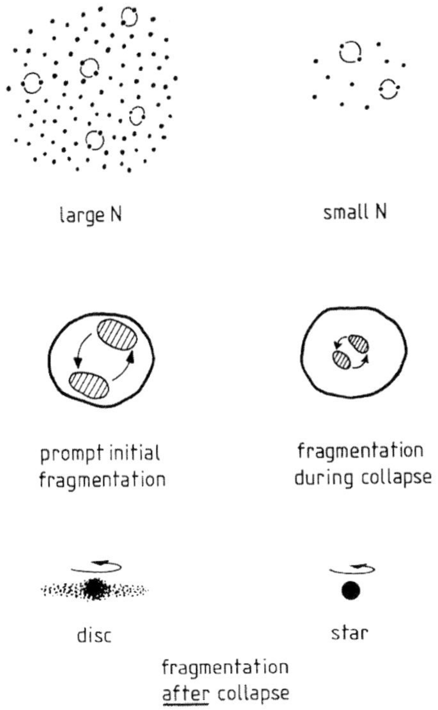

Figure 16.1 Schematic of binary formation scenarios (Clarke, 1992).

in 1991. The upper line depicts scenarios in which stars pair up as binaries within the context of a natal cluster of membership number N. If we consider encounters with a relative velocity of order to the virialised velocity dispersion of the cluster ($v \sim (GmN/R)^{0.5}$ where R is a measure of the cluster radius and m is a typical stellar mass), then the only way in which two stars can become bound in the absence of dissipation is if energy is removed from the relative orbit and transferred to the kinetic energy of a third star in the vicinity. This transfer needs to be mediated by a gravitationally focused three-body encounter, which requires, for so-called dynamical capture, that three stars are present within a size scale $\sim Gm/v^2 \sim R/N$. This immediately implies that dynamical capture in a large N cluster requires regions where the local density exceeds the mean cluster density by a factor of order N^2, consistent with the fact that dynamical simulations typically create only a handful of such binaries and that all of these are formed within the cluster core (see light grey points in Figure 16.2). On the other hand, in a small N cluster this condition can be frequently satisfied: although the number of binaries created by purely N-body interactions in small N clusters is typically ~ 1 per cluster, this can evidently imply a significant binary fraction if N is sufficiently small (van Albada, 1968; McDonald and Clarke, 1993).

There are, however, several additional binary formation channels within cluster environments. First of all, there is the situation where close encounters are associated with in dissolving clusters described earlier. On the negative side, more recent work on the whole disfavours the two categories in the bottom line of Figure 16.1 (e.g., Tohline, 2002). However, the lack of new mechanisms does *not* imply that the field has stagnated; indeed, we will see that it has been transformed by the capability of performing large-scale hydrodynamical simulations, and it is to these that we now turn.

16.3 The Era of Binary Mass Production in Hydrodynamical Simulations

Since 2002, there have been a number of landmark simulations that – instead of studying the fragmentation properties of individual dense cores within molecular clouds (e.g., Delgado-Donate et al., 2004a; Delgado et al, 2004b; Goodwin et al., 2004a,a, 2006) – have instead attempted to model larger complexes containing many hundreds of solar masses (e.g., Bate et al., 2002a; Bate, Bonnell and Bromm, 2002, 2003; Bate, 2009a, 2012; Myers et al., 2013). Such studies are typically initiated with 'turbulent' initial conditions, i.e., are seeded with a velocity field with a power spectrum consistent with the observed size linewidth relation within molecular clouds (Larson, 1981). Shocks resulting from such supersonic 'turbulence' produce a complex filamentary density field whose inhomogeneities are amplied by self-gravity in the gas. These latter regions undergo localised collapse, typically at the intersection of dense cloud filaments. Crucially, given the focus on multiple formation and the discussion in Section 16.2, the typical entity for star formation is the small N cluster.

Within these small N groupings there occur the dynamical effects discussed previously: energy is exchanged within small nonhierarchical groupings, resulting in the formation of bound pairs plus ejected singles. This process occurs in a gas-rich environment, both in the form of distributed gas (often channelled into the star-forming centre along filaments) and circumstellar discs. Further complexity is introduced by formation of additional companions within discs and also by the fact that separate small N star-forming regions, which share in the collapse of gas along filaments, can interact before they have completed their internal dynamical reconfiguration. Needless to say, these simulations are highly chaotic in the sense that the fate of individual objects can depend sensitively on initial conditions. However, the *statistical* properties of the multiple systems created in such simulations are remarkably consistent across simulations.

Perhaps more remarkably, these statistical properties agree pretty well with the statistics of observed multiple systems as illustrated in Figures 16.3 and 16.4. For example, the greater robustness of systems containing more massive components leads to an increase in binary fraction with stellar mass, in broad agreement with that observed. Similarly, the simulations do a good job of generating a wide range of separations since a binary's ultimate separation – rather than being a simple consequence of the spin of its natal cloud core – is shaped by a history of interaction with both gas and other stars in the small N grouping. The rich local dynamical environment results in the plentiful production of higher-order multiples, again in reasonable agreement with observations. A full analysis of the extent to which such simulations match observations is contained in Bate (2009a).

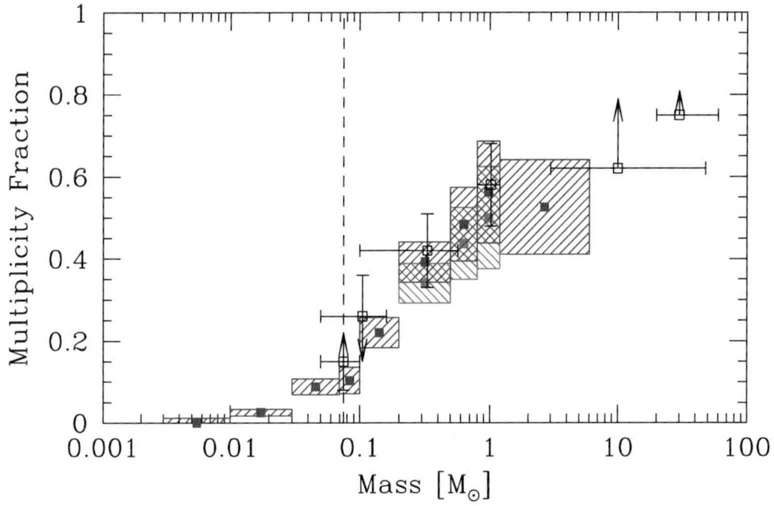

Figure 16.3 The multiplicity fraction as a function of primary mass from the simulation of Bate (2009a) with filled squares and hatched regions representing Poissonian errors. The open symbols represent observed values and their uncertainties (see Bate, 2009a, for the sources of observational data).

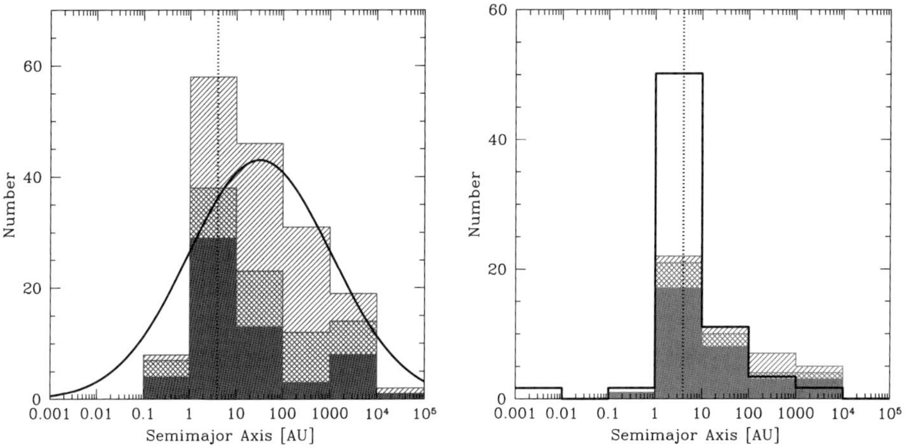

Figure 16.4 The separation distribution of binaries formed in the simulations of Bate (2009a) in the case of solar type and and very low-mass stars compared with the observational compilations of Duquennoy and Mayor (1991) and N. Siegler at http://vlmbinaries.org/ (last updated on 4 February 2008). The solid, single-hatched and double-hatched histograms denote contributions from binaries, triples and quadruples.

While the success of such studies was encouraging, there were several features of the original simulations that required correction. Despite their success in reproducing binary statistics, the original generation of simulations overproduced low-mass stars and brown dwarfs, an effect that could plausibly be traced to the fact that in the simulations the gas the ubiquity of binaries and protostellar discs is well attested observationally, this suggests that there is some aspect of reality that is not captured by the simulations.

Attempts to resolve this paradox have centred either on modification of the initial conditions or on also including nonideal MHD effects. In the former category, disc formation is relatively favoured if the magnetic field is initially misaligned with respect to the rotation axis (Ciardi and Hennebelle, 2010; Joos et al., 2012), while Seifried et al. (2012) instead emphasised that turbulent motions in the environment external to the nascent disc suppress the development of toroidal fields and associated magnetic braking. Misalignment of magnetic fields and turbulent conditions are both to be expected in the context of the fragmentation of cores that belong to a larger-scale turbulent cloud complex and early cloud-scale MHD simulations showed some success in producing binaries (e.g., Price and Bate, 2007). Nevertheless, in general, large-scale cloud simulations employing ideal MHD find that fragmentation is suppressed by magnetic fields (e.g., Hennebelle et al., 2011; Lewis and Bate, 2017); this is particularly true when both magnetic fields *and* radiative feedback are included (Commercon, Hennebelle and Henning, 2011; Myers et al., 2013): suppression of fragmentation leads to the formation of a single central object to which material is funnelled along filaments (see Figure 16.5), a situation that is obviously unsuitable for multiple star formation. On the other hand, inclusion of nonideal effects has not much relieved the situation (Wurster et al., 2017). While the fragmentation properties of cloud regions with a supercritical mass-to-flux ratio are not much affected by relaxing the assumption of flux freezing, nonideal effects do allow the collapse of regions with subcritical mass-to-flux ratio that would not have otherwise collapsed. This does not, however, assist the binary fragmentation crisis: cores that have slowly expelled their magnetic fields as a result of

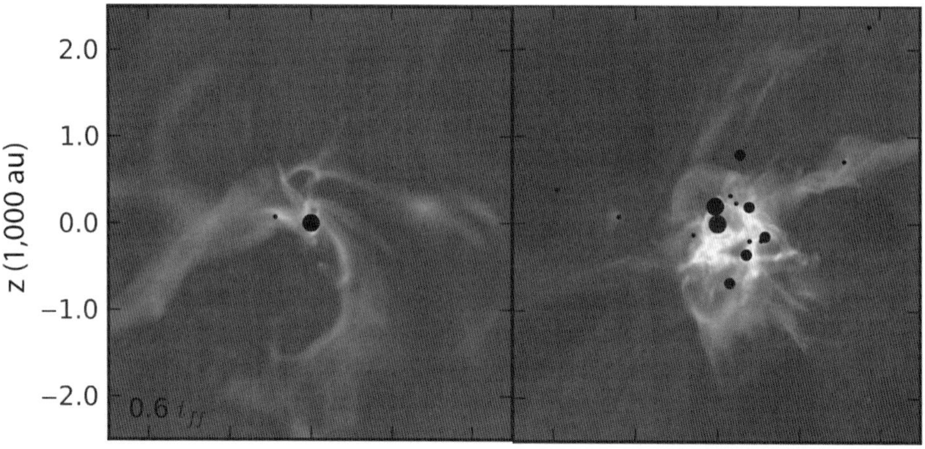

Figure 16.5 Zoom-in of two star-forming simulations with (left) and without (right) inclusion of magnetic fields. Clearly whereas the right-hand simulation undergoes multiple fragmentation, the left-hand case forms a single massive object. From Myers et al. (2013). Reproduced by permission of the AAS.

ambipolar diffusion are already centrally concentrated at the onset of collapse, an initial condition that has long been known to be unfavourable to multiple star formation (Boss, 1987). Subcritical cores in these simulations thus tend only to produce single stars.

In contrast to the case of radiative feedback previously reviewed, where episodic accretion can mitigate the effects suppressing binary formation, there is no ready solution to the magnetic field problem. It would be very hard to argue on the basis of existing observational evidence that magnetic fields are *not* in fact of dynamical significance. It is also hard to argue that the fragmentation crisis is a numerical artefact since both Eulerian and Lagrangian MHD codes with a variety of implementations of magnetic fields are in broad agreement. This impasse places us in an unfortunate place where we either hone the observational predictions of models that manifestly neglect important physics or else perform simulations that clearly disagree with observations by making binary formation a rare outcome of star formation.

16.5 Clustering at Birth: Hints from Higher-Order Multiple Statistics

Although the 'fragmentation crisis' described in the preceding section means that we do not have a good end-to-end picture of binary star formation, we can use observations to make some progress on understanding the nature of the process. Laying aside for now the problem of how initial fragmentation is achieved, the simulations that have been successful in reproducing observed binary statistics are those (see Section 16.3) that involve dynamical interactions within small N groupings. Before looking at how such initial conditions may relate to higher-order multiplicity statistics, we take a brief look at *direct* observational evidence that the small N group is the fundamental unit of star formation.

A number of statistical techniques have been used to classify subclustering in observed star-forming regions (Larson, 1995; Bate et al., 1998; Cartwright and Whitworth, 2004). In general, substructure is more readily picked out by eye in relatively low-density regions such as Taurus-Auriga, which is readily explicable in that the timescale for the dynamical erasure of subclustered initial conditions is short in high-density regions such as the Orion Nebula Cluster (Scally and Clarke, 2002; Fellhauer, Wilkinson and Kroupa, 2009; Parker et al., 2014). In Taurus, it is, however, possible to identify distinct groupings (Gomez et al., 1993) and to seek correlations between binary properties and their local environment at an age of $\sim$2 Myr: Joncour et al. (2017), for example, have shown that binaries are more likely to possess neighbours within 10^4 au than single stars. More direct evidence of binary interactions within their natal groups is provided by Herschel imaging of the environs of the two young stars, HV Tau and DO Tau (Howard et al., 2013). Figure 16.6 shows a 160 μm map from Howard et al. (2013) of the two pre–main-sequence systems HV Tau and DO Tau. Although these stars are currently unbound and at a separation of $> 10^4$ au there is clear evidence for large-scale tidal tails surrounding the system, compatible with the two systems having undergone a prior close star–disc encounter (Winter et al., 2018). Interestingly, HV Tau is itself a triple system, so that the present configuration is the outcome of a complex dynamical interaction involving at least four stars, at an age when these interactions would

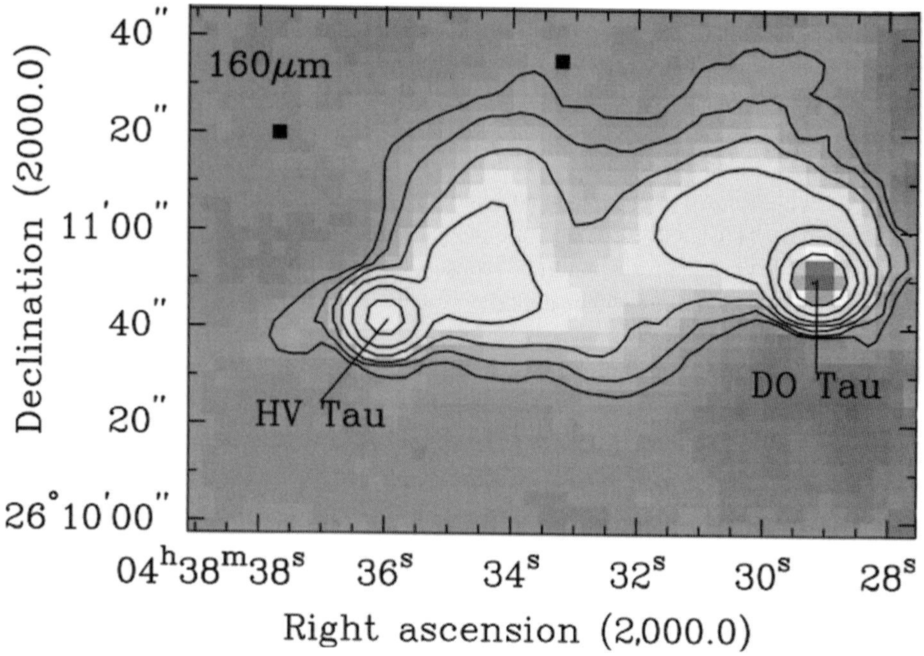

Figure 16.6 160μm map of the two pre–main-sequence systems HV Tau and DO Tau. Although separated by $>10^4$ au on the sky, the 'tidal' morphology of the extended dust emission suggests a prior dynamical encounter between the triple HV Tau and the single DO Tau. From Howard et al. (2013). Reproduced by permission of the AAS.

be strongly mediated by circumstellar gas. Although the HV/DO Tau system is unusual in betraying evidence of prior interactions, there is more widespread evidence that multiplicity levels are higher among pre–main-sequence stars than older systems (Chen et al., 2013) and also that wider components are more abundant at young ages (Connelley, Reipurth and Tokunaga, 2008; Tobin et al., 2016a).

We thus have direct evidence that young stars commonly reside in environments where they are bound to more than one companion. Evidently only a subset of these will end up in configurations that are dynamically stable $N > 2$ systems, with orbital reconfiguration being effected by both internal (Delgado-Donate et al., 2004a; Reipurth and Mikkola, 2012) and external (Moeckel and Bate, 2010) interactions. The survivors of this processing (the higher-order multiples that are found in main-sequence populations) are becoming increasingly well characterised thanks to recent large-scale surveys (e.g., Tokovinin, 2014; Halbwachs et al., 2017). While the interpretation of these statistics is still a matter for debate, it is worth setting out here a few notable properties of higher-order multiple systems (here restricting ourselves to solar-type systems, since these are the most fully characterised).

Although higher-order multiples are less abundant in main-sequence populations than in younger stars, the numbers are not negligible, with around 8% and 4% of systems being triples and quadruples respectively (Tokovinin, 2014). Remarkably, the periods of inner and outer pairs in these systems fill the entire region of phase space that is dynamically stable, a clear sign that there are likely to have been many systems that strayed beyond these limits and dynamically decayed. A possible exception to such phase space filling is in the case of the closest pairs, where there is potentially revealing evidence for a deficit of close triple companions (periods of hundreds of days: see Borkovits et al., 2016). Secondly, quadruple systems with a 'planetary' architecture are extremely rare (Tokovinin, 2014): the normal configuration for a quadruple consists of a pair of close binaries. Thirdly, the distributions of mass ratios within multiple systems appear to be insensitive to whether pairs are pure binaries or are binary components within higher-order multiple systems. Finally, although the overall fraction of systems that are triples is not high, there are separation regimes where membership of a triple appears to be the norm. In young systems, ultrawide pairs are more abundant than they are in main-sequence populations (Connelley, Reipurth and Tokunaga, 2008; Tobin et al., 2016a), and here it is found that the binary fraction of the individual wide components is very high (Elliott and Bayo, 2016). This contrasts with the situation in mature populations, where the binary fraction for the individual components of wide systems is no higher than in the general population (Law et al., 2010). This suggests that many young triples decay by losing their outlying components; very wide main-sequence systems may then largely arise through the dissolving cluster capture mechanism discussed in Section 16.2, which leads to random association of pairs and singles. The other region of parameter space where triples are very common is in the case of the closest pairs, where – also on the main sequence – the frequency of more distant tertiary companions is high (Lohr et al., 2015): at a period of around three days, 60% of pairs have such tertiary companions, a figure that drops steeply at a period of around 10 days (Tokovinin, 2014). This is consistent with a picture in which the closest pairs are driven into such orbits by a combination of the Kozai–Lidov mechanism and tidal dissipation (see Naoz and Fabrycky, 2014, for detailed modelling of this scenario). Given that the Kozai–Lidov mechanism relies on misalignment between the orbital planes of the two binaries, it is notable that there is now substantial observational evidence that the orbital planes within multiple systems can be significantly misaligned, both in the case of main-sequence (Tokovinin, 2018) and pre–main-sequence multiples (see the following section).

16.6 An Observational Frontier: Direct Observations of Binary Star Formation from ALMA and the VLA

In parallel with continued efforts to improve numerical modelling of binary formation, fresh insights are also to be expected from high-resolution imaging data, from ALMA and the VLA, which target the youngest protostars. Already such studies are highlighting the high frequency of binaries and higher-order multiples in these samples, together with the existence of nonhierarchical groupings at the youngest ages (Rodriguez et al., 2014; Tobin et al., 2016a,b). These observations allow detailed modelling of the relationship between

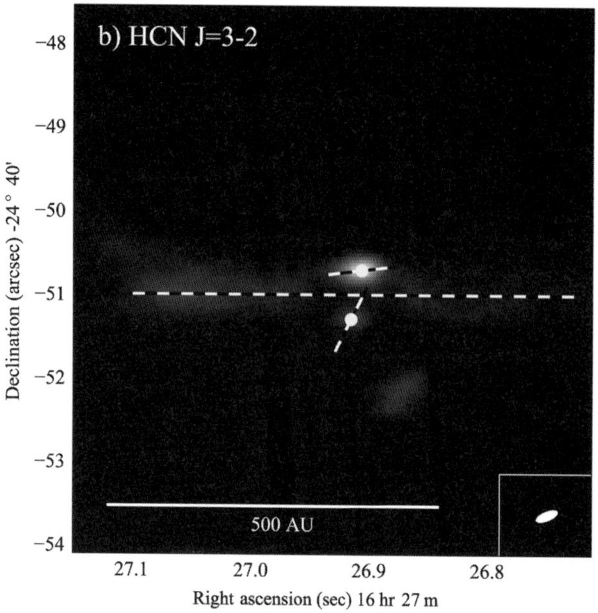

Figure 16.7 Image of the protobinary system IRS 43 in HCN J=3-2 showing the misalignment between the binary orbital plane, that of the circumbinary disc and those of each of the circumstellar discs. From Brinch et al. (2016). Reproduced by permission of the AAS.

protobinaries and their associated circumstellar and circumbinary discs (e.g., Takakuwa et al., 2017) and are providing evidence of the geometrical complexity involved. For example, Figure 16.7 shows an image of the protobinary system IRS 43, where the orbital plane of the binary and of its circumbinary disc and of each of the individual circumbinary discs are all mutually misaligned (Brinch et al., 2016).

Currently, this information is available for wide binaries (i.e., those with separations exceeding several hundred au). From the perspective of the current volume, there is at least equal interest in understanding the protobinary environment of much closer pairs (and this also connects to the issue of forming the circumbinary planets inferred from Kepler data; Armstrong et al., 2014). Until recently, however, the sample of close (sub-au) pairs among pre–main-sequence stars has been small, and so it is currently unknown what quantities of gas typically reside in circumbinary discs in such systems. Whereas such systems are too compact to allow discs to be resolved – even with ALMA – the recent identification of sizeable samples of pre–main-sequence spectroscopic binaries (Kounkel et al., 2016; Fernandez et al., 2017) means that it will soon be possible to measure the circumbinary mass reservoir in these systems. Since, in simulations, discs play an important role in driving binaries to small separations, circumbinary disc mass measurements will be usefully constraining of the origin of close binaries.

References

Armstrong, D. J., Osborn, H. P., Brown, D. J. A. et al. 2014. On the abundance of circumbinary planets. *MNRAS*, **444**, 1873–1883

Bate, M. 2009a, Stellar, brown dwarf and multiple star properties from hydrodynamical simulations of star cluster formation. *MNRAS*, **392**, 590–616

Bate, M. 2009b, The importance of radiative feedback for the stellar initial mass function. *MNRAS*, **392**, 1363–1380

Bate, M. 2012. Stellar, brown dwarf and multiple star properties from a radiation hydrodynamical simulation of star cluster formation. *MNRAS*, **419**, 3115–3146

Bate, M., Bonnell, I., and Bromm, V. 2002a. The formation mechanism of brown dwarfs. *MNRAS*, **332**, L65–L68

Bate, M., Bonnell, I., and Bromm, V. 2002b. The formation of close binary systems by dynamical interactions and orbital decay. *MNRAS*, **336**, 705–713

Bate, M., Bonnell, I., and Bromm, V. 2003. The formation of a star cluster: predicting the properties of stars and brown dwarfs. *MNRAS*, **339**, 577–599

Bate, M., Clarke, C., and McCaughrean, M. 1998. Interpreting the mean surface density of companions in star-forming regions. *MNRAS*, **297**, 1163–1181

Bonnell, I., Arcoragi, J.-P., Martel, H., et al. 1992. Fragmentation of elongated cylindrical clouds. IV. Clouds with solid-body rotation about an arbitrary axis. *ApJ*, **400**, 579–594

Bonnell, I., Martel, H., Bastien, P., et al. 1991. Fragmentation of elongated cylindrical clouds. III. Formation of binary and multiple systems. *ApJ*, **377**, 553–558

Borkovits, T., Hajdu, T., Sztakovics, J, et al. 2016. A comprehensive study of the Kepler triples via eclipse timing. *MNRAS*, **455**, 4136–4165

Boss, A. 1986. Protostellar formation in rotating interstellar clouds. V. Nonisothermal collapse and fragmentation. *ApJS*, **62**, 519–552

Boss, A. 1987. Protostellar formation in rotating interstellar clouds. VI. Nonuniform initial conditions. *ApJ*, **319**, 149–161

Boss, A. 1991. Formation of hierarchical multiple protostellar cores. *Nature*, **351**, 298

Boss, A., and Bodenheimer, P. 1979. Fragmentation in a rotating protostar: a comparison of two three-dimensional computer codes. *ApJ*, **284**, 289–295

Boss, A., and Keiser, S. 2014. Collapse and fragmentation of magnetic molecular cloud cores with the Enzo AMR MHD code. II. Prolate and oblate cores. *ApJ*, **794**, 44

Brinch, C., Joergensen, J., Hogerheijde, M., et al. 2016. Misaligned disks in the binary protostar IRS 43. *ApJ*, **830**, L16

Cartwright, A., and Whitworth, A. 2004. The statistical analysis of star clusters. *MNRAS*, **348**, 589–598

Chapman, S., Pongracic, H., Disney, M., et al. 1992. The formation of binary and multiple star systems. *Nature*, **359**, 207–210

Chen, X., et al. 2013. SMA observations of class 0 protostars: a high angular resolution survey of protostellar binary systems. *ApJ*, **768**, 110

Ciardi, B. and Hennebelle, P. 2010. Outflows and mass accretion in collapsing dense cores with misaligned rotation axis and magnetic field. *MNRAS*, **409**, L39–L43

Clarke, C. 1992. The formation of binary stars. in *Workshop on Binaries as Tracers of Star Formation*, eds. A. Duquennoy and M. Mayor, 38–51. Cambridge: Cambridge University Press.

Clarke, C. and Pringle, J.,1991a. Star–disc interactions and binary star formation *MNRAS*, **249**, 584–587

Clarke, C. and Pringle, J. 1991b. The role of discs in the formation of binary and multiple star systems. *MNRAS*, **249**, 588–595

Commercon, B., Hennebelle, P. and Henning, Th. 2011. Collapse of massive magnetized dense cores using radiation magnetohydrodynamics: early fragmentation inhibition. *ApJ*, **742**, L9

Connelley, M., Reipurth, B. and Tokunaga, A. 2008. The evolution of the multiplicity of embedded protostars. II. Binary separation distribution and analysis *AJ*, **135**, 2526–2536

Crutcher, R. 2012. Magnetic fields in molecular clouds. *ARAA*, **50**, 29–63

Dale, J. and Bonnell, I. 2008. The effect of stellar winds on the formation of a protocluster. *MNRAS*, **391**, 2–13

Dale, J., Ercolano, B. and Bonnell, I. 2012. Ionizing feedback from massive stars in massive clusters. II. Disruption of bound clusters by photoionization. *MNRAS*, **424**, 377–392

Delgado-Donate, E., Clarke, C. and Bate, M. 2004a. Accretion and dynamical interactions in small-N star-forming clusters: N= 5. *MNRAS*, **342**, 926–938

Delgado-Donate, E., Clarke, C., Bate, M. and Hodgkin, S. 2004b. On the properties of young multiple stars. *MNRAS*, **351**, 617–629

Duquennoy, A. and Mayor, M. 1991. Multiplicity among solar-type stars in the solar neighbourhood. II. Distribution of the orbital elements in an unbiased sample. *A&A*, **248**, 485–524

Elliott, P. and Bayo, A. 2016. The crucial role of higher order multiplicity in wide binary formation: a case study using the -Pictoris moving group. *MNRAS*, **459**, 4499–4507

Fabian, A., Pringle, J. and Rees, M. 1975. Tidal capture formation of binary systems and X-ray sources in globular clusters. *MNRAS*, **172**, 15–18

Fellhauer, M., Wilkinson. M. and Kroupa, P. 2009. Merging time-scales of stellar subclumps in young star-forming regions. *MNRAS*, **397**, 954–962

Fernandez, M. A., Covey, K. R., De Lee, N. et al. 2017. IN-SYNC VI. Identification and radial velocity extraction for 100+ double-lined spectroscopic binaries in the APOGEE/IN-SYNC fields. *PASP*, **129**, 8

Forgan, D., Rice, K., Stamatellos, D. and Whitworth, A. 2009. Introducing a hybrid radiative transfer method for smoothed particle hydrodynamics. *MNRAS*, **394**, 882–891

Gomez, M., Hartmann, L., Kenyon, S. and, Hewett, R. 1993. On the spatial distribution of pre–main-sequence stars in Taurus. *AJ*, **105**, 1927–1937

Goodwin, S., Whitworth, A. and Ward-Thompson, D. 2004a. Simulating star formation in molecular cloud cores. I. The influence of low levels of turbulence on fragmentation and multiplicity. *A&A*, **414**, 633–650

Goodwin, S., Whitworth, A. and Ward-Thompson, D. 2004b. Simulating star formation in molecular cores. II. The effects of different levels of turbulence. *A&A* **423**, 169–182

Goodwin, S., Whitworth, A., and Ward-Thompson, D. 2006. Star formation in molecular cores. III. The effect of the turbulent power spectrum. *A&A*, **452**, 487–492

Halbwachs, J., Mayor, M. and Udry, S, 2017. Double stars with wide separations in the AGK3. II. The wide binaries and the multiple systems. *MNRAS*, **464**, 4966–4976

Heggie, D. 1975. Binary evolution in stellar dynamics. *MNRAS*, **173**, 729–787

Hennebelle, P., and Teyssier, R. 2008. Magnetic processes in a collapsing dense core. II. Fragmentation. Is there a fragmentation crisis? *A&A*, **477**, 25–34

Hennebelle, P., Commercon, B., Joos, M. et al. 2011. Collapse, outflows and fragmentation of massive, turbulent and magnetized prestellar barotropic cores. *A&A*, **528**, A72

Howard, C. D., Sandell, G., Vacca, W. D. et al. 2013. Herschel/PACS survey of protoplanetary disks in Taurus/Auriga: observations of [O I] and [C II], and far-infrared continuum. *ApJ*, **776**, 21

Joncour, I., Duchene, G. and Moraux, E. 2017. Multiplicity and clustering in Taurus star-forming region. I. Unexpected ultra-wide pairs of high-order multiplicity in Taurus. *A&A*, **599**, A14

Joos, M., Hennebelle, P. and Ciardi, A. 2012. Protostellar disk formation and transport of angular momentum during magnetized core collapse. *A&A*, **543**, A128

Kenyon, S., Hartmann, L., Strom, K. and Stro, S. 1990. An IRAS survey of the Taurus-Auriga molecular cloud. *AJ*, **99**, 869–887

Kounkel, M., Hartmann, L., Tobin, J., Mateo, M., Bailey, M. and Spencer, M. 2016. Spectroscopic binaries in the Orion Nebula Cluster and NGC 2264. *ApJ*, **821**, 8

Kouwenhoven, M., Goodwin, S., Parker, R. et al. 2010. The formation of very wide binaries during the star cluster dissolution phase. *MNRAS*, **404**, 1835–1848

Kozai, Y. 1962. Secular perturbations of asteroids with high inclination and eccentricity. *AJ*, **67**, 591

Krumholz, M., Klein, R. and McKee, C. 2011. Radiation-hydrodynamic simulations of the formation of Orion-like star clusters I. Implications for the origin of the initial mass function. *ApJ*, **740**, 74

Krumholz, M., Klein, R. and McKee, C. 2012. Radiation-hydrodynamic simulations of the formation of Orion-like star clusters II. The initial mass function from winds, turbulence and radiation. *ApJ*, **754**, 71

Larson, R. 1978. Calulations of three-dimensional collapse and fragmentation. *MNRAS*, **184**, 69–85

Larson, R. 1981. Turbulence and star formation in molecular clouds. *MNRAS*, **194**, 809–826

Larson, R. 1995. Star formation in groups. *MNRAS*, **272**, 213–220

Law, N., Dhital, S., Kraus, A. et al. 2010, The higher order multiplicity of unusually wide M-dwarf binaries: eleven new triple and quadruple systems. *ApJ*, **720**, 1727–1737

Lewis, B. and Bate, M. 2017. The dependence of protostar formation on the geometry and strength of the initial magnetic field. *MNRAS*, **467**, 3324–3337

Lidov, M. 1962. The evolution of orbits of artificial satellites of planets under the action of gravitational perturbations of external bodies. *Planetary & Space Science*, **9**, 719–759

Lohr, M., Norton, A., Payne, S. et al. 2015. Orbital period changes and the higher-order multiplicity fraction amongst SuperWASP eclipsing binaries. *A&A*, **578**, A136

Lomax, O., Whitworth, A., Hubber, D. et al. 2015. Simulations of star formation in Ophiuchus. II. Multiplicity. *MNRAS*, **447**, 1550–1564

Machida, M., Tomisaka, K., Matsumoto, T. et al. 2008. Formation scenario for wide and close binary systems. *ApJ*, **677**, 327–347

McDonald, J. and Clarke, C. 1993. Dynamical biasing in binary star formation: implications for brown dwarfs in binaries. *MNRAS*, **262**, 800–804

Meyer, D., Kuiper, R., Kley, W. et al. 2017. Forming spectroscopic massive proto-binaries by disk fragmentation. *MNRAS*, **404**, 3615–3637

Moe, M., and Kratter, K., 2018. Dynamical formation of close binaries during the pre–main-sequence phase. *ApJ*, **854**, 44

Moeckel, N. and Bate, M. 2010, On the evolution of a star cluster and its multiple stellar systems following gas dispersal. *MNRAS*, **404**, 721

Moeckel, N., and Clarke, C. 2011. The formation of permanent soft binaries in dispersing clusters. *MNRAS*, **415**, 1179–1187

Mouschovias, T. and Spitzer, L. 1976. Note on the collapse of magnetic interstellar clouds. *ApJ*, **201**, 326

Myers, A., McKee, C., Cunnigham, A. et al. 2013. The fragmentation of magnetized massive star forming cores with radiative feedback. *ApJ*, **766**, 97

Naoz, S. and Fabrycky, D. 2014. Mergers and obliquities in stellar triples. *ApJ*, **793**, 137

Offner, S., Klein, R., McKee, C. and Krumholz, M. 2009. The effects of radiative transfer on low mass star formation. *ApJ*, **703**, 131–149

Offner, S., Kratter, K., Matzner, C. et al. 2010. The formation of low-mass binary star systems via turbulent fragmentation. *ApJ*, **725**, 1485–1494

Parker, R., Wright, N., Goodwin, S. and Meyer, M. 2014. Dynamical evolution of star-forming regions. *MNRAS*, **438**, 620–638

Price, D. and Bate, M. 2007. The impact of magnetic fields on single and binary star formation. *MNRAS*, **377**, 77–90

Pringle, J. 1989. On the formation of binary stars. *MNRAS*, **239**, 361–370

Raghavan, D. McAlister, H. A., Henry, T. J. et al. 2010. A survey of stellar families: multiplicity of solar-type stars. *ApJS*, **190**, 1–42

Reipurth, B. and Mikkola, S. 2012. Formation of the widest binaries from dynamical unfolding of triple systems. *Nature*, **492**, 221

Rodriguez,L., Reipurth, B. and Chiang, H.-F. 2014. Radio continuum sources associated with the HH 92 and HH 34 jets. *RMxAA*, **50**, 285–291

Scally, A. and Clarke, C. 2002. Primordial substructure in the Orion Nebula Cluster. *MNRAS* **334**, 156

Seifried, D., Banerjee, R., Pudritz, R. and Klessen, R. 2012. Disc formation in turbulent massive cores: circumventing the magnetic braking catastrophe. *MNRAS*, **423**, L40–L44

Stamatellos, D., Whitworth, A., Bisbas, T. and Goodwin, S. 2007. Radiative transfer and the energy equation in SPH simulations of star formation. *A&A*, **475**, 37–49

Stamatellos, D., Whitworth, A. P. and Hubber, D. A. 2011. Episodic accretion, protostellar radiative feedback, and their role in low-mass star formation. *ApJ*, **730**, 32

Takakuwa, S., Saigo, K., Matsumoto, T. et al. 2017. Spiral arms, infall, and misalignment of the circumbinary disk from the circumstellar disks in the protostellar binary system L1551 NE. *ApJ*, **837**, 86

Tobin, J. et al. 2016a. The VLA nascent disk and multiplicity survey of Perseus protostars (VANDAM). II. Multiplicity of protostars in the Perseus Molecular Cloud. *ApJ*, **818**, 73

Tobin, J. et al. 2016b. A triple protostar system formed via fragmentation of a gravitationally unstable disk. *Nature*, **538**, 483–486.

Tohline, J. 2002. The origin of binary stars. *ARAA*, **40**, 349–385

Tokovinin, A. 2014. From binaries to multiples. II. Hierarchical multiplicity of F and G dwarfs. *AJ*, **147**, 87

Tokovinin, A. 2018. The updated multiple star catalog. *ApJS*, **235**, 6

van Albada, T. 1968. Numerical integrations of the N-body problem. *Bull. Astron. Inst. Neth.*, **19**, 479

Whitehouse, S. and Bate, M. 2004. Smoothed particle hydrodynamics with radiative transfer in the flux-limited diffusion approximation. *MNRAS*, **353**, 1078–1094

Winter, A. J., Clarke, C. J., Rosotti, G. et al. 2018, Protoplanetary disc truncation mechanisms in stellar clusters: comparing external photoevaporation and tidal encounters. *MNRAS*, 478(2), 2700–2722
Wurster, J., Price, D. and Bate, M. 2017. The impact of non-ideal magnetohydrodynamics on binary star formation. *MNRAS*, **466**, 1788
Zhu, Z, Hartmann, L. and Gammie, C. 2009. Nonsteady accretion in protostars. *ApJ*, **694**, 1045–1055.
Zinnecker, H. and Yorke, H. 2007. Towards understanding massive star formation. *ARAA*, **45**, 481–563.

17

The Maxwell's Demon of Star Clusters

MICHELA MAPELLI[†]

17.1 Binaries as Sources of Energy

For a long time, we have known that binaries are an energy reservoir for star clusters. The internal energy of a binary can be expressed as

$$E_{\rm int} = \frac{1}{2}\,\mu\, v^2 - \frac{G\,m_1\,m_2}{r}, \tag{17.1}$$

where $\mu = m_1\,m_2/(m_1+m_2)$ is the reduced mass, m_1 and m_2 are the masses of the primary and secondary component of the binary, G is the gravitational constant and v and r are the relative velocity and the distance between the two components of the binary.

Part of this energy can be exchanged with single stars if they approach the binary in a close encounter (with minimum distance of the order of a few times the orbital separation of the binary). In particular, if the binary transfers a fraction of its internal energy to the single star, then the single star's kinetic energy increases, while the binary shrinks (i.e., its binding energy increases). In this case, we say that the binary *hardens*. Vice versa, if the binary acquires kinetic energy from the single star, then its semimajor axis increases and we say that the binary *softens*. The binary might even be *ionised*, i.e., its binding energy may become zero.

In a star cluster, we usually define as *hard* (*soft*) binaries those binaries with binding energy $E_{\rm b} > \frac{1}{2}\langle m\rangle\sigma^2$ ($E_{\rm b} < \frac{1}{2}\langle m\rangle\sigma^2$), where $\langle m\rangle$ is the average mass of a star in the star cluster and σ is the velocity dispersion. According to Heggie (1975), which is nothing less than the bible of three-body encounters, hard binaries tend to harden as a consequence of three-body encounters, while soft binaries tend to soften. This statement has been commonly referred to as Heggie's law. Under the impulse approximation, the hardening rate of a binary does not depend on the mass of the binary and can be expressed as

$$\frac{d}{dt}\left(\frac{1}{a}\right) = 2\pi\,\xi\,\frac{G\,\rho}{\sigma}, \tag{17.2}$$

[†] MM acknowledges financial support from the Italian Ministry of Education, University and Research (MIUR) through grant FIRB 2012 RBFR12PM1F, from the National Institute for Astrophysics (INAF) through grant PRIN-2014-14, and from the Mobilising European Research in Astrophysics and Cosmology (MERAC) Foundation.

where a is the semimajor axis of the binary, $\xi \sim 0.2 - 3$ is a dimensionless efficiency parameter (Hills, 1983; Quinlan, 1996; Miller and Hamilton, 2002), ρ is the local mass density of stars, and σ is the local velocity dispersion. Based on Equation 17.2, we can say that a hard binary hardens at a constant rate, depending only on the environment (i.e., the local stellar density and velocity dispersion). However, a slight dependence of the hardening rate on the properties of the binary is hidden in the ξ parameter, which can be calibrated only from simulations.

Exchanges are another possible outcome of three-body encounters. During exchanges, one of the members of the binary is kicked off the binary and replaced by the intruder. Exchanges are most likely when the intruder mass is larger than the mass of one of the members of the binary, leading to the formation of more and more massive binaries (Hills and Fullerton, 1980).

Figure 17.1 shows some examples of simulated three-body encounters: a *fly-by* (during which energy is exchanged between the binary and the single stars but the binary does

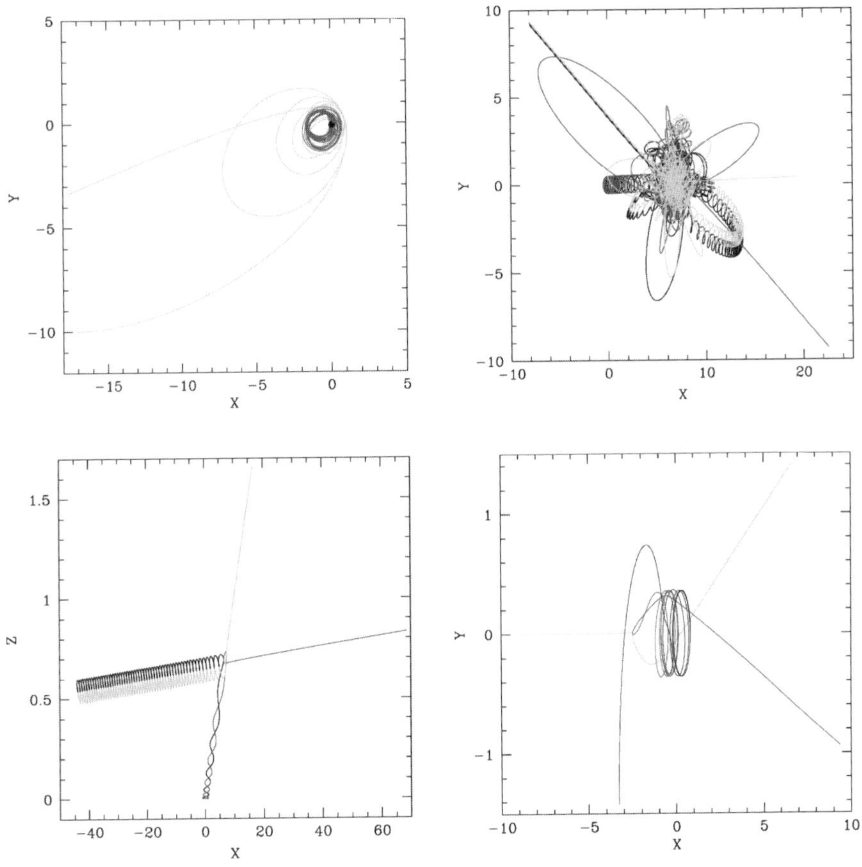

Figure 17.1 Simulations of three-body encounters. Top-left: dynamical flyby; top-right: resonant exchange; bottom-left: prompt exchange; bottom-right: ionisation. Axes are in N-body units.

not change its components), two exchanges (during which the binary exchanges one of its members with the intruder) and a ionisation (during which the binary breaks).

Via three-body encounters, binaries can pump kinetic energy (from their internal energy) in the star cluster, changing its dynamical state completely. In this sense, binaries act like the Maxwell's demon (Maxwell, 1872), which can change the entropy state of a system in statistical mechanics without any external energy input. The main difference between Maxwell's demons and binaries is that binary systems are definitely real and even rather common astrophysical systems, while Maxwell's demons are just a *Gedankenexperiment*.

17.2 Impact of Binaries on Core Collapse

One of the most important and well-studied effects of three-body encounters in star clusters is the reverse of core collapse.

Isolated star clusters composed of a dense core surrounded by a low-density halo are expected to undergo core collapse because of their negative heat capacity (Spitzer, 1987). In other words, two-body relaxation is particularly efficient in the core of star clusters, leading to the evaporation of the fastest stars from the core. This leads to a decrease of both the kinetic energy (K) and the modulus of the potential energy ($|W|$) of the cluster core, but the decrease of K is more significant than the decrease of $|W|$ because only the fastest stars leave the core. Thus, the core contracts (because potential energy is no longer balanced by kinetic energy), decreasing its radius and increasing its density. If the density increases, two-body encounters are even more efficient (because the two-body relaxation timescale scales as $t_{\rm rlx} \sim \rho^{-1}\,\sigma^3$, where ρ is the density and σ the velocity dispersion; Spitzer and Hart, 1971), leading to an even faster rate of evaporations. This induces a further contraction, the core radius decreases and the density increases even more, enhancing evaporation, and so on.

This runaway mechanism, called core collapse, would lead to an infinite core density and a null core radius. Something is needed to break this loop and reverse core collapse, since we do not observe star clusters with pathologically high central densities. A Maxwell's demon is needed to break the loop: something that changes the energetic state of the star cluster.

There are essentially two ways of reversing core collapse: by removing potential energy $|W|$ from the core without removing much kinetic energy and by injecting fresh kinetic energy into the core. Hénon (1961) suggested for the first time that binaries might be the Maxwell's demon of the situation, able to change the dynamical state of a cluster (see also Elson et al., 1987, for a review). When the core density is low, binaries do not exchange significant energy with the other stars, because three-body encounters are rare. During core collapse, when the core density rises by orders of magnitude, three-body encounters become more and more frequent. Moreover, additional binaries can form dynamically during core collapse, because of the higher stellar density. Hard binaries harden, transferring kinetic energy to the other stars. If the intruder stars are not ejected, they increase the kinetic energy of the core. If they are ejected, their kinetic energy is lost but does not trigger a

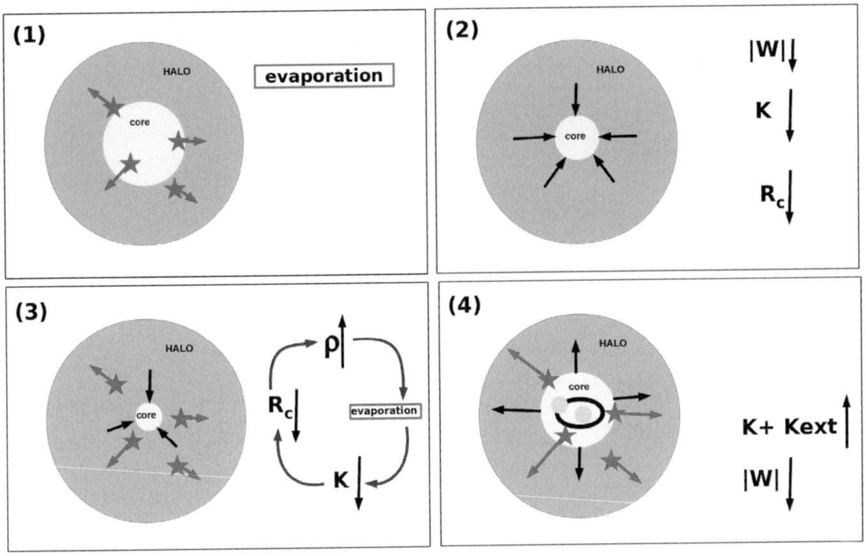

Figure 17.2 Cartoon of a core collapse and its reverse. The star cluster is represented as a dense core (white) embedded into a low-density halo (grey). In panel (1), stars evaporate from the core because of two-body encounters. The evaporation removes both kinetic energy (K) and potential energy (W) from the core. Since the fastest stars are more likely to evaporate, K decreases faster than $|W|$: the core contracts (2), because potential energy is no longer balanced by kinetic energy, and the radius of the core R_c decreases. The decrease of R_c increases the density ρ of the core. Thus, two-body encounters become more likely and lead to more efficient evaporation. This further reduces K and the core contracts again, increasing its density (3). The result is a loop (or runaway process): the core keeps contracting forever unless a new source of kinetic energy is switched on in the star cluster. The source of kinetic energy is represented by hard binaries (4), which release kinetic energy (K_{ext}) into the core by means of three-body encounters. As an effect of kinetic energy injection, the core expands and the core collapse is reversed.

further collapse because these stars were not necessarily the fastest ones before interacting with the binary. The net result is that the kinetic energy of the core increases and the core expands. Figure 17.2 is a schematic representation of core collapse reversal by binaries.

Several studies (Angeletti and Giannone, 1977, 1980; Chernoff and Weinberg, 1990; Schulman et al., 2012; Downing, 2012) suggest that mass loss by massive stars could be another mechanism able to reverse the core collapse, by removing a large fraction of core's potential energy. However, Mapelli and Bressan (2013) and Trani et al. (2014) highlight that mass loss of massive stars is a viable mechanism to reverse core collapse only if the collapse happens within the lifespan of massive stars, which might be the case only for dense young massive star clusters. In fact, the lifetime of a $\sim$30 $M_\odot$ star is $\sim$6 Myr, and core collapse happens on a shorter timescale only in star clusters with two-body relaxation timescale $t_{rlx} \lesssim 50\,\mathrm{Myr}\,(M/10^5 M_\odot)^{1/2}\,(R/1\,\mathrm{pc})^{3/2}$, assuming that core collapse occurs at $t_{rlx} \sim 0.1 - 0.2\,t_{rlx}$ (see Fujii and Portegies Zwart, 2014).

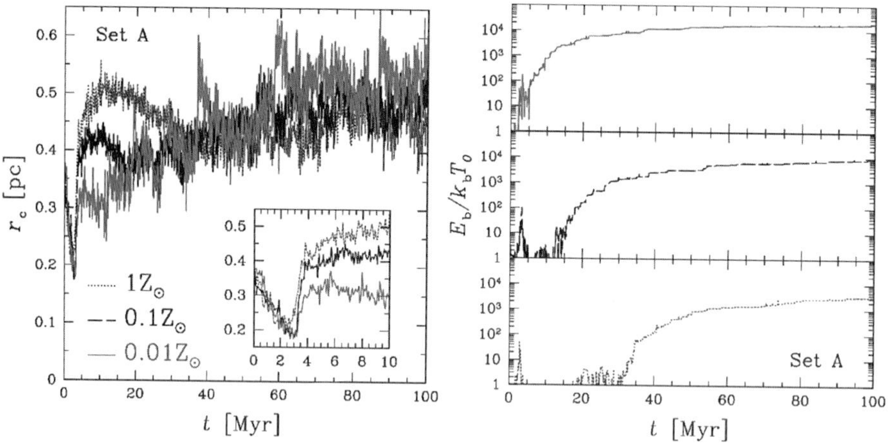

Figure 17.3 Core radius (left-hand panel) as a function of time for simulated star clusters with three different metallicities $Z = 1,\ 0.1,\ 0.01\,Z_{\odot}$. In the inset: zoom of the first 10 Myr. Right-hand panel: total internal energy of the binary content of the simulated star clusters as a function of time, normalised to the initial $k_{\rm b}\,T_0 = 1/3\,\langle K\rangle|_{t=0}$, where $\langle K\rangle$ is the average kinetic energy of a star, $k_{\rm b}$ is the Boltzmann constant and T_0 the initial average kinetic temperature of the star cluster. Solid grey line: $Z = 0.01\ Z_{\odot}$; dashed black line: $Z = 0.1\ Z_{\odot}$; dotted grey line: $Z = 1\ Z_{\odot}$. Each line is the median value of 10 N-body realisations of star clusters, modelled as King models (King, 1966) with initial virial radius $r_{\rm vir} = 1$ pc and dimensionless central potential $W_0 = 5$. These simulations start without primordial binaries: all binaries form from dynamical interactions. Adapted from figure 3 of Trani et al. (2014).

The left-hand panel of Figure 17.3 shows the evolution of the core radius in simulations of young, dense star clusters with different metallicity. The collapse is apparent at $t \sim 3$ Myr, followed by a fast reverse. In metal-rich clusters, mass loss by stellar winds is particularly efficient in reversing core collapse, as it is apparent from the right-hand panel of Figure 17.3. This figure shows the total binding energy stored in binaries as a function of time. While at low metallicity the binding energy of binaries grows rapidly during and after core collapse, at solar metallicity we must wait for $t \sim 30$ Myr before the first hard binaries appear. This difference can be easily explained because mass loss by stellar winds depends on metallicity as $\dot{M} \propto Z^{\alpha}$ with $\alpha \sim 0.85$ (Vink et al., 2001).

17.3 Impact of Binaries on Equipartition

Core collapse reverse is the best-known effect of binaries on the global evolution of star clusters, but there are also other crucial, although more subtle, effects that are still matters of investigation.

For example, hard binaries might play a crucial role in the onset of Spitzer's instability. From the equipartition theorem of statistical mechanics (Boltzmann, 1876), we know that in gas systems at thermal equilibrium energy is shared equally by all particles. For analogy with gas, we expect that in a two-body relaxed stellar system the kinetic energy of a star i is *l*ocally the same as that of the star j ($m_i v_i^2 \sim m_j v_j^2$). Thus, if a stellar system is in equipartition, the local velocity of a star[1] should depend on its mass as $v(m) \propto m^{-0.5}$. If the velocities of all stars are initially drawn from the same distribution, massive stars are then expected to transfer kinetic energy to lighter stars and slow down until they reach equipartition.

Spitzer (1969) already predicted that there are star clusters that cannot reach equipartition. According to Spitzer's analytic model, an idealised star cluster composed of two classes of stars (the stars in the first and the second class having mass m_1 and m_2, respectively, with $m_1 << m_2$) cannot reach equipartition if $M_2 > 0.16 M_1 (m_2/m_1)^{3/2}$, where M_1 and M_2 are the total mass of class 1 and 2, respectively. This has a concrete physical interpretation: if there are too many massive stars in a star cluster, then the light stars are not able to absorb all the excess of kinetic energy that the massive stars need to transfer to them in order to reach equipartition. The light stars are ejected on larger orbits, while the massive stars decouple dynamically from the light stars and tend to interact between each other, forming binary systems and eventually ejecting each other through three-body encounters (Bonnell and Davies, 1998; Allison et al., 2009; Portegies Zwart et al., 2010). This process takes the name of Spitzer's instability.

While the onset of Spitzer's instability has been demonstrated analytically only for rather idealised systems (Spitzer, 1969; Vishniac, 1978; Merritt, 1981; Miocchi, 2006), Monte Carlo (Bianchini et al., 2016) and N-body simulations (Trenti and van der Marel, 2013; Parker et al., 2016; Spera et al., 2016) show that Spitzer's instability is rather common, if not universal, in star clusters (both open clusters and globular clusters).

The left-hand panel of Figure 17.4 is an example of the onset of Spitzer's instability in star clusters. The massive stars initially transfer kinetic energy to the light stars and their velocity dispersion decreases, approaching the trend expected for equipartition. However, at time $\gtrsim 30$ Myr, massive stars become dynamically hotter than lower-mass stars, and the process of reaching equipartition is interrupted: Spitzer's instability develops. The right-hand panel of Figure 17.4 shows that there is an interesting correlation between the formation of binary systems and the onset of Spitzer's instability. The simulations shown in Figure 17.4 start from zero binary systems. Thus, all binaries in the clusters have dynamical origins. It is apparent that the formation of binaries starts exactly at $\sim$30 Myr, when Spitzer instability also manifests itself, and then the number of binaries and their total binding energy grow very fast. A possible explanation of this link between binary formation/hardening and the onset of Spitzer's instability is the following. Spitzer's instability implies that massive stars decouple dynamically from light stars; thus massive stars can interact only between each other, forming binaries and undergoing three-body encounters that increase their velocity dispersion and make them hotter.

[1] It is important to stress that the equipartition theorem is a local theorem: it does not apply to the star cluster as a whole.

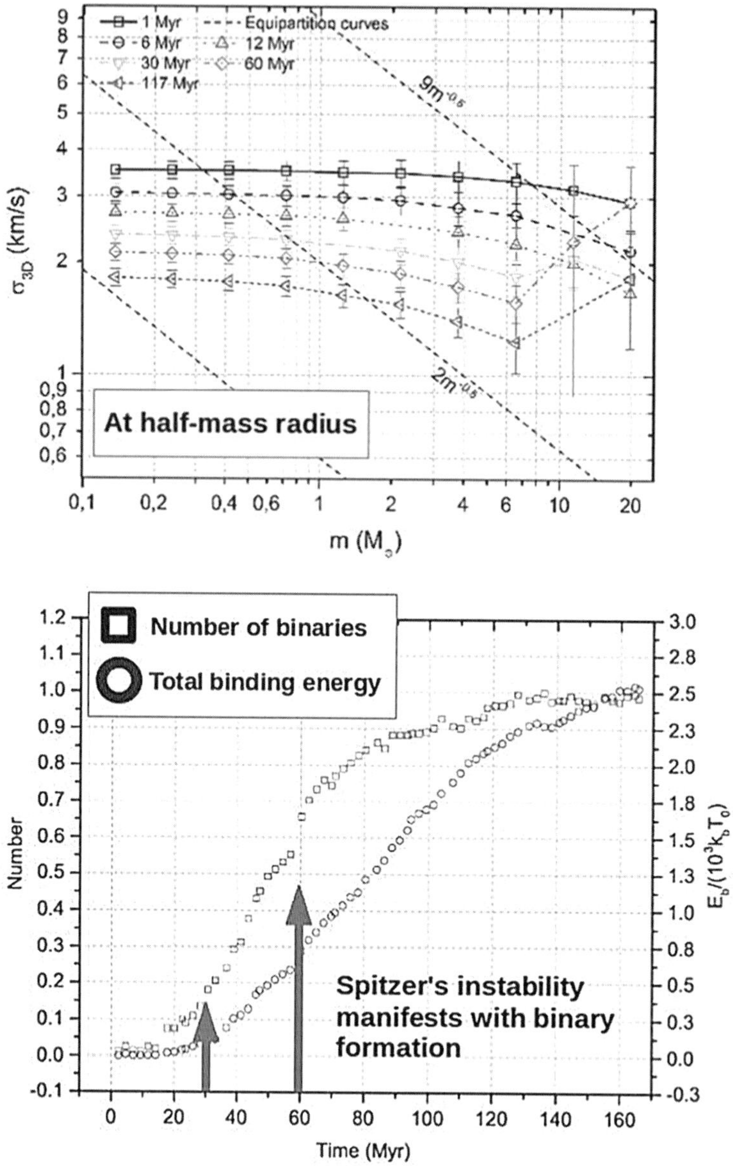

Figure 17.4 Left-hand panel: three-dimensional velocity dispersion σ_{3D} as a function of the stellar mass in 200 realisations of star clusters with $N = 10^4$ and two-body relaxation timescale $t \sim 27$ Myr. The lines show the median values of σ_{3D}, the error bars represent the standard deviation over 200 different Monte Carlo realisations. Different lines correspond to a different time since the beginning of the simulation. Solid dark-grey line with open squares: $t = 1$ Myr; dashed dark-grey lines with open circles: $t = 6$ Myr; dotted grey line with open triangles: $t = 12$ Myr; dash-dot light-grey line with downward triangles: $t = 30$ Myr; dash-double-dot grey line with diamonds: $t = 60$ Myr; short dashed grey line with leftward open triangles: $t = 117$ Myr. Dashed black diagonal lines correspond to the trend expected from equipartition ($\sigma_{3D} \propto m^{-0.5}$). Only stars inside half-mass radius are considered. Right-hand panel: total number of binaries (open black squares and left y axis) and total binding energy (open black circles and right y axis) as a function of time in the same simulations as shown in the left-hand panel. Vertical arrows correspond to $t = 30$ and $t = 60$ Myr. These simulations start without primordial binaries: all binaries form from dynamical interactions. Adapted from figures 4 and 11 of Spera et al. (2016).

17.4 Stellar Exotica

The dynamics of binaries in star clusters is essential to generate the so-called *stellar exotica*, i.e., stellarlike objects that cannot be produced (or can hardly be produced) by 'simple' stellar evolution (Davies, 2002). Famous examples of stellar exotica are blue straggler stars (BSSs) and intermediate-mass black holes (IMBHs, i.e., black holes with mass $\sim 10^2 - 10^5$ $M_\odot$), but even massive stellar black hole (BH) binaries, such as the ones observed by the LIGO–Virgo collaboration (Abbott et al., 2016a), might be part of this category.

17.4.1 Blue Straggler Stars

Lying on the continuation of the main sequence, but above the turnoff in the colour-magnitude diagram of a star cluster, BSSs appear to be core hydrogen-burning stars which are more massive than the turnoff. This has been interpreted as the effect of stellar rejuvenation, which might come from mass transfer in a binary or from the collision (and merger) between two main-sequence stars, likely triggered by a three-body encounter. In both cases, BSSs require a binary system to form.

Because of their formation involving the dynamics of binaries, BSSs are thought to be a good indicator of the dynamical evolution of a star cluster. As described in detail in Chapter 19 (see also Mapelli et al., 2004, 2006; Beccari et al., 2006), the radial distribution of BSSs, normalised to the average light distribution in a star cluster, is suggestive of the dynamical evolution of a star cluster: BSSs are among the most massive objects in a star cluster; thus, they sink to the centre by dynamical friction. The presence of a minimum in the normalised distribution of BSSs indicates the maximum distance from the centre below which dynamical friction was efficient in sweeping BSSs towards the core.

In this scenario, BSSs would be a powerful indicator of the dynamical friction timescale, and thus also of the two-body relaxation timescale of a star cluster[2]. In this sense, BSSs would be a genuine indicator of the dynamical age of a star cluster: a 'dynamical clock', as coined by Ferraro et al. (2012).

Have we understood BSSs, then? No. Our picture of BSSs is still far from optimal. For example, Hypki and Giersz (2017) simulate the dynamical evolution of BSSs in a globular cluster through the MOCCA Monte Carlo code. They produce the normalised distribution of BSSs from their simulations as a function of time, finding that the position of the minimum and the shape of the curve shift much faster than a two-body relaxation timescale. This suggests that we still do not have a complete understanding of BSSs and we should keep investigating on them, from both the theoretical and the observational point of view. Chapters 18 and 19 offer a detailed description and observational characterisation of the population of the BSS in open clusters and globular clusters respectively.

2 The two-body relaxation timescale $t_{\rm rlx}$ and the dynamical friction timescale $t_{\rm DF}(M)$ of a star with mass M are linked by the following relation: $t_{\rm DF}(M) = \frac{\langle m \rangle}{M}\, t_{\rm rlx}$, where $\langle m \rangle$ is the average mass of a star in the star cluster.

17.4.2 Massive Black Hole Binaries

On 14 September 2015, the two LIGO interferometers captured the first gravitational wave signal from a merging BH binary (Abbott et al., 2016a). GW150914 has been interpreted as the merger of two BHs with mass $m_{\rm BH} = 36.2^{+5.2}_{-3.8}$, $29.1^{+3.7}_{-4.4}$ $M_\odot$ (Abbott et al., 2016b). Since then, the LIGO–Virgo collaboration has detected five BH mergers, three of them containing BHs more massive than ~ 30 $M_\odot$ (Abbott et al., 2016b, 2017a,b,c; LIGO Scientific Collaboration et al., 2017).

The formation channel of these massive BH binaries is still an open question. Two main channels have been proposed in the last few years: isolated binary evolution and dynamical formation in star clusters.

In the isolated binary evolution scenario, two massive stars form gravitationally bound, from the same cloud, and then they evolve through several processes (e.g., Roche-lobe overflow, common envelope, etc.) until they become two BHs and eventually merge. The two progenitor stars should be relatively metal poor ($Z \lesssim 0.3$ $Z_\odot$) in order to end up as two massive BHs (Mapelli et al., 2009, 2017; Belczynski et al., 2010, 2016; Spera et al., 2015; Giacobbo et al., 2018).

Alternatively, BH binaries can form from dynamical processes in star clusters. For example, a dynamical exchange can lead to the formation of a double BH binary in the core of a dense star cluster. Ziosi et al. (2014) show that more than $\sim$90% of the double BH binaries in star clusters form through dynamical exchanges. BHs are particularly efficient in capturing companions through dynamical exchanges, because they are more massive than the vast majority of the other stars in star clusters. As we mentioned in Section 17.1, exchanges are more likely if the mass of the intruder is larger than the mass of one of the members of the binary. Thus, exchanges favour the formation of more and more massive binaries, similar to the ones observed by LIGO and Virgo. Moreover, binaries produced by exchanges have initially large eccentricities. A third crucial feature of BH binaries formed by exchanges is the distribution of BH spins: exchanges tend to randomise spins, and we expect a nearly isotropic distribution of BH spins in dynamically formed binaries.

Moreover, BH binaries are expected to be relatively hard binaries; thus, three-body encounters can harden them even more, possibly enhancing the merger rate of BH binaries. Finally, recoil kicks induced by three-body encounters can also eject BH binaries from their parent star clusters (Mapelli et al., 2013). Figure 17.5 is a summary of the main effects of star cluster dynamics on BH binaries.

Hardening by three-body encounters is thus a crucial mechanism to enhance the merger rate of BH binaries. It is even possible to make a simple analytic estimate of the evolution of the semimajor axis a of a double BH binary that is affected by three-body encounters and by gravitational wave emission (see equation 9 of Colpi et al., 2003):

$$\frac{da}{dt} = -2\pi\,\xi\,\frac{G\rho}{\sigma}\,a^2 - \frac{64}{5}\,\frac{G^3\,m_1\,m_2\,(m_1+m_2)}{c^5\,(1-e^2)^{7/2}}\,a^{-3}, \tag{17.3}$$

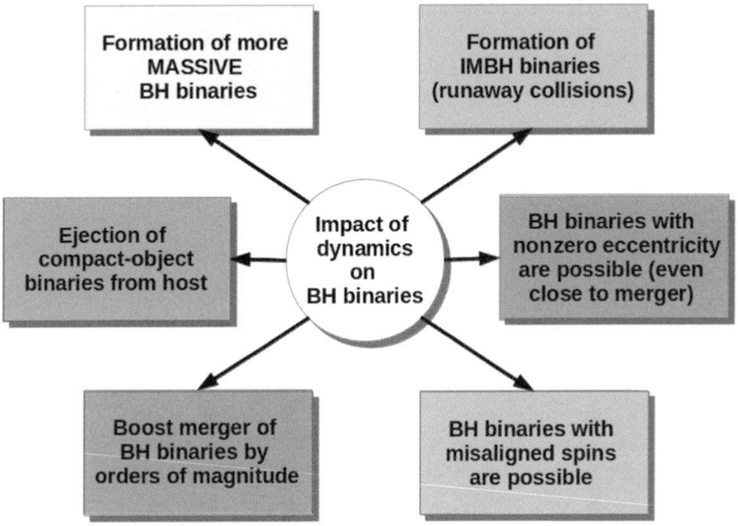

Figure 17.5 Summary of the main effects of star cluster dynamics on the formation and evolution of BH binaries.

where $\xi \sim 0.2 - 3$ is the same dimensionless parameter as defined in Equation 17.2, ρ is the local mass density of stars, σ is the local velocity dispersion, c is the light speed and e is the eccentricity of the binary. The first part of the right-hand term of Equation 17.3 accounts for the effect of three-body hardening on the semimajor axis (from Equation 17.2). It scales as $da/dt \propto a^2$, indicating that the larger the binary is, the more effective the hardening. This can be easily understood considering that the geometric cross section for three-body interactions with a binary scales as a^2.

The second part of the right-hand term of Equation 17.3 accounts for energy loss by gravitational wave emission. It is the first-order approximation of the calculation by Peters (1964). It scales as $da/dt \propto a^{-3}$, indicating that gravitational wave emission becomes efficient only when the two BHs are very close to each other.

In Figure 17.6, we solve Equation 17.3 numerically for three double BH binaries with different mass. All binaries evolve through (i) a first phase in which hardening by three-body encounters dominates the evolution of the binary; (ii) a second phase in which the semimajor axis stalls because three-body encounters become less efficient as the semimajor axis shrink, but the binary is still too large for gravitational-wave emission to become efficient; and (iii) a third phase in which the semimajor axis drops because the binary enters the regime where gravitational wave emission is efficient.

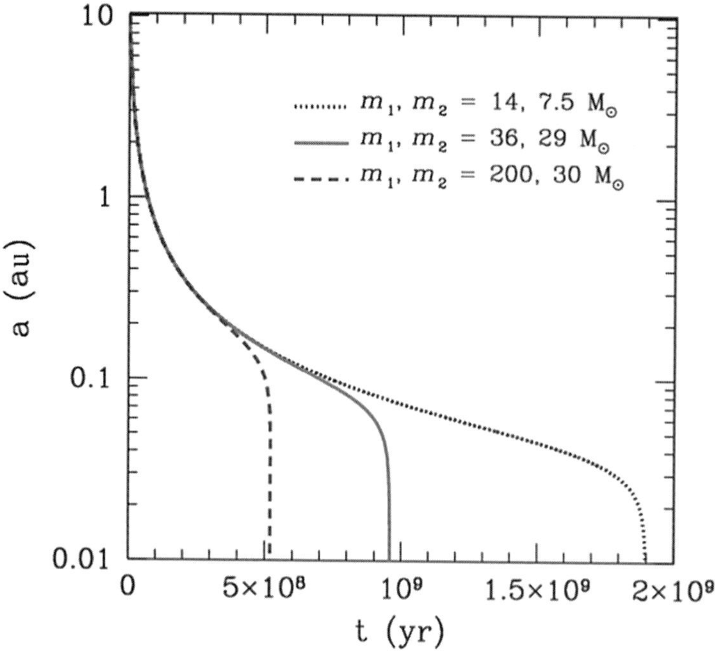

Figure 17.6 Time evolution of the semimajor axis of three BH binaries estimated from Equation 17.3. Black dashed line: BH binary with masses $m_1 = 200\ \mathrm{M}_\odot$, $m_2 = 30\ \mathrm{M}_\odot$; grey solid line: $m_1 = 36\ \mathrm{M}_\odot$, $m_2 = 29\ \mathrm{M}_\odot$; black dotted line: $m_1 = 14\ \mathrm{M}_\odot$, $m_2 = 7.5\ \mathrm{M}_\odot$. For all BH binaries: $\xi = 1$, $\rho = 10^5\ \mathrm{M}_\odot\ \mathrm{pc}^{-3}$, $\sigma = 10\ \mathrm{km\ s}^{-1}$, initial semimajor axis of the BH binary $a_\mathrm{i} = 10$ au, eccentricity $e = 0$.

17.4.3 Intermediate-Mass Black Holes

Our understanding of IMBHs is remarkably poor. We are not even sure they exist, because the observational constraints are rather scanty:

- ESO 243-49 HLX-1 is an IMBH candidate (Farrell et al., 2009; Servillat et al., 2011) with mass $\sim 10^4\ \mathrm{M}_\odot$. It is associated with a hyperluminous X-ray source (peak luminosity $\sim 10^{42}\ \mathrm{erg\ s}^{-1}$) whose light curve shows some periodicity and a fast-rise exponential decay behaviour.
- Kızıltan et al. (2017) claim that the motion of pulsars is suggestive of a $2,300^{+1500}_{-850}\ \mathrm{M}_\odot$ IMBH in the centre of 47 Tucanae.
- Other IMBH candidates are supported by the combination of spectra and photometry in globular clusters (e.g., G1; Gebhardt et al. 2005).
- LIGO and Virgo provide an upper limit to the IMBH merger rate of $R_\mathrm{IMBH} < 0.93\ \mathrm{Gpc}^{-3}\ \mathrm{yr}^{-1}$ in the range 100 to 500 $\mathrm{M}_\odot$ (Abbott et al., 2017a).

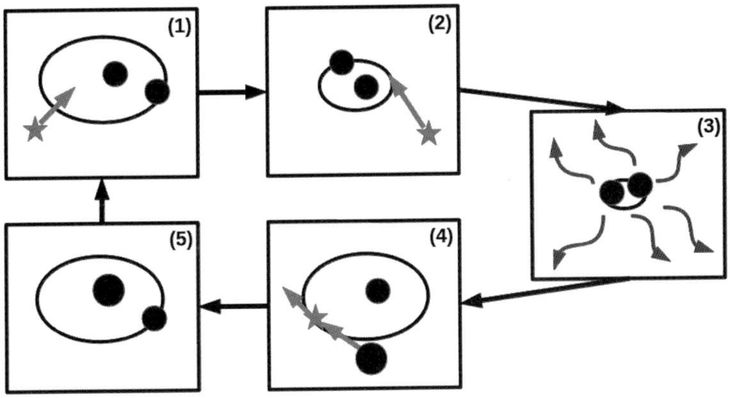

Figure 17.7 Cartoon of the repeated merger scenario in old star clusters (see, e.g. Miller and Hamilton 2002; Giersz et al. 2015). From top to bottom and from left to right: (1) a BH binary undergoes three-body encounters in a star cluster; (2) three-body encounters harden the BH binary, shrinking its semimajor axis; (3) the BH binary hardens by three-body encounters until it enters the regime where gravitational wave emission is efficient: the binary semimajor axis decays by gravitational wave emission and the binary merges; (4) a single bigger BH forms as result of the merger, which may acquire a new companion by dynamical exchange (if it is not ejected by gravitational wave recoil); (5) the new binary containing the bigger BH starts shrinking again by three-body encounters. This loop may be repeated several times until the main BH becomes an IMBH.

From the theoretical point of view, the formation channels of IMBHs are deeply connected with the dynamics of binaries. In this section, we will describe the repeated merger and the runaway collision scenario, which are two of the most popular theoretical models for the formation of IMBHs.

In the *repeated merger* scenario (Miller and Hamilton, 2002; Giersz et al., 2015), a stellar-mass BH binary in the core of a globular cluster undergoes repeated three-body encounters, hardening until it can merge by gravitational wave emission (if it is not ejected by a dynamical kick). This leads to the formation of a more massive BH, which, if it is not ejected by gravitational wave recoil, can enter a new binary by exchange. The new binary hardens by three-body encounters until it can merge by gravitational wave emission, and so on (see Figure 17.7).

The main issue of the repeated merger scenario is that BHs might be ejected by three-body kicks or by gravitational wave recoils before accreting enough mass to become IMBHs. Moreover, the assembly by repeated mergers is rather inefficient.

In contrast, the *runaway collision* scenario is expected to happen only in very young massive dense star clusters (such as R136 in the Large Magellanic Cloud), where the dynamical friction timescale for very massive stars is shorter than the time for the first supernovae to explode (Portegies Zwart and McMillan, 2002; Portegies Zwart et al., 2004). If this condition is satisfied, the most massive stars sink to the centre of the star cluster before they explode as supernovae, enhancing the central density of massive objects. Once

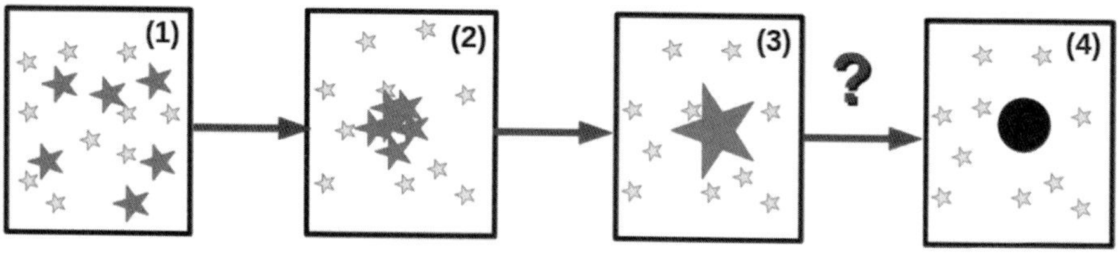

Figure 17.8 Cartoon of the runaway collision scenario in dense young star clusters (see, e.g., Portegies Zwart and McMillan, 2002). From left to right: (1) the massive stars (dark-grey big stars) and the low-mass stars (grey small stars) follow the same initial spatial distribution; (2) dynamical friction leads the massive stars to sink to the core of the cluster, where they start colliding between each other; (3) a very massive star ($\gg 100\ M_\odot$) forms as a consequence of the runaway collisions; (4) this massive star might be able to directly collapse into a BH.

in the core, the massive stars are likely to form binaries and to collide with each other by three-body encounters. This leads to the formation of a very massive star, as shown in Figure 17.8. The fate of the massive star represents the main uncertainty of the runaway collision scenario. In fact, the star might either collapse to an IMBH directly or produce a smaller remnant via a supernova explosion.

Recent studies (Mapelli, 2016; Spera and Mapelli, 2017) suggest that the formation of an IMBH is inhibited at solar metallicity, because of the strong stellar winds. At low metallicity ($\lesssim 0.1\ Z_\odot$), about 10 to 20% of the runaway collision products might produce IMBHs with mass ~ 100 to $500\ M_\odot$. Most runaway collision products are expected to form double BH binaries by exchanges, with important implications for gravitational wave detections (Mapelli, 2016).

17.5 Summary

It is hence fully demonstrated now the important role that binaries play for the dynamical evolution of a star cluster. The internal energy of binaries represents the main energy reservoir of a star cluster. When they interact via three-body and multibody encounters, binaries can exchange a fraction of their internal energy with the other stars in the cluster. In particular, hard binaries (i.e., binaries whose binding energy is larger than the average kinetic energy of a star in the star cluster) tend to transfer a part of their kinetic energy to the other stars in the star cluster (Heggie, 1975).

From this perspective, binaries act as ‘Maxwell’s demons’ of star clusters, because they change the dynamical state of a star cluster without any external energy input, just by exchanging their internal energy with stars. Three-body encounters with hard binaries are also called ‘superelastic’ encounters, because they increase the translational kinetic energy of the involved bodies.

The importance of binaries as energy reservoirs is apparent from the study of core collapse: star clusters are subject to gravothermal catastrophe because of their negative heat capacity. This leads their core to contract in a runaway sense. Three-body encounters are the most effective process in reversing core collapse, because they pump fresh kinetic energy into the core (extracted from the internal energy of hard binaries) and stop the collapse (see Spitzer, 1987, and references therein).

While the reverse of core collapse is the best-known example of the importance of binaries for star clusters, binaries play a crucial role in many other dynamical processes. For example, N-body simulations (Spera et al., 2016) show that binary hardening is a clear indicator of the onset of Spitzer's instability (Spitzer, 1969).

Finally, binaries are necessary to explain the formation of stellar exotica, i.e., stellarlike objects that cannot be produced (or can hardly be produced) by 'simple' stellar evolution. For example, blue straggler stars are interpreted as the product of mass transfer in binaries or of stellar collisions, triggered by three-body encounters.

Even BH binaries are efficiently produced by three-body encounters, especially by dynamical exchanges (Ziosi et al., 2014). Exchanges tend to produce more massive BH binaries than isolated binary evolution, with higher eccentricity and with a nearly isotropic spin distribution. Moreover, dynamical hardening is expected to enhance the merger rate of BH binaries (Rodriguez et al., 2016; Askar et al., 2017; Banerjee, 2017). Finally, dynamical processes such as the runaway collision (Portegies Zwart et al., 2004; Mapelli, 2016) and the repeated merger scenario (Miller and Hamilton, 2002; Giersz et al., 2015) are among the most popular models to explain the formation of IMBHs. Thus, the dynamics of binaries in star clusters is a crucial ingredient to shed light on the formation channels of gravitational wave events (Abbott et al., 2016b).

References

Abbott, B. P., Abbott, R., Abbott, T. D., et al. 2016a. Observation of Gravitational Waves from a Binary Black Hole Merger. *Physical Review Letters*, **116**, 061102.

Abbott, B. P., Abbott, R., Abbott, T. D., et al. 2016b. Binary Black Hole Mergers in the First Advanced LIGO Observing Run. *Physical Review X*, **6**, 041015.

Abbott, B. P., Abbott, R., Abbott, T. D., et al. 2017a. Search for Intermediate Mass Black Hole Binaries in the First Observing Run of Advanced LIGO. *PRD*, **96**, 022001.

Abbott, B. P., Abbott, R., Abbott, T. D., et al. 2017b. GW170104: Observation of a 50-Solar-Mass Binary Black Hole Coalescence at Redshift 0.2. *Physical Review Letters*, **118**, 221101.

Abbott, B. P., Abbott, R., Abbott, T. D., et al. 2017c. GW170814: A Three-Detector Observation of Gravitational Waves from a Binary Black Hole Coalescence. *Physical Review Letters*, **119**, 141101.

Allison, R. J., Goodwin, S. P., Parker, R. J., et al. 2009. Dynamical Mass Segregation on a Very Short Timescale. *ApJ L*, **700**, L99.

Angeletti, L., and Giannone, P. 1977. Dynamical Evolution of Clusters with Stellar Mass Loss. *A&A*, **58**, 363.

Angeletti, L., and Giannone, P. 1980. *A&A*, Dynamical Evolution of Cluster Models with a Continuous Stellar Mass Loss. *A&A*, **85**, 113.

Askar, A., Szkudlarek, M., Gondek-Rosińska, D., et al. 2017. MOCCA-SURVEY Database. I. Coalescing Binary Black Holes Originating from Globular Clusters. *MNRAS*, **464**, L36.

Banerjee, S. 2017. Stellar-Mass Black Holes in Young Massive and Open Stellar Clusters and Their Role in Gravitational-Wave Generation. *MNRAS*, **467**, 524.

Beccari, G., Ferraro, F. R., Possenti, A., et al. 2006. The Dynamical State and Blue Straggler Population of the Globular Cluster NGC 6266 (M62). *AJ*, **131**, 2551.

Belczynski, K., Bulik, T., Fryer, C. L., et al. 2010. On the Maximum Mass of Stellar Black Holes. *ApJ*, **714**, 1217.

Belczynski, K., Holz, D. E., Bulik, T., et al. 2016. The First Gravitational-Wave Source from the Isolated Evolution of Two Stars in the 40–100 Solar Mass Range. *Nature*, **534**, 512.

Bianchini, P., van de Ven, G., Norris, M. A., et al. 2016. A Novel Look at Energy Equipartition in Globular Clusters. *MNRAS*, **458**, 3644.

Boltzmann, L. 1876. Über die Natur der Gasmoleküle (On the Nature of Gas Molecules). Wiener Berichte.

Bonnell, I. A., and Davies, M. B. 1998. Mass Segregation in Young Stellar Clusters. *MNRAS*, **295**, 691.

Chernoff, D. F., and Weinberg, M. D. 1990. Evolution of Globular Clusters in the Galaxy. *ApJ*, **351**, 121.

Colpi, M., Mapelli, M. and Possenti, A. 2003. Probing the Presence of a Single or Binary Black Hole in the Globular Cluster NGC 6752 with Pulsar Dynamics. *ApJ*, **599**, 1260.

Davies, M. B. 2002. Omega Centauri, a Unique Window into Astrophysics. ASP Vol. **265**, 215. http://adsabs.harvard.edu/abs/2002ASPC..265..215D.

Downing, J. M. B. 2012. Is there a size difference between red and blue globular clusters? *MNRAS*, **425**, 2234.

Elson, R., Hut, P. and Inagaki, S. 1987. Dynamical Evolution of Globular Clusters. *ARA&A*, **25**, 565.

Farrell, S. A., Webb, N. A., Barret, D., Godet, O. and Rodrigues, J. M. 2009. An Intermediate-Mass Black Hole of over 500 Solar Masses in the Galaxy ESO243-49. *Nature*, **460**, 73.

Ferraro, F. R., Lanzoni, B., Dalessandro, E., et al. 2012. Dynamical Age Differences among Coeval Star Clusters as Revealed by Blue Stragglers. *Nature*, **492**, 393.

Fujii, M. S. and Portegies Zwart, S. 2014. The Moment of Core Collapse in Star Clusters with a Mass Function. *MNRAS*, **439**, 1003.

Gebhardt, K., Rich, R. M. and Ho, L. C. 2005. An Intermediate-Mass Black Hole in the Globular Cluster G1. *ApJ*, **634**, 1093.

Giacobbo, N., Mapelli, M., and Spera, M. 2018. Merging Black Hole Binaries: The Effects of Progenitor's Metallicity, Mass-Loss Rate and Eddington Factor. *MNRAS*, **474**, 2959.

Giersz, M., Leigh, N., Hypki, A., et al. 2015. MOCCA Code for Star Cluster Simulations: IV. A New Scenario for Intermediate Mass Black Hole Formation in Globular Clusters. *MNRAS*, **454**, 3150.

Heggie, D. C. 1975. Binary Evolution in Stellar Dynamics. *MNRAS*, **173**, 729.

Hénon, M. 1961. Sur l'évolution dynamique des amas globulaires. *Annales d'Astrophysique*, **24**, 369.

Hills, J. G. 1983. The Effect of Low-Velocity, Low-Mass Intruders (Collisionless Gas) on the Dynamical Evolution of a Binary System. *AJ*, **88**, 1269.

Hills, J. G. and Fullerton, L. W. 1980. Computer Simulations of Close Encounters between Single Stars and Hard Binaries. *AJ*, **85**, 1281.

Hypki, A., and Giersz, M. 2017. MOCCA Code for Star Cluster Simulations. VI. Bimodal Spatial Distribution of Blue Stragglers. *MNRAS*, **471**, 2537.

King, I. R. 1966. The Structure of Star Clusters. III. Some Simple Dynamical Models. *AJ*, **71**, 64.

Kızıltan, B., Baumgardt, H. and Loeb, A. 2017. An Intermediate-Mass Black Hole in the Centre of the Globular Cluster 47 Tuc. *Nature*, **542**, 203.

LIGO Scientific Collaboration, the Virgo Collaboration, Abbott, B. P., et al. 2017. GW170608: Observation of a 19 Solar-Mass Binary Black Hole Coalescence. *ApJ L*, **851**, L35.

Mapelli, M. 2016. Massive Black Hole Binaries from Runaway Collisions: The Impact of Metallicity. *MNRAS*, **459**, 3432.

Mapelli, M. and Bressan, A. 2013. Impact of Metallicity on the Evolution of Young Star Clusters. *MNRAS*, **430**, 3120.

Mapelli, M., Colpi, M. and Zampieri, L. 2009. Low Metallicity and Ultra-Luminous X-Ray Sources in the Cartwheel Galaxy. *MNRAS*, **395**, L71.

Mapelli, M., Giacobbo, N., Ripamonti, E., et al. 2017. The Cosmic Merger Rate of Stellar Black Hole Binaries from the Illustris Simulation. *MNRAS*, **472**, 2422.

Mapelli, M., Sigurdsson, S., Colpi, M., et al. 2004. The Contribution of Primordial Binaries to the Blue Straggler Population in 47 Tuc. *ApJ L*, **605**, L29.

Mapelli, M., Sigurdsson, S., Ferraro, F. R., et al. 2006. The Radial Distribution of Blue Straggler Stars and the Nature of Their Progenitors. *MNRAS*, **373**, 361.

Mapelli, M., Zampieri, L., Ripamonti, E., et al. 2013. Dynamics of Stellar Black Holes in Young Star Clusters with Different Metallicities. I. Implications for X-Ray Binaries. *MNRAS*, **429**, 2298.

Maxwell, J. C. 1872. *Theory of Heat* (New York: D. Appleton and Co.).

Merritt, D. 1981. Two-Component Stellar Systems in Thermal and Dynamical Equilibrium. *AJ*, **86**, 318.

Miller, M. C. and Hamilton, D. P. 2002. Production of Intermediate-Mass Black Holes in Globular Clusters. *MNRAS*, 330, 232.

Miocchi, P. 2006. Central Energy Equipartition in Multimass Models of Globular Clusters. *MNRAS*, **366**, 227.

Parker, R. J., Goodwin, S. P., Wright, N. J., et al. 2016. Mass Segregation in Star Clusters Is Not Energy Equipartition. *MNRAS*, **459**, L119.

Peters, P. C. 1964. Gravitational Radiation and the Motion of Two Point Masses. *Physical Review*, **136**, 1224.

Portegies Zwart, S. F. and McMillan, S. L. W. 2002. The Runaway Growth of Intermediate-Mass Black Holes in Dense Star Clusters. *ApJ*, **576**, 899.

Portegies Zwart, S. F., Baumgardt, H., Hut, P., et al. 2004. Formation of Massive Black Holes through Runaway Collisions in Dense Young Star Clusters. *Nature*, **428**, 724.

Portegies Zwart, S. F., McMillan, S. L. W. and Gieles, M. 2010. Young Massive Star Clusters. *ARA&A*, **48**, 431.

Quinlan, G. D. 1996. The Time-Scale for Core Collapse in Spherical Star Clusters. *New Astronomy*, **1**, 35.

Rodriguez, C. L., Chatterjee, S. and Rasio, F. A. 2016. Binary Black Hole Mergers from Globular Clusters: Masses, Merger Rates, and the Impact of Stellar Evolution. *PRD*, **93**, 084029.

Schulman, R. D., Glebbeek, E. and Sills, A. 2012. The Effect of Stellar Metallicity on the Sizes of Star Clusters. *MNRAS*, **420**, 651.

Servillat, M., Farrell, S. A., Lin, D., et al. 2011. X-Ray Variability and Hardness of ESO 243-49 HLX-1: Clear Evidence for Spectral State Transitions. *ApJ*, **743**, 6.

Spera, M., and Mapelli, M. 2017. Very Massive Stars, Pair-Instability Supernovae and Intermediate-Mass Black Holes with the SEVN Code. *MNRAS*, **470**, 4739.

Spera, M., Mapelli, M., and Bressan, A. 2015. The Mass Spectrum of Compact Remnants from the PARSEC Stellar Evolution Tracks. *MNRAS*, **451**, 4086.

Spera, M., Mapelli, M., and Jeffries, R. D. 2016. Do Open Star Clusters Evolve towards Energy Equipartition? *MNRAS*, **460**, 317.

Spitzer, L., Jr. 1969. Equipartition and the Formation of Compact Nuclei in Spherical Stellar Systems. *ApJ L*, **158**, L139.

Spitzer, L. 1987. *Dynamical Evolution of Globular Clusters*. (Princeton: Princeton University Press).

Spitzer, L., Jr. and Hart, M. H. 1971. Random Gravitational Encounters and the Evolution of Spherical Systems. I. Method. *ApJ*, **164**, 399.

Trani, A. A., Mapelli, M. and Bressan, A. 2014. The Impact of Metallicity-Dependent Mass-Loss versus Dynamical Heating on the Early Evolution of Star Clusters. *MNRAS*, **445**, 1967.

Trenti, M. and van der Marel, R. 2013. No Energy Equipartition in Globular Clusters. *MNRAS*, **435**, 3272.

Vink, J. S., de Koter, A. and Lamers, H. J. G. L. M. 2001. Mass-Loss Predictions for O and B Stars as a Function of Metallicity. *A&A*, **369**, 574.

Vishniac, E. T. 1978. A Necessary Condition for Equilibrium in Stellar Systems with a Continuous Mass Spectrum. *ApJ*, **223**, 986

Ziosi, B. M., Mapelli, M., Branchesi, M., et al. 2014. Dynamics of Stellar Black Holes in Young Star Clusters with Different Metallicities. II. Black Hole–Black Hole Binaries. *MNRAS*, **441**, 3703.

18

Alternative Stellar Evolution Pathways

ROBERT D. MATHIEU AND EMILY M. LEINER

18.1 Introduction

Our understanding of the evolution of single stars is one of the great intellectual accomplishments of the past century. This accomplishment relied heavily on star clusters: for the delineation of sequences in the properties of stars, as ongoing testbeds for ever more sophisticated models and as exemplars for education.

Nonetheless, from their beginnings – e.g., the Johnson and Sandage (1955) colour-magnitude diagram of M67 – cluster colour-magnitude diagrams (CMDs) have shown many stars that do not lie along their main sequences and giant branches. The most famous of these are the blue stragglers, but there are numerous others as well. As research focused on the evolution of single stars, consideration of these stars as anomalies became the norm.

In fact, a quarter of the evolved[1] *stars in older open clusters do not fall on single-star isochrones.* In Figure 18.1, we show a modern CMD of M67, including only three-dimensional kinematic cluster members. Fully 25% of the evolved stars do not fall on the subgiant or giant branches of an M67 isochrone.[2] Rather than being anomalies, these stars are on frequently travelled alternative pathways of stellar evolution. Indeed, these are necessary evolutionary pathways of the many solarlike stars in M67 that are in binaries with periods less than 10^4 days (Geller et al., 2015; Geller et al., in preparation).

The overarching goal of our team's research program has been to define and understand the products of these alternative stellar evolution pathways, using open clusters as well-populated, accessible laboratories. These narratives will start from the large populations of hard binary stars in open clusters. Their accounts will include both evolutionary processes internal to these binaries and external stellar dynamical influences.

Blue stragglers are the best-known examples of alternative stellar evolution pathways, but the M67 CMD shows an array of cases reflecting a variety of such paths. For example:

- S1082 is a remarkable system including two blue stragglers: one is the 2.5 $M_\odot$ primary of a very close eclipsing binary and the other is a (likely) tertiary companion (Sandquist et al., 2003; van den Berg et al., 2001). The secondary of the eclipsing binary – while photometrically located on the upper main sequence – has a mass of 1.6 $M_\odot$, in excess of

[1] We use 'evolved' broadly to mean stars that photometrically are not on the single- or binary-star main sequence.

[2] Possible photometric binaries with subgiant or giant primaries are counted as consistent with the isochrone.

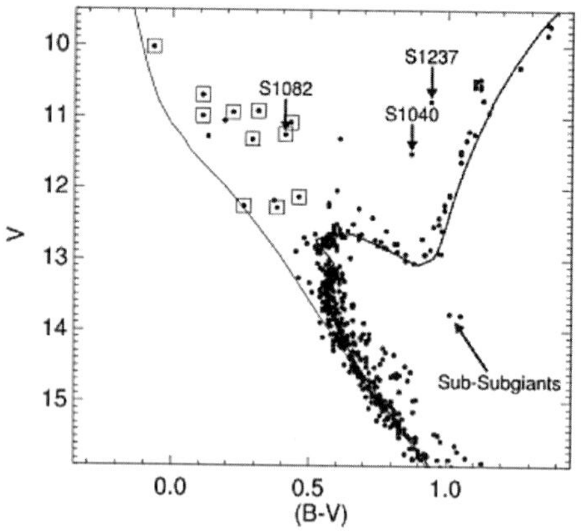

Figure 18.1 CMD of three-dimensional kinematic members of M67 (4 Gyr). Squares identify spectroscopic binaries among the blue stragglers. The stars identified with arrows also cannot be explained by standard single-star stellar evolution models and are described in the text. The curves are the zero-age main sequence (ZAMS) and a 4 Gyr isochrone. (Adapted from Geller, Latham and Mathieu, 2015.)

the turnoff mass of 1.3 $M_\odot$. Combined, these masses ensure that formation of this system involved dynamical encounters of multiple binary stars, with some combination of mass transfer, mergers and collisions. S1082 shows that alternative evolution pathways can lie at the interface of stellar evolution and stellar dynamics.

- The two sub-subgiants are binaries with periods of 2.8^d and 18.4^d (Mathieu et al., 2003). We and others (e.g., Albrow et al., 2001; Hurley et al., 2005; Leiner et al., 2017) have conjectured that sub-subgiant evolution paths involve mass-transfer or encounter processes that rapidly reduce the radii of the primaries. Alternatively, they may be the consequence of tidally induced rapid rotation and consequent magnetic reduction of convective efficiency and luminosity (Leiner et al., 2017).
- S1040 is a 40^d-period yellow-giant binary whose position blueward of the giant branch cannot be explained by light from the faint secondary (Mathieu et al., 1990). The insightful application of tidal circularisation theory by Verbunt and Phinney (1995) suggested that the system is a mass-transfer product including a white dwarf companion, as then found by Landsman et al. (1997). Landsman et al. proposed a still-untested alternative evolutionary path in which the current primary star is an evolved blue straggler in a post–mass-transfer system.
- S1237 is also a yellow-giant binary, with an orbital period of 698^d. The frequency analysis of *Kepler* K2 light curves yielded a radius of 9.2 ± 0.2 $R_\odot$ and a mass of 2.9 ± 0.2 $M_\odot$ (Leiner et al., 2016). This result would suggest S1237 to be an evolved blue straggler with a formation history requiring multiple stars. Observed X-ray emission from this star has not been explained, given the small projected rotation velocity of the primary. Perhaps there is a rapidly rotating secondary, which would be unusual at the age of this cluster and a clue to the system formation.

The M67 cluster is an exemplar, but not unique. Our team has under way an integrated program of observational and theoretical study to define and understand these alternative pathways of stellar evolution. We have characterised in detail the products of alternative stellar evolution pathways in five older open clusters (t > 1.6 Gyr; 1.1 $M_\odot$ < $M_{turnoff}$ < 1.7 $M_\odot$); we describe our discoveries in NGC 188 and NGC 6819 in the next section. We have also taken first steps in modelling the mass transfer and the dynamical processes that yield these alternative stellar evolution products (Geller et al., 2013; Gosnell et al., 2014, 2015; Leiner et al., 2017).

Crucially, we also have characterised in these open clusters the populations of hard binaries, the sites of alternative stellar evolution (e.g., Geller and Mathieu, 2012; Milliman et al., 2014, 2016; Geller et al., 2015; Leiner et al., 2015, Geller et al., in preparation). One-third of cluster solar-like stars are in fact the primary stars of binaries with periods of less than 10^4 days; stars in binaries with such periods, and somewhat longer, will interact as they evolve. Frequent stellar evolution along alternative pathways is inevitable.

18.2 Alternative Stellar Evolution Pathways: The Blue Stragglers of NGC 188 and NGC 6819

Recent work on blue straggler formation has focused on several possible formation pathways: mass transfer on the red giant branch (RGB) or asymptotic giant branch (AGB) (Chen and Han, 2008), collisions during dynamical encounters (Knigge et al., 2009; Leigh and Sills, 2011), mergers due to angular momentum loss (Andronov et al., 2006) and mergers in hierarchical triples (Perets and Fabrycky, 2009). An observational challenge has been to securely discriminate among these channels.

We have done an intensive study of the rich blue straggler population (Figure 18.2a) in NGC 188 (7 Gyr, turnoff mass 1.1 $M_\odot$, [Fe/H]=0.0), yielding a wealth of discoveries and astrophysical insight into blue straggler formation. Specifically, we found the NGC 188 blue stragglers to be characterised by the following:

- A binary frequency of 80% for orbital periods less than 10^4 days (Figure 18.2a); the main-sequence binary frequency is 29%.
- Typical orbital periods around 1,000 days (Figure 18.2b).
- Typical secondary-star masses of 0.5 $M_\odot$ (Figure 18.2c), indicative of white dwarf companions.
- Modestly rapid rotation, upwards to 100 km s^{-1}.
- A bimodal radial spatial distribution.
- Dynamical masses greater than standard stellar evolution masses (based on close binaries).

These findings are reviewed in Mathieu and Geller (2015), with complete references there. Several of these properties are found in the M67 and the field blue stragglers as well (Carney et al., 2001; Latham, 2007).

All of these observations strongly, but indirectly, indicated that many long-period blue straggler binaries – which dominate the NGC 188 blue straggler population – are formed by AGB or occasionally RGB mass transfer onto main-sequence stars.

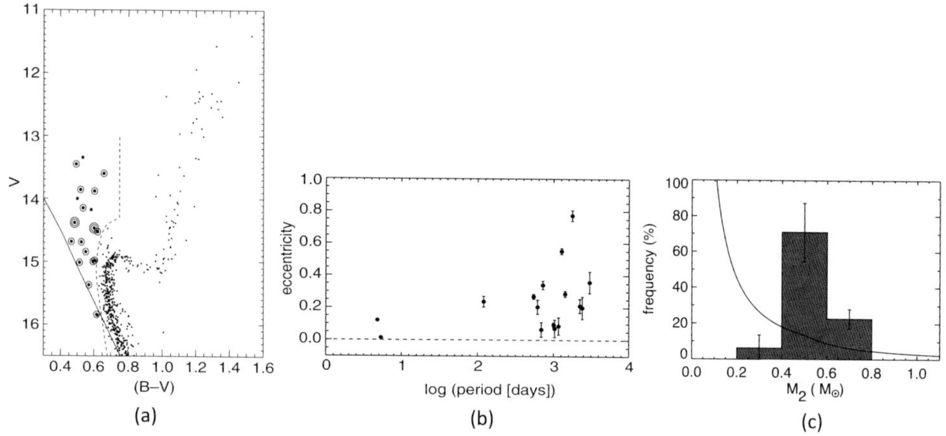

Figure 18.2 (a) CMD of the 7 Gyr cluster NGC 188, with binary blue stragglers circled (double-lined binaries have double circles); Mathieu and Geller, 2015; (b) blue straggler period-eccentricity distribution (Mathieu and Geller, 2009); (c) statistical blue straggler secondary-mass distribution, compared to the Kroupa (2001) initial mass function (adapted from Geller and Mathieu, 2011).

Our subsequent *HST* ACS/SBC photometry immediately revealed hot white dwarf (WD) companions to three long-period blue stragglers in NGC 188 (Gosnell et al., 2014). These WDs were conclusive evidence that these blue stragglers formed through mass transfer, the WDs being the cores of the evolved primaries of the progenitor binaries. Furthermore, combined with cooling models the WD temperatures directly provide the time since mass transfer ended. With effective temperatures higher than 12,000 K, all three WDs formed within the past 300 Myr, extremely recent in comparison to the 7 Gyr age of NGC 188. These three blue stragglers formed 'yesterday'.

These very short times since the end of mass transfer place strong constraints on the primary-star mass and the orbital period of the progenitor binary, and provided the first opportunity to empirically describe in some detail the formation story of specific blue stragglers. Interestingly, these three blue stragglers may have formed through three qualitatively different mass-transfer histories (Gosnell et al., 2014): Case B with an RGB star in Roche Lobe overflow (RLO); Case C with an AGB star in Roche-lobe overflow; and efficient wind accretion from an AGB star, sometimes referred to as Case D mass transfer.

Further analysis of the individual ultraviolet fluxes from the entire NGC 188 blue straggler population produced another direct detection of a WD companion with a temperature of 13,500 K and statistically three additional cooler WD companions with effective temperatures between 11,000 K and 12,000 K (Gosnell et al., 2015). The presence of seven WD companions sets a lower limit on the frequency of mass-transfer formation of blue stragglers at 33%. The remaining eight long-period blue straggler binaries likely also include mass-transfer cases, but with WDs currently too cool to detect photometrically. Statistical models and the frequency of long-period orbits indicate a mass-transfer formation frequency of $\approx$67%.

The formation paths for the rest of the blue stragglers are not as clearly defined. However, the two short-period ($P < 6^d$; Figure 18.2b) double-lined blue straggler binaries again point to the influence of dynamical encounters. One of these binaries contains two blue stragglers and is assuredly the result of a binary encounter in the cluster, perhaps placing together two mass-transfer blue stragglers or leading to at least one collisional-origin blue straggler (Mathieu and Geller, 2009; Geller and Mathieu, 2012). Finally, our N-body models showed that the few blue stragglers without detected velocity variation could be formed from collisions in resonant binary encounters or mergers in triples (Geller et al., 2013). Such formation processes are also suggested by their observed rotation distributions (discussed in the next section).

Finally, we note that the distance of the NGC 188 blue stragglers from the ZAMS is not necessarily aligned with age since formation. Since WD temperature correlates with age, Gosnell et al. (2015) used the CMD location of blue stragglers with detected WD companions to probe relationships between blue straggler age and optical luminosity and colour. In Figure 18.3, several of the youngest blue stragglers (dark blue diamonds) are on the ZAMS. Blue straggler binaries along the ZAMS include two of the blue stragglers with hot WD companions, two of the blue stragglers with cool WD companions and an additional binary with a 10% probability of having a WD companion. Based on the detected WD temperatures, the spread of ages along the ZAMS is 77 to 360 Myr (Holberg and Bergeron, 2006; Tremblay et al., 2011), and may extend to older ages if the additional binary has a cold WD companion.

At the same time, two of the youngest mass-transfer blue stragglers lie well off the ZAMS. These results suggest that blue straggler products along the ZAMS may be among the youngest in the population, but distance from the ZAMS is not necessarily equivalent with age since formation. One must use caution when inferring the age of blue stragglers from single-star isochrones, especially for blue stragglers that lie away from the ZAMS.

Interestingly, three of the four non-velocity variable blue stragglers are also among the most optically luminous of the population (Gosnell et al., 2015). If these blue stragglers formed through collisions, they would likely be among the most massive blue straggler stars (BSSs), perhaps accounting for their high optical luminosity (Leigh and Sills, 2011). And yet, they are not more luminous than the most luminous mass transfer–formed blue stragglers. There is not a clear colour or luminosity separation between mass-transfer blue stragglers and possible collision BSSs in the optical CMD.

The blue stragglers of the intermediate-age (2.5 Gyr) cluster NGC 6819 have effective temperatures of 6,500 K to 8,500 K, compared to 6,200 K to 6,750 K in NGC 188. With a turnoff mass of 1.4 $M_\odot$, NGC 6819 allows exploration of alternative evolution pathways for stars with substantially different internal structures than the turnoff stars of NGC 188.

For example, AGB stars in NGC 6819 have thermal pulses that lead to the production of s-process elements that can be tracers of mass transfer. In Figure 18.4 (left panel), we show the discovery of barium excesses among five – but not all – of the NGC 6819 blue

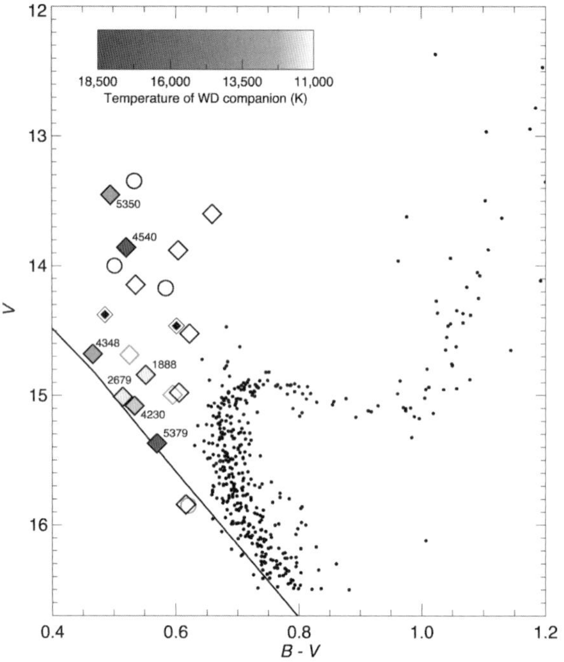

Figure 18.3 Optical CMD of NGC 188 cluster members with the blue straggler population highlighted according to binarity and the measured temperature of WD companions. The solid black line is the ZAMS for NGC 188. Binary blue stragglers are shown as diamonds, and nonvelocity variable blue stragglers are shown as large circles. The blue stragglers with WD detections are shown with a colour from dark grey to light grey representing the temperature of the WD companion, as indicated with the colour bar. The sources outlined in grey (two SB1 binaries, one single blue straggler) are the three sources without a 3σ ultraviolet detection. The outlined solid black diamonds are the two double-lined blue straggler binaries. (From Gosnell et al., 2015. Reproduced by permission of the AAS.)

stragglers (Milliman et al., 2015). We have not found similar barium excesses among blue stragglers in the older open clusters (including those of definitive mass-transfer origin; Milliman, 2016), nor have they been found in globular cluster blue stragglers. We argue that the discovery of barium excesses only in an intermediate-age cluster is the result of s-process formation happening only in AGB stars with masses above 1.3 $M_{\odot}$ (e.g., Cristallo et al., 2011), and that their discovery extends the narrative of mass-transfer origins to these higher-mass stars.

Remarkably, though, these barium-excess stars are not in long-period binaries as is typical for barium stars (Jorissen et al., 1998) and as expected from the mass-transfer formation scenario. Also surprising is that the excesses are only found within a specific domain of the CMD (Figure 18.4, right panel). Understanding the implications of these new abundance results for mass-transfer evolution pathways is an outstanding goal for theoretical work.

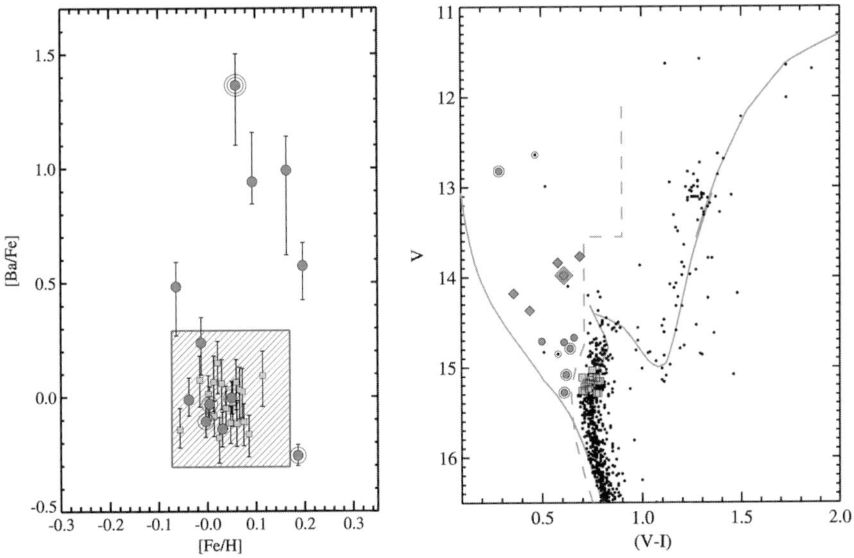

Figure 18.4 Left: [Ba/Fe] as a function of [Fe/H] for 18 main-sequence stars (light grey squares) and 12 blue stragglers (grey circles) in NGC 6819. The box is the region within 3 σ of the average values for the main-sequence stars. Five blue stragglers are seen to have significant barium enhancement. Right: NGC 6819 CMD using the same symbols, but with barium enhanced blue stragglers shown with light grey diamonds. Also shown are the ZAMS and a 2.5 Gyr isochrone (Milliman et al., 2015).

18.3 Alternative Stellar Evolution Pathways: Across the CMD: Yellow Giants and Sub-Subgiants

18.3.1 Yellow Giants

The name *yellow giants* has a varied history in the literature. Here we define yellow giants as stars to the blue of the giant branch of an open cluster. In some cases, such stars are simply photometric binaries combining a giant and a main-sequence star (or a blue straggler). However, there is increasing evidence that many yellow giants are on alternative evolutionary pathways, likely as evolved blue stragglers.

That high-level statement belies their complex evolutionary pathways, as yet not fully modelled in even one case. As one example, to explain S1040 in M67 (Figure 18.1), Landsman et al. (1997) propose a progenitor binary comprised of 1.24 $M_{\odot}$ and 0.82 $M_{\odot}$ stars and a period of 1.86 days. As the primary evolves, stable mass transfer yields a 42-day-period binary with a 1.48 $M_{\odot}$ primary and a 0.25 $M_{\odot}$ helium WD. Notably, based on the WD cooling model, the blue straggler primary needs to evolve into a yellow giant only 75 Myr from the end of mass transfer. This time interval is very short, alternatively suggesting the progenitor secondary was very close to the turnoff at formation and the hydrogen-depleted core was not rotationally remixed as some blue straggler models suggest (see, for example, Sills et al., 2001). This system calls out for an updated analysis of its evolution.

Another yellow giant in M67 is S1237, which Leiner et al. (2016) argue is a helium-burning descendent of a blue straggler. Leiner et al. find a radius of 9.2 $\pm$ 0.2 $R_\odot$ and a mass of 2.9 $\pm$ 0.2 $M_\odot$ from asteroseismic scaling relations. Given the 1.3 $M_\odot$ turnoff mass in M67, such a star requires a collision or merger of at least three cluster stars to form. The star's photometry, however, is more consistent with the evolutionary track of a 2.3 to 2.4 $M_\odot$, 3 σ below the scaling relation mass. Subsequent frequency modelling of the star has now revised the asteroseismic mass down to 2.4 $M_\odot$, consistent with this photometry and possibly compatible with a formation history involving just two stars (Leiner et al., in preparation).

Finally, no successful model has been put forward for S1072, a 1,000^d-period binary 1.2 mag above the cluster turnoff (Figure 18.1; Mathieu and Latham, 1986; Nissen et al., 1987). Unexpectedly, this star lies in the Hertzsprung gap.

The yellow giants of M67 are exemplars of an ever-increasing number of over-massive giants being discovered with the *Kepler* satellite. Corsaro et al. (2012) identified at least two overmassive giants in NGC 6819 and one in NGC 6791 from the first 19 months of *Kepler* data. A subsequent individual frequency analysis in NGC 6819 using the full *Kepler* time series revealed eight of 50 giants studied were more massive than typical for the cluster (Handberg et al., 2017). In addition, they discovered one undermassive giant hypothesised to have gone through an episode of enhanced mass loss. An updated analysis under way in NGC 6791 using the full *Kepler* time series and improved kinematic membership information yields four more overmassive giants that are kinematic cluster members. These detections indicate yellow giants are not the only evolved blue stragglers; many more exist in clusters that photometrically blend with the red giant branch. Asteroseismology can allow us to detect these populations.

18.3.2 Sub-Subgiants

Sub-subgiants were first recognised in M67 (Figure 18.1; Belloni et al., 1998; Mathieu et al., 2003) and 47 Tuc (Albrow et al., 2001, where they were called red stragglers). Geller et al. (2017a; see also Milliman et al., 2016) compile nine more sub-subgiants in the open clusters NGC 188, NGC 6791 and NGC 6819. Eight of these sub-subgiants are binaries; six have orbital solutions, with periods less than $\approx 10^d$. Another in NGC 188 is a W UMa binary.

Geller et al. (2017a) recently assembled the first catalog of sub-subgiants from the literature, including over 70 stars in both open and globular clusters. The ever-increasing identifications of sub-subgiants in open and globular clusters, and possibly in the field (Huber et al., 2014), point towards sub-subgiants as a class of alternative stellar evolution pathways.

Broadly, these stars share several characteristics: X-ray emission; Hα emission; photometric variability; and when radial-velocity information is available, binary companions with orbital periods less than 20^d. These similarities strongly suggest sub-subgiants are a newly discovered class of objects resulting from binary interaction processes. Furthermore, sub-subgiants resemble RS CVn binaries (Hall, 1976). These are binary systems with subgiant or giant primaries that have been tidally spun up to large rotation velocities by a close binary companion. This rapid rotation gives rise to strong magnetic fields, yielding X-ray emission, Hα emission and large starspots. In a few cases (e.g. Hummel et al., 2017), RS CVn stars have been suggested to be underluminous.

Leiner et al. (2017) developed and studied three hypotheses of sub-subgiant formation. These fall into two domains: either sub-subgiants have lost substantial envelope mass, or internal magnetic fields create cooler, underluminous giants.

The luminosities of the sub-subgiants of M67 and NGC 6791 could correspond to giants a few 10ths of a solar mass below the mass of a standard cluster giant. Roche-lobe overflow in a binary is a candidate channel to create such undermassive subgiant or giant stars. However, the original secondary star has to be of low enough mass so as to not outshine the sub-subgiant after receiving the transferred mass. Leiner et al. (2017) find that stable mass transfer with such lower-mass secondary stars can only occur early on the subgiant track. Models predict the resulting sub-subgiant systems to be in substantially closer orbits than observed. Furthermore, only a very small number of progenitor binary systems are in such close orbits, so this mechanism is unable to produce the observed numbers of sub-subgiants in M67 and NGC 6791. These challenges can be reduced if significant mass loss from the primary star occurs well before the onset of RLO, for example via a tidally enhanced wind (Tout and Eggleton, 1988). Such mass loss would lower the mass ratio and permit stable mass transfer in wider subgiant binaries. However, a tidally enhanced wind would have to be much more efficient than the standard prescription for donors well within their Roche lobes.

Alternatively, stripping of stellar envelopes on dynamical timescales yields a rapid decrease in luminosity, and subsequent evolution below the current subgiant branch to a lower-mass red giant branch. A possible mechanism is a close dynamical encounter in which a giant is rapidly stripped of envelope material by tidal forces but avoids a full collision. The incoming star would have to be captured as the sub-subgiant binary companion. Leiner et al. (2017) find the frequency of such encounters in open clusters to be very low. However, in denser globular clusters this may be a viable mechanism (Geller et al., 2017b). Indeed, encounters with black holes or neutron stars in globular clusters may produce such stars (Ivanova et al., 2017). There may be other ways for a subgiant to rapidly lose substantial envelope mass. For example, if a binary system with a subgiant primary went through a period of common-envelope evolution, ejecting some but not all of a subgiant's envelope material before the mass transfer stabilised, the remaining system might resemble a sub-subgiant. Hurley et al. (2005) create a sub-subgiant in an N-body simulation of M67 using a similar mechanism.

Sub-subgiant stars may also be of normal mass but have inflated radii, cooler surface temperatures and lower luminosities due to strong magnetic fields inhibiting convection. Such phenomena have been observed among low-mass eclipsing binaries, although whether this is a magnetic effect is still a matter of debate (Torres et al., 2006; Morales et al., 2008; Feiden and Chaboyer, 2013). Leiner et al. (2017) use lowered mixing-length models to mimic the effects of magnetic fields on stellar evolutionary tracks (following Chabrier et al. 2007). While at most points along standard evolutionary tracks the luminosity is unchanged regardless of mixing length, lowered mixing-length models are significantly underluminous at the base of the red giant branch. This is the observed location of the known sub-subgiants in M67 and NGC 6791. While first results are intriguing, significant advances in magnetic stellar evolution modelling are needed to produce more accurate physical models of these stars.

18.4 Alternative Stellar Evolution Pathways: Within the Main Sequence

Binary mass transfer, mergers and collisions can also create stellar products with masses below the main-sequence turnoff (e.g., Chen and Han, 2008; Geller et al., 2013). These lower-luminosity counterparts to the classical blue stragglers would blend photometrically with the cluster main sequence, making them difficult to detect.

A way to identify these stars can be through rotation signatures. Interaction products should be spun up to large rotation velocities at formation (Sills et al., 2005). Indeed, blue stragglers are sometimes observed with projected rotational velocities over 100 km s^{-1} (e.g., Shara et al., 1997; De Marco et al., 2004). Field main-sequence stars with young white dwarf companions (implying recent mass-transfer formation) have also been observed to have large rotation rates (Jeffries and Stevens, 1996). In old open clusters, solar-type main-sequence stars typically rotate quite slowly (e.g., $P_{rot} \sim 25^d$ days for stars in M67 at 4 Gyr; Barnes et al., 2016). Main-sequence stars with significantly faster rotation rates, but absent close tidally locked companions, could be products of recent stellar interactions.

We have piloted this approach in M67 using *Kepler* K2 campaign data to V = 16.5 (Leiner et al. 2019, in preparation), yielding 11 rapidly rotating single stars on or near the cluster main sequence (Figure 18.5). These stars all have rotation periods under 8 days, compared to the 20 to 30 days typical for main-sequence cluster members. A few of these 13 stars are found slightly bluer or brighter than the main-sequence locus. Geller et al.

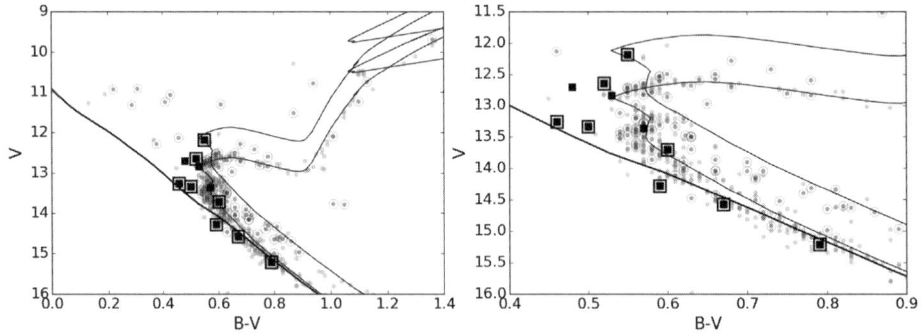

Figure 18.5 Left: CMD of three-dimensional kinematic members of M67 (Geller et al., 2015). Light grey points indicate rapidly rotating main-sequence stars. Binary systems are circled. Right: Same as left, but rescaled to better show the M67 main sequence.

(2015) do not classify these borderline cases as blue stragglers because they are either very close to the blue hook or reside in a region above the turnoff that can to contain binaries composed of two normal main-sequence turnoff stars. The detection of rapid rotation in these cases suggests that these stars are in fact an extension of the blue straggler sequence, and indeed show a continuum of blue stragglers from the classical population down into the main sequence.

This population shares several noteworthy characteristics with the classical blue straggler population. Nine of the 13 are in binary systems with long-period orbits ($P_{orb} > 100^d$). This binary fraction is much higher than the overall M67 main-sequence binary frequency of 25% ($P < 10^4$ days; Geller et al., 2015), and is comparable to that found among blue straggler populations of older open clusters (Mathieu and Geller, 2009). Two of these systems are in long-period circular binaries, very unusual among standard main-sequence binaries but explainable given mass transfer from an RGB or an AGB star in a binary system. While most of the primary stars are too hot to permit photometric detection of >100 Myr WD companions in GALEX data, one of the coolest systems has a significant far-UV excess in GALEX consistent with a ∼300 Myr WD companion (N. Gosnell, private communication).

Stellar merger and mass-transfer products have been identified with spectral signatures such as barium, carbon, oxygen or lithium abundance anomalies (e.g., Jorissen et al., 1998; Ferraro et al., 2006; Aoki et al., 2008; Milliman et al., 2015; Shetrone and Sandquist, 2000). An abundance survey of these stars (in low vsin *i* cases), and indeed the entire main-sequence population searching for additional candidates, would be well taken.

With 13 candidates, this newly discovered population is nearly as numerous as the classical blue straggler population of M67, suggesting photometric studies may have been blind to as many as half the mass-transfer, merger and collision population. In the era of high-precision, time-series photometric surveys, photometric rotation periods may be a key to identifying extensions of blue straggler populations into main sequences.

18.5 Alternative Stellar Evolution Pathways: Angular Momentum Evolution

Observations of blue stragglers reveal that many are rotating rapidly, with vsin *i* over 100 $\mathrm{km\,s^{-1}}$ (e.g., Shara et al., 1997; De Marco et al., 2004; Lovisi et al., 2010). Likely these stars have been spun up to large rotational velocities by mass transfer, mergers or collisions. In general, such spin-up is consistent with theoretical models (Packet, 1981; Sills et al., 2001), though very little work has been done to compare rotational distributions among blue straggler populations to these models. Indeed, angular momentum evolution is a long-standing problem in binary evolution physics, one very much in need of new observational constraints. The extensive radial-velocity archive of the WIYN Open Cluster Study (WOCS) can provide vsin *i* measurements for the blue straggler populations of several open clusters down to a limiting vsin *i* of 10 km s (Geller et al., 2008), providing an ideal data set to study blue straggler rotational evolution.

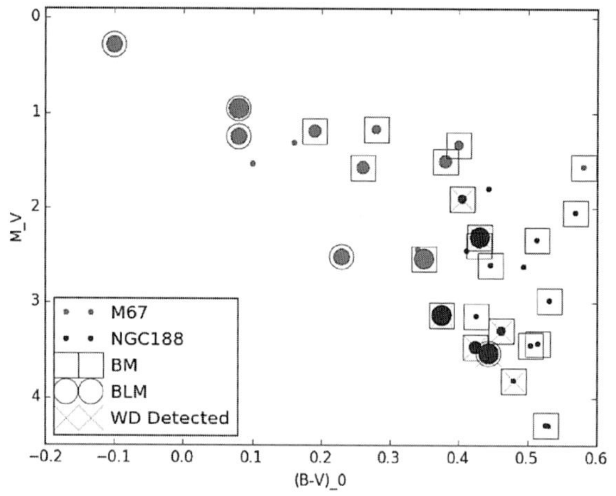

Figure 18.6 CMD of blue stragglers in M67 (grey) and NGC 188 (dark grey). The point size indicates rotation rates from largest (vsin i >100 km s^{-1}) to smallest (vsin i < 10 km s^{-1}). Binary members or likely members are boxed or circled, respectively. NGC 188 blue stragglers with hot white dwarf companions are indicated with Xs (Gosnell et al., 2015).

In Figure 18.6, we show a CMD of the blue straggler populations of the old open cluster M67 and NGC 188, with projected rotation velocity also indicated. Several interesting features are apparent. In both clusters, binary blue stragglers (boxed or circled points) are rotating more rapidly than single blue stragglers. This may suggest that merger or collisional blue stragglers (more likely to be single stars) either are not spun up to large velocities at formation or are able to shed their excess angular momentum quickly (e.g., due to disk locking or outflowing winds; Sills et al., 2005). In contrast, the binary blue stragglers (more likely of mass-transfer origin) are often found to be rapidly rotating with velocities in excess of 10 km s^{-1}, implying significant angular momentum accretion at formation and perhaps slower spin-down.

In NGC 188, seven binary blue stragglers have photometrically detected WD companions, implying ages < 400 Myr. The other binary blue stragglers do not have detectable WDs; if WD companions are present, they have ages of > 400 Myr. The young blue stragglers are rapid rotators, while the older blue stragglers display rotation rates below the WOCS vsin i measurement limit of 10 km s^{-1}. These results directly support the picture of significant spin-up during mass transfer, followed by a slow spin-down. Based on the ensemble of young blue stragglers, this spin-down is consistent with rotation models developed for single main-sequence stars (e.g., Gallet and Bouvier, 2013, Leiner et al., 2018). As noted earlier, the youngest and most rapidly rotating blue stragglers are not necessarily on the ZAMS. Perhaps rotation provides a more reliable diagnostic for age than CMD position.

18.6 Closing Thoughts

One of the most notable developments of the past year has been the recognition of similarities across a wide range of objects not evidently associated at first glance – blue stragglers, barium stars, carbon stars, carbon-enhanced metal-poor stars (CEMP-s), post-AGB stars and planetary nebula cores. (See Chapters 20, 7 and 8, respectively). Perhaps most striking has been their very similar e-log P diagrams (Figure 18.2b). Very likely, all of these classes of stars harken back to mass transfer on to main-sequence stars.

Blue stragglers in open clusters have the potential to yield deep insight into the processes forming all of these stars. For the NGC 188 blue stragglers with WD detections in particular, we know the progenitor donor mass, the progenitor core mass and evolutionary state, the progenitor donor radius and the progenitor Roche lobe (providing constraints on the progenitor orbit) and have limits on the progenitor accretor mass (less than turnoff mass, greater than the blue straggler mass and less than the donor envelope mass). We also know the final orbital parameters, the final accretor luminosity and effective temperature, the WD mass (from *HST* COS spectra) and temperature and the formation time. This is a tremendous amount of information and constraints on mass-transfer modelling, which in turn will reflect back on mass-transfer physics.

Finally, detailed analysis of vsin *i* and photometric period distributions of blue stragglers in WOCS clusters will reveal very soon new insights into binary angular momentum evolution. Considering the many improvements to single-star angular momentum models in the past decade, the time has come for new investigations into the rotational evolution of these mass-transfer, merger and collision products.

18.7 Acknowledgements

The findings reported here represent the contributions of many people who began as graduate students at Wisconsin. We want to especially acknowledge the intellectual leadership of Aaron Geller, Natalie Gosnell and Katelyn Milliman. We also want to thank the many Wisconsin undergraduate and graduate students and the WIYN Observatory staff who have helped acquire the 100,000 WIYN/Hydra spectra on which this work is based. Lastly, we acknowledge with deep appreciation the funding support from the National Science Foundation NASA, Sigma Xi Grants-in-Aid of Research and the University of Wisconsin–Madison.

References

Albrow, M. D., Gilliland, R. L., Brown, et al. 2001. The Frequency of Binary Stars in the Core of 47 Tucanae. *ApJ*, **559**, 1060.

Andronov, N., Pinsonneault, M. H. and Terndrup, D. M. 2006. Mergers of Close Primordial Binaries. *ApJ*, **646**, 1160.

Aoki, W., Beers, T. C., Sivarani, et al. 2008. Carbon-Enhanced Metal-Poor Stars. III. Main-Sequence Turnoff Stars from the SDSS SEGUE Sample. *ApJ*, **678**, 1351.

Barnes, S. A., Weingrill, J., Fritzewski, et al. 2016. Rotation Periods for Cool Stars in the 4 Gyr Old Open Cluster M67, the Solar-Stellar Connection, and the Applicability of Gyrochronology to at Least Solar Age. *ApJ*, **823**, 16.

Belloni, T., Verbunt, F. and Mathieu, R. D. 1998. X-Rays from Old Open Clusters: M 67 and NGC 188. *A&A*, **339**, 431.

Carney, B. W., Latham, D. W., Laird, J. B., et al. 2001. A Survey of Proper-Motion Stars. XIV. Spectroscopic Binaries among Metal-Poor Field Blue Stragglers. *AJ*, **122**, 3419.

Chabrier, G., Gallardo, J. and Baraffe, I. 2007. Evolution of Low-Mass Star and Brown Dwarf Eclipsing Binaries. *A&A*, **472**, L17.

Chen, X. and Han, Z. 2008. Mass Transfer from a Giant Star to a Main-Sequence Companion and Its Contribution to Long-Orbital-Period Blue Stragglers. *MNRAS*, **387**, 1416.

Corsaro, E., Stello, D., Huber, D., et al. 2012. Asteroseismology of the Open Clusters NGC 6791, NGC 6811, and NGC 6819 from 19 Months of Kepler Photometry. *ApJ*, **757**, 190.

Cristallo, S., Piersanti, L., Straniero, O., et al. 2011. Evolution, Nucleosynthesis, and Yields of Low-Mass Asymptotic Giant Branch Stars at Different Metallicities. II. The FRUITY Database. *ApJ S*, **197**, 17.

De Marco, O., Lanz, T., Ouellette, J. A., et al. 2004. First Evidence of Circumstellar Disks around Blue Straggler Stars. *ApJ L*, **606**, L151

Feiden, G. A., and Chaboyer, B. 2013. Magnetic Inhibition of Convection and the Fundamental Properties of Low-Mass Stars. I. Stars with a Radiative Core. *ApJ*, **779**, 183.

Ferraro, F. R., Sabbi, E., Gratton, R., et al. 2006. Discovery of Carbon/Oxygen-Depleted Blue Straggler Stars in 47 Tucanae: The Chemical Signature of a Mass Transfer Formation Process. *ApJ L*, **647**, L53.

Gallet, F. and Bouvier, J. 2013. Improved Angular Momentum Evolution Model for Solar-Like Stars. *A&A*, **556**, A36.

Geller, A. M. and Mathieu, R. D. 2011. A Mass Transfer Origin for Blue Stragglers in NGC 188 as Revealed by Half-Solar-Mass Companions. *Nature*, **478**, 356.

Geller, A. M., and Mathieu, R. D. 2012. WIYN Open Cluster Study. XLVIII. The Hard-Binary Population of NGC 188. *AJ*, **144**, 54.

Geller, A. M., Hurley, J. R. and Mathieu, R. D. 2013. Direct N-Body Modeling of the Old Open Cluster NGC 188: A Detailed Comparison of Theoretical and Observed Binary Star and Blue Straggler Populations. *AJ*, **145**, 8.

Geller, A. M., Latham, D. W. and Mathieu, R. D. 2015. Stellar Radial Velocities in the Old Open Cluster M67 (NGC 2682). I. Memberships, Binaries, and Kinematics. *AJ*, **150**, 97.

Geller, A. M., Leiner, E. M., Bellini, A., et al. 2017a. On the Origin of Sub-Subgiant Stars. I. Demographics. *ApJ*, **840**, 66.

Geller, A. M., Leiner, E. M., Chatterjee, S., et al. 2017b. On the Origin of Sub-Subgiant Stars. III. Formation Frequencies. *ApJ*, **842**, 1.

Geller, A. M., Mathieu, R. D., Harris, H. C., et al. 2008. WIYN Open Cluster Study. XXXII. Stellar Radial Velocities in the Old Open Cluster NGC 188. *AJ*, **135**, 2264.

Gosnell, N. M., Mathieu, R. D., Geller, A. M., et al. 2015. Implications for the Formation of Blue Straggler Stars from HST Ultraviolet Observations of NGC 188. *ApJ*, **814**, 163.

Gosnell, N. M., Mathieu, R. D., Geller, A. M., et al. 2014. Detection of White Dwarf Companions to Blue Stragglers in the Open Cluster NGC 188: Direct Evidence for Recent Mass Transfer. *ApJ L*, **783**, L8.

Hall, D. S. 1976. The RS CVn Binaries and Binaries with Similar Properties. In *Multiple Periodic Variable Stars*, Astrophysics and Space Science Library. Vol. 60, IAU Colloq. 29, ed. W. S. Ftch, 287. Springer.

Handberg, R., Brogaard, K., Miglio, A., et al. 2017. NGC 6819: Testing the Asteroseismic Mass Scale, Mass Loss and Evidence for Products of Non-Standard Evolution. *MNRAS*, **472**, 979.

Holberg, J. B. and Bergeron, P. 2006. Calibration of Synthetic Photometry Using DA White Dwarfs. *AJ*, **132**, 1221.

Huber, D., Silva Aguirre, V., Matthews, J. M., et al. 2014. Revised Stellar Properties of Kepler Targets for the Quarter 1-16 Transit Detection Run. *ApJ S*, **211**, 2.

Hummel, C. A., Monnier, J. D., Roettenbacher, R. M., et al. 2017. Orbital Elements and Stellar Parameters of the Active Binary UX Arietis. *ApJ*, **844**, 115.

Hurley, J. R., Pols, O. R., Aarseth, S. J., et al. 2005. A Complete N-Body Model of the Old Open Cluster M67. *MNRAS*, **363**, 293.

Ivanova, N., da Rocha, C. A., Van, K. X., et al. 2017. Formation of Black Hole X-Ray Binaries with Non-Degenerate Donors in Globular Clusters. *ApJ L*, **843**, L30.

Jeffries, R. D. and Stevens, I. R. 1996. Wind-Accretion Induced Rapid Rotation and a New Class of Active Star. *MNRAS*, **279**, 180.

Johnson, H. L. and Sandage, A. R. 1955. The Galactic Cluster M67 and Its Significance for Stellar Evolution. *ApJ*, **121**, 616.

Jorissen, A., Van Eck, S., Mayor, M. and Udry, S. 1998. Insights into the Formation of Barium and Tc-Poor S Stars from an Extended Sample of Orbital Elements. *A&A*, **332**, 877.

Knigge, C., Leigh, N. and Sills, A. 2009. A Binary Origin for 'Blue Stragglers' in Globular Clusters. *Nature*, **457**, 288.

Kroupa, P. 2001. On the Variation of the Initial Mass Function. *MNRAS*, **322**, 231.

Landsman, W., Aparicio, J., Bergeron, P., et al. 1997. S1040 in M67: A Post–Mass Transfer Binary with a Helium Core White Dwarf. *ApJ L*, **481**, L93.

Latham, D. W. 2007. Spectroscopic Binaries in M 67: Highlights of Astronomy, **14**, 444.

Leigh, N. and Sills, A. 2011. An Analytic Technique for Constraining the Dynamical Origins of Multiple Star Systems Containing Merger Products. *MNRAS*, **410**, 2370.

Leiner, E. M., Mathieu, R. D. and Geller, A. M. 2017. On the Origin of Sub-Subgiant Stars. II. Binary Mass Transfer, Envelope Stripping, and Magnetic Activity. *ApJ*, **840**, 67.

Leiner, E. M., Mathieu, R. D., Gosnell, N. M. and Sills, A. 2018. Observations of Spin-down in Post-Mass-Transfer Stars and the Possibility for Blue Straggler Gyrochronology. *ApJ L*, **869**, L29.

Leiner, E. M., Mathieu, R. D., Stello, D., et al. 2016. The K2 M67 Study: An Evolved Blue Straggler in M67 from K2 Mission Asteroseismology. *ApJ L*, **832**, L13.

Leiner, E. M., Mathieu, R. D., Gosnell, N. M., et al. 2015. WIYN Open Cluster Study. LXVI. Spectroscopic Binary Orbits in the Young Open Cluster M35 (NGC 2168). *AJ*, **150**, 10.

Lovisi, L., Mucciarelli, A., Ferraro, F. R., et al. 2010. Fast Rotating Blue Stragglers in the Globular Cluster M4. *ApJ L*, **719**, L121.

Mathieu, R. D. and Geller, A. M. 2009. A Binary Star Fraction of 76 Per Cent and Unusual Orbit Parameters for the Blue Stragglers of NGC 188. *Nature*, **462**, 1032.

Mathieu, R. D. and Geller, A. M. 2015. The Blue Stragglers of the Old Open Cluster NGC 188. in *Ecology of Blue Straggler Stars*, ed. H. M. J. Boffin, G. Carraro and G. Beccari, 29. Springer.

Mathieu, R. D. and Latham, D. W. 1986. The Spatial Distribution of Spectroscopic Binaries and Blue Stragglers in the Open Cluster M67. *AJ*, **92**, 1364.

Mathieu, R. D., Latham, D. W. and Griffin, R. F. 1990. Orbits of 22 Spectroscopic Binaries in the Open Cluster M67. *AJ*, **100**, 1859.

Mathieu, R. D., van den Berg, M., Torres, G., et al. 2003. Sub-Subgiants in the Old Open Cluster M67? *AJ*, **125**, 246.

Milliman, K. E., Leiner, E. L., Mathieu, R. D., et al. 2016. Spectroscopic Binary Orbits of Sub-Subgiants in NGC6791. *AJ*, arXiv:1408.0239.

Milliman, K. E., Mathieu, R. D. and Schuler, S. C. 2015. Barium Surface Abundances of Blue Stragglers in the Open Cluster NGC 6819. *AJ*, **150**, 84.

Milliman, K. E., Mathieu, R. D., Geller, A. M., et al. 2014. WIYN Open Cluster Study. LX. Spectroscopic Binary Orbits in NGC 6819. *AJ*, **148**, 38.

Morales, J. C., Ribas, I. and Jordi, C. 2008. The Effect of Activity on Stellar Temperatures and Radii. *A&A*, **478**, 507.

Nissen, P. E., Twarog, B. A. and Crawford, D. L. 1987. UvbyH-beta Photometry of Main-Sequence Stars in M67. *AJ*, **93**, 634.

Packet, W. 1981. On the Spin-Up of the Mass Accreting Component in a Close Binary System. *A&A*, **102**, 17.

Perets, H. B. and Fabrycky, D. C. 2009. On the Triple Origin of Blue Stragglers. *ApJ*, **697**, 1048.

Sandquist, E. L., Latham, D. W., Shetrone, M. D., et al. 2003. The Blue Straggler RS Canum Venaticorum Star S1082 in M67: A Detailed Light Curve and the Possibility of a Triple. *AJ*, **125**, 810.

Shara, M. M., Saffer, R. A. and Livio, M. 1997. The First Direct Measurement of the Mass of a Blue Straggler in the Core of a Globular Cluster: BSS 19 in 47 Tucanae. *ApJ L*, **489**, L59.

Shetrone, M. D. and Sandquist, E. L. 2000. Spectroscopy of Blue Stragglers and Turnoff Stars in M67 (NGC 2682). *AJ*, **120**, 1913.

Sills, A., Adams, T. and Davies, M. B. 2005. Blue Stragglers as Stellar Collision Products: The Angular Momentum Question. *MNRAS*, **358**, 716.

Sills, A., Faber, J. A., Lombardi, Jr., et al. 2001. Evolution of Stellar Collision Products in Globular Clusters. II. Off-Axis Collisions. *ApJ*, **548**, 323.

Torres, G., Lacy, C. H., Marschall, L. A., et al. 2006. The Eclipsing Binary V1061 Cygni: Confronting Stellar Evolution Models for Active and Inactive Solar-Type Stars. *ApJ*, **640**, 1018.

Tout, C. A. and Eggleton, P. P. 1988. Tidal Enhancement by a Binary Companion of Stellar Winds from Cool Giants. *MNRAS*, **231**, 823.

Tremblay, P.-E., Bergeron, P. and Gianninas, A. 2011. An Improved Spectroscopic Analysis of DA White Dwarfs from the Sloan Digital Sky Survey Data Release 4. *ApJ*, **730**, 128.

van den Berg, M., Orosz, J., Verbunt, F., et al. 2001. The Blue Straggler S 1082: A Triple System in the Old open cluster M 67. *A&A*, **375**, 375.

Verbunt, F. and Phinney, E. S. 1995. Tidal Circularization and the Eccentricity of Binaries Containing Giant Stars. *A&A*, **296**, 709.

19

Clocks and Scales: Playing with the Physics of Blue Stragglers

FRANCESCO R. FERRARO AND BARBARA LANZONI

19.1 The BSS–Binaries Connection

According to their first detection (Sandage, 1953) in the external region of the Galactic Globular Cluster (GC) M3, blue straggler stars (BSSs) are commonly defined as stars brighter and bluer than the main-sequence (MS) turnoff (TO) point in the colour-magnitude diagram (CMD) of the host stellar cluster (e.g., Sandage, 1953; Ferraro et al., 1992, 1993, 1997a, 2004, 2006a; Lanzoni et al., 2007a,b, Leigh et al., 2007; Dalessandro et al., 2008b; Moretti et al., 2008; Beccari et al., 2011, 2013; Parada et al., 2016; Simunovic and Puzia, 2016; see Figure 19.1).

Although the details of their origin are not completely understood yet, it is widely accepted that their formation process must involve binary systems. In fact, their location in the CMD along the bright extrapolation of the MS suggests that they are hydrogen-burning stars more massive than the MS-TO objects (Shara et al., 1997; Gilliland et al., 1998; Fiorentino et al., 2014). However, since GCs are completely devoid of gas, any recent star formation event can be realistically ruled out. Hence the origin of BSSs has to be searched in some mechanisms able to increase the initial mass of a single star and bring new hydrogen into the core, thus prolonging the MS stage in a sort of *rejuvenation process*. Two main mechanisms are currently favoured: (1) mass transfer (MT) in binary systems (McCrea, 1964) possibly up to the complete coalescence of the two stars, and (2) direct stellar collisions (COL; Hills and Day 1976). Both of these processes can have an efficiency that depends on the local environment (Fusi Pecci et al., 1992; Ferraro et al., 1993, 2003; Davies et al., 2004; Dalessandro et al., 2008a,b; Knigge et al., 2009), and they can act within the same cluster simultaneosly (Bailyn, 1995; Ferraro et al., 2009; Dalessandro et al., 2013a; Xin et al., 2015).

With the final aim of identifying the main BSS formation channel in GCs, many recent works in the literature have been devoted to search for correlations between the BSS specific frequency (or number) and the parameters tracing the cluster structure (as the total luminosity, mass, central density, etc.; e.g., Davies et al. 2004; Sollima et al. 2008; Chen and Han 2009; Ferraro et al. 2009; Knigge et al. 2009; Chatterjee et al. 2013; Leigh et al., 2013; Sills et al. 2013; Xin et al. 2015). Though not conclusive, this approach has provided a number of interesting results:

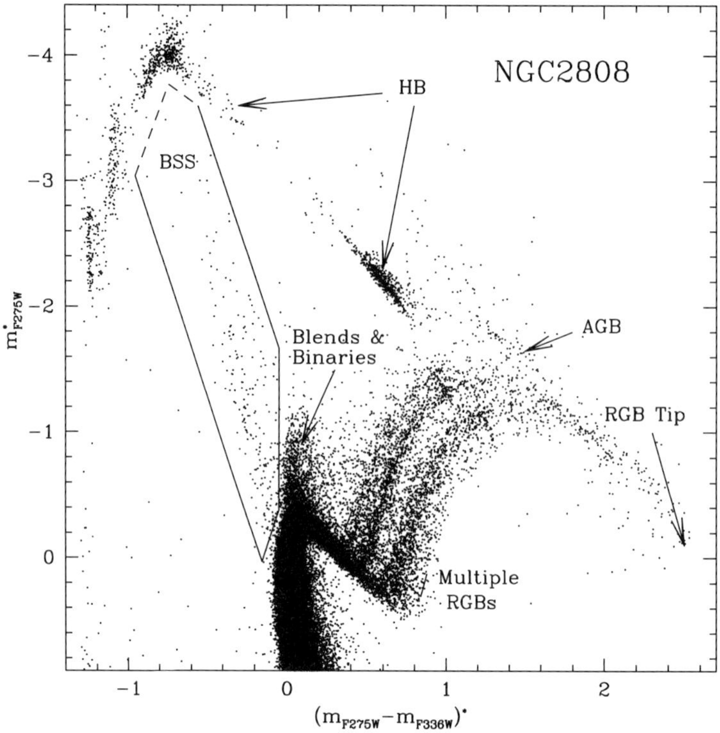

Figure 19.1 'Normalized' ultraviolet (UV) CMD of NGC 2808: magnitudes and colours have been arbitrarily shifted to locate the MS-TO at $m^*_{F275W}=0$ and $(m_{F275W}-m_{F336W})^*=0$. The location of the main evolutionary sequences is labelled. The brightest objects in this diagram are hot horizontal branch (HB) stars and BSSs, while cool giants, as red giant branch (RGB) and asymptotic giant branch (AGB) stars, are significantly less luminous. The box highlights the BSS population.

1. No correlation has been found with the collisional parameter (e.g., Davies et al., 2004; Piotto et al., 2004; Leigh et al., 2007, 2013).
2. A correlation has been revealed between the number of BSSs in cluster cores and the core mass (Knigge et al., 2009; Leigh et al., 2013).
3. A correlation between the binary fraction (estimated from the distribution of stars along the 'secondary' MS) and the BSS frequency has been found in a sample of 13 low-density GCs (with central luminosity $\log \nu_0 < 3$ in units of $L_\odot/\mathrm{pc}^3$; Sollima et al., 2008; see Figure 19.2).
4. A correlation between the BSS specific frequency and the core binary fraction has been found for a subsample of 24 (out of 59) GCs for which the MS binary fraction has been measured (Milone et al., 2012). This has been confirmed also by Leigh et al. (2013), who, however, obtain a much stronger correlation between the number of core BSSs and the cluster core mass.

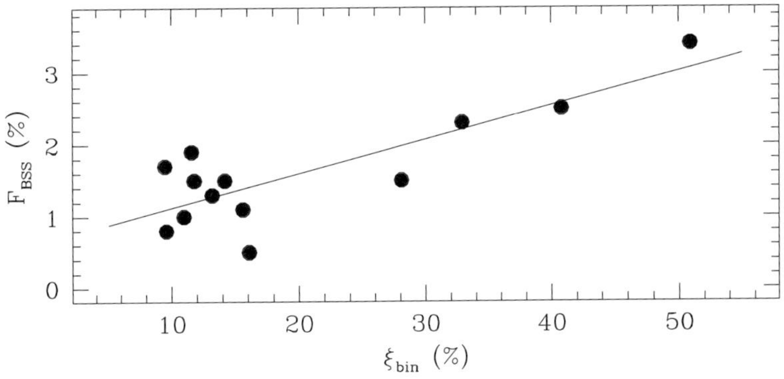

Figure 19.2 BSS specific frequency (i.e, ratio between the number of BSSs and that of MS stars) as a function of the core binary fraction measured in a sample of low-density GCs. The best-fit linear correlation is also plotted (solid line). From Sollima et al. (2008). Reproduced by permission of the AAS.

Item (3) represents the cleanest evidence that the unperturbed evolution of primordial binaries is the dominant BSS formation process in low-density environments (this is also consistent with thes results obtained in open clusters; e.g., Mathieu and Geller, 2009). On the other hand, items (1), (2) and (4) have been interpreted as the evidence that the dominant BSS formation channel is the evolution of binary systems, independently of the central density and the dynamical state of the parent cluster (Knigge et al., 2009). However, it is interesting to note that post–core collapse (PCC) clusters (Djorgovski & King, 1986) lie well outside the correlation between the binary and the BSS fraction found by Milone et al. (2012). This likely indicates that the role of internal dynamics in determining the BSS content in GC cores cannot be neglected. Indeed, the internal dynamical evolution of star clusters is expected to play a crucial role in this game: GC are 'not dormant systems' (Meylan and Heggie, 1997). They are dynamically active stellar systems, where processes as dynamical friction and mass segregation continuously change the radial distribution of stars according to their mass. Hence it is quite possible that the BSSs that we currently observe in the core of a cluster did not all form there, but they (or their binary progenitors) sank over the time into the core because of dynamical friction, thus joining the population of BSSs (or proto-BSSs) already present. Moreover, in high-density environments such a mixed population is expected to be subject to frequent dynamical interactions with other binaries, single stars and even multiple systems. These interactions can either bring stellar collisions or significantly alter the physical properties of binaries, even promoting mass-transfer activity. Hence, internal dynamical processes can promote and activate both the MT and the COL formation channels by operating on the overall population of binaries (locally formed in the core or sank there from the outer regions). Thus, after all, it is not surprising that the two formation channels cannot be clearly distinguished just on the basis of the observed BSS/binary fraction, neglecting the dynamical evolutionary stage of the parent cluster.

19.1.1 Another Puzzling Feature: The Double BSS Sequence

One of the most promising observables for the possible distinction between the two BSS formation channels was the double BSS sequence identified in the Galactic GC M30 (Ferraro et al., 2009). Using an exceptional set of 44 high-resolution images secured with the HST-WFPC2, Ferraro et al. (2009) obtained a high-precision CMD of the central region of M30, revealing the existence of two well-separated and almost parallel sequences of BSSs (Figure 19.3). This was the very first time that such a feature was detected in any stellar system, and it opened the possibility to distinguish between the two BSS formation processes.

The comparison with evolutionary models of BSSs formed by direct collisions of two MS stars (Sills et al., 2009) shows that the blue-BSS sequence is well fit by collisional isochrones with ages of ~ 2 Gyr (see Figure 19.3; dark-grey and grey points represent the blue- and red-BSSs, respectively). Instead, the red-BSS population is far too red to be reproduced by collisional isochrones of any age, and its origin should therefore be different. Unfortunately, the comparison with MT binary models does not bring an analogously univocal answer. In fact, Xin et al. (2015) showed that the colour-magnitude distribution of synthetic MT-BSSs nicely matches the observed red-BSS sequence (thus providing strong support to an MT origin for these stars), and it never reaches the observed location of the blue-BSS sequence (thus reinforcing the hypothesis that the latter formed through a

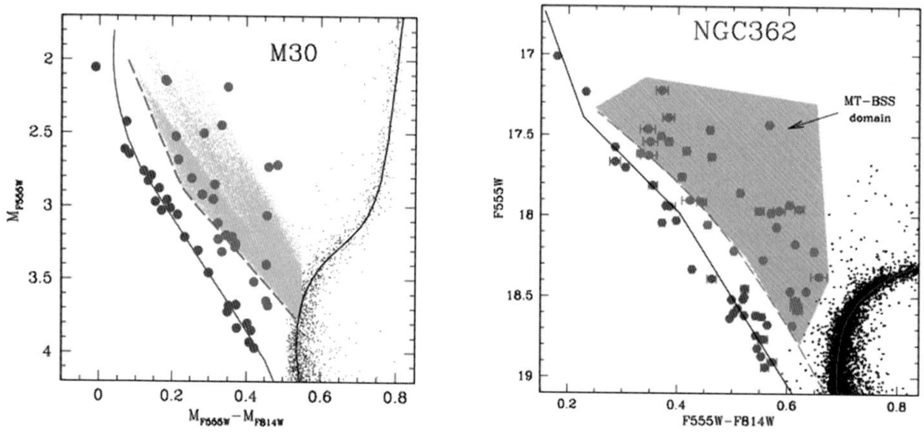

Figure 19.3 Left panel: The *double* BSS sequence discovered in the post–core collapse cluster M30 (blue BSS in dark-grey and red BSS in grey circles; from Ferraro et al., 2009). The red and the blue BSS sequences are similarly populated (consisting of 21 and 24 stars, respectively), and they are separated in magnitude by $\Delta\,(F555W) \simeq 0.4$ mag and in colour by $\Delta\,(F555W - F814W) \simeq 0.12$ mag. The solid dark-grey line is a 2 Gyr collisional isochrone (Sills et al., 2009). The light-grey dots correspond to 40,000 models of synthetic MT-BSSs (from Xin et al., 2015), and the dashed grey line marks the 'low-luminosity boundary' of their distribution. Clearly, the blue sequence is well reproduced by the COL-BSS model, while the position of the red sequence fully agrees with the distribution of MT-BSSs. Right panel: The double BSS sequence detected in NGC 362 (Dalessandro et al., 2013a).

different channel, likely collisions). Conversely, more recent binary models (Jiang et al., 2017) show that the blue-BSS sequence can be populated by BSSs that have ended the mass-transfer process and have a cool white dwarf companion.

However, it is not clear from the paper how well the simulations reproduce the observational data of M30. One problem is that the vast majority (60–70%) of the simulated BSSs populate the blue sequence, while the corresponding fraction observed in M30 is smaller than 50%. Moreover, the authors conclude that the simulated BSSs all form from standard binary evolution, which is a continuous (instead of a short-lived) process. This clearly opens the question of why, then, two well-separated BSS sequences are not observed in *all* GCs. Jiang et al. (2017) propose that such nondetection could be due to large photometric errors or differential reddening and distance variations among the BSSs populating a cluster. However, it should be kept in mind that many GCs have very low reddening values (hence they are essentially free from differential reddening problems), but show no evidence of double BSS sequences. Morever, for a typical Galactic GC (with average distance $d \sim 10$ kpc and half-mass radius $r_h < 5$ pc; Harris, 1996) the difference in distance modulus between a BSS physically located in the core and a BSS orbiting at r_h is only 0.001 mags. This difference increases at most to 0.01 mag between a core BSS and a BSS at the tidal radius (typically $r_t < 50$ pc). Such magnitude differences are so small that their possible effect on the BSS location in the CMD is completely negligible, and they certainly cannot explain why two BSS sequences separated by an empty gap are not observed in all GCs. Hence only large photometric errors can be advocated to explain the nonuniversal detection of double BSS sequences. While this is in principle possible, it appears quite unlikely since even the most recent and high S/N photometric surveys of GCs apparently fail to detect significant numbers of double BSS sequences (while new evidence is emerging in another post–core collapse cluster: M15; Beccari et al., 2018). In any case, the CMD distributions of simulated stars shown in figures 5 and 6 of Jiang et al. (2017) do not seem to properly reproduce the observations: while the highest-density distribution found at the bluest colours is well aligned with the observed blue-BSS sequence, neither the observed red sequence nor the gap devoid of stars that is observed between the two chains in M30 seem to be reproduced by the simulations. However, beside the details, the most intriguing result from the simulations of Jiang et al. (2017) is that also BSSs originated from MT in binary systems can populate the bluest side of the BSS sequence. Thus, if this result is confirmed by future investigations, both COLL- and MT-BSSs can in principle coexist along the blue sequence of M30, and the key feature to focus our attention on is the overall morphology of the two sequences, i.e., the existence of a clear gap between them.

Indeed, the presence of this gap suggests that at least the blue-BSS population has been *generated by a recent and short-lived event (instead of a continuous phenomenon, as in the Jiang et al., 2017 models).* In fact, due to standard stellar evolution, in a few billion years the progeny of blue-BSSs will populate the region between the two observed sequences and fill the gap. Quite interestingly, M30 is classified as a PCC cluster (see the steep power-law cusp in the innermost $5'' - 6''$, or ~ 0.2 pc, of its star density profile in figure 3 of Ferraro et al., 2009). Hence the possibility that the enhanced rate of gravitational interactions characterising the cluster core collapse phase could be at the origin of the blue-BSS sequence is still valid (note that the collisional parameter scales as $\Gamma \propto \rho_0^{1.5} r_c^2$, where r_c is the core radius). In fact, these can trigger the formation of new BSSs, both via direct stellar collisions and via MT activity in dynamically shrunk binary systems. These considerations support a scenario where both the observed BSS populations have been affected by core collapse: most of the BSSs lying along the blue sequence arose from the enhanced stellar collision activity, while the red BSSs are the result of the evolution of binary systems that first sank into the cluster centre because of dynamical friction (or they were already present in the core), and then have been brought to an active MT phase either by secular evolution or by hardening processes due to the enhanced gravitational interaction rate. According to this scenario, *the existence of a well-defined gap suggests that all BSSs along the blue sequence formed simultaneously and the extension in luminosity of this sequence dates the occurrence of the core collapse event back to 1 to 2 Gyr ago.*

Interestingly, the double BSS feature has been detected also in two additional clusters: NGC 362 (Dalessandro et al., 2013a) and NGC 1261 (Simunovic et al., 2014). Preliminary indications suggest that it could be present also in the core of M15 (Beccari et al., 2018). Detailed spectroscopic investigations are certainly worth performing to obtain a complete characterisation of the BSS properties (orbital periods, rotation velocities, etc.) along the two sequences. In particular, following the scenario presented by Jiang et al. (2017), the detection of radial velocity variations due to the presence of a white dwarf companion would allow to estimate the percentage of MT-BSSs populating the blue sequence, since no such signal is expected if these stars formed through collisions.

19.2 Clocks: Defining and Refining the Dynamical Clock for Star Clusters

Ferraro et al. (2012, hereafter F12) showed that the observed shape of the BSS normalised radial distribution[1] (hereafter BSS-nRD) can be used to set a *dynamical clock* able to measure the level of dynamical evolution reached by the host system. In fact, chronologically old and coeval GCs can be grouped in at least three distinct families on the basis of the shape of their BSS-nRD (see Figure 19.4):

- Family I, for which the BSS-nRD is flat (i.e., the radial distribution of BSSs within the cluster is fully compatible with that of the light)
- Family II, for which the BSS-nRD is bimodal (with a high peak in the cluster centre, a dip at an intermediate radius $r_{\rm min}$ and a rising branch in the external regions)

[1] With 'BSS normalised radial distribution' here, we indicate the double normalised ratio first introduced by Ferraro et al. (1993), defined as the ratio between the fraction of BSSs sampled in a radial bin and the fraction of cluster light sampled in the same bin. Thus, $R_{\rm BSS}(r) = [N_{\rm BSS}(r)/N_{\rm BSS,tot}]/[L_{\rm samp}(r)/L_{\rm samp,tot}]$, where $N_{\rm BSS}(r)$ is the number of BSSs measured in any given radial bin, $N_{\rm BSS,tot}$ is the total number of such stars and $L_{\rm samp}(r)$ and $L_{\rm samp,tot}$ are the analogous quantities for the sampled luminosity.

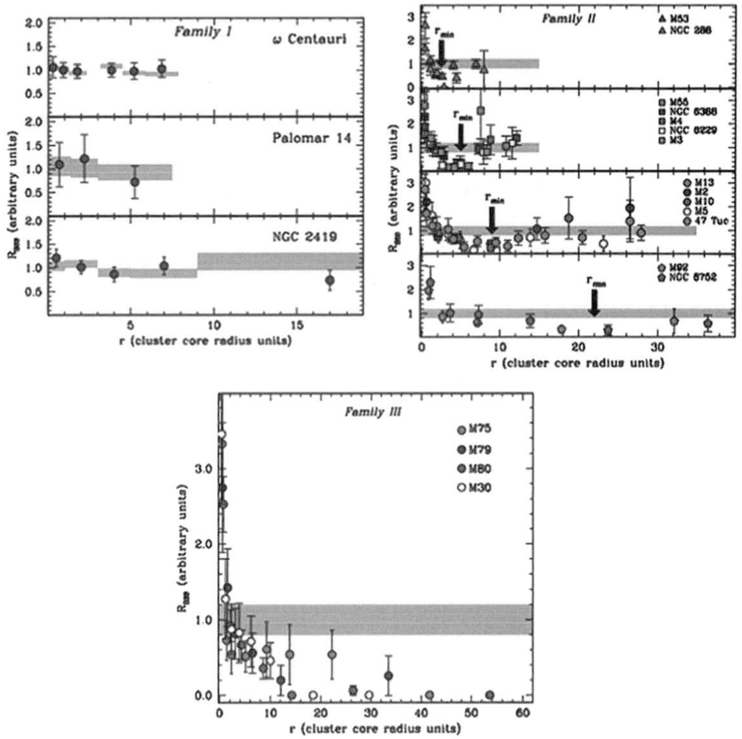

Figure 19.4 Normalised radial distribution of BSSs (grey-scaled symbols) compared to that of normal cluster stars taken as reference (grey regions) in dynamically young clusters (Family I, left panel), in systems with intermediate dynamical age (Family II, central panel) and in dynamically old systems (Family III, right panel), for a sample of GCs in our Galaxy. The width of the grey bands is representative of the uncertainty. By definition, the normalised radial distribution is equal to unity for any population with a radial distribution that follows that of the cluster integrated luminosity (such as the reference stars). From Ferraro et al., 2012.

- Family III, for which the BSS-nRD shows only a central peak followed by a monotonically decreasing trend.

Since BSSs (and their binary progenitors) have masses larger than normal cluster stars, these behaviours have been interpreted as the manifestation of the progressive effect of dynamical friction, which drives the objects more massive than the average towards the cluster centre, with an efficiency that mainly depends on the local star density (i.e., it decreases for increasing radial distance). Hence, a flat BSS-nRD indicates that dynamical friction has not played a major role yet in affecting the BSS population (not even in the innermost regions) and GCs belonging to Family I therefore are ‘dynamically young’. In more evolved clusters (Family II), dynamical friction has been more and more efficient, affecting regions at increasingly larger distances from the centre, as marked by the observed position of r_{min}. In 'dynamically old' systems, we expect that dynamical friction affected also the most remote BSSs, thus producing a monotonic BSS-nRD with a central peak only (as observed in Family III GCs).

In spite of many and independent observational confirmations, the current generation of N-body and Monte Carlo simulations of GCs seems to be unable to reproduce the details of the BSS-nRD evolution, possibly indicating a still insufficient level of realism. In fact, while the peak of the distribution is seen to form in simulations, and the final stage of the synthetic BSS-nRD is indeed monotonic and single-peaked (Miocchi et al., 2015), a bimodal shape appears to be a temporary and unstable feature (see Miocchi et al., 2015; Hypki and Giersz, 2017). This is at odds with the observations, which instead indicate that the BSS-nRD is bimodal in the vast majority of the investigated clusters (even in extra-Galactic environments; see the case of Hodge 11 discussed in Li et al., 2013). On the other hand, when dealing with discrete and not numerous populations (as BSSs are), exactly locating the minimum of the radial distribution is not an easy task. The correct identification of the minimum is critically dependent on the choice of the radial binning, since adopting too wide intervals tends to clean out the feature, while too narrow bins generate noisy distributions because of the decreasing number of sampled BSSs. Thus it is not surprising that 'blind' automatic searches for r_{min} in simulated BSS radial distributions do not succeed in detecting a stable minimum (see, e.g., Hypki and Giersz, 2017). Given these difficulties, we searched for additional and 'more easily detectable' observables able to describe the BSS segregation process.

In Alessandrini et al. (2016), we used 10^5 particle N-body simulations to investigate the effect of dynamical friction on BSSs in GCs, and we proposed a new parameter (A^+) to measure the level of BSS central sedimentation. The parameter is defined as the area enclosed between the cumulative radial distribution of BSSs, $\phi_{BSS}(x)$ and that of a reference (lighter) population, $\phi_{REF}(x)$:

$$A^+(x) = \int_{x_0}^{x} \phi_{BSS}(x') - \phi_{REF}(x')\, dx', \tag{19.1}$$

where x_0 and x correspond to the innermost and the outermost distances from the cluster centre sampled by the observations in units of the half-mass radius rh: $x = \log(r/rh)$. Clustercentric distances are expressed in logarithmic units to maximise the sensitivity of A^+ to the effects of dynamical friction (which is most efficient in the central cluster regions). N-body simulations demonstrate that this parameter increases significantly as a function of time, following the dynamical evolution of the cluster and, more specifically, tracking the process of BSS segregation relative to the reference population (see figure 5 in Alessandrini et al., 2016). At initial times, when all stars are perfectly mixed regardless of their mass, BSSs and the REF population have the same cumulative radial distributions and $A^+ = 0$. As mass segregation proceeds, BSSs migrate towards the centre of the system, and the two curves start to separate, thus providing a progressively increasing value of the A^+ parameter.

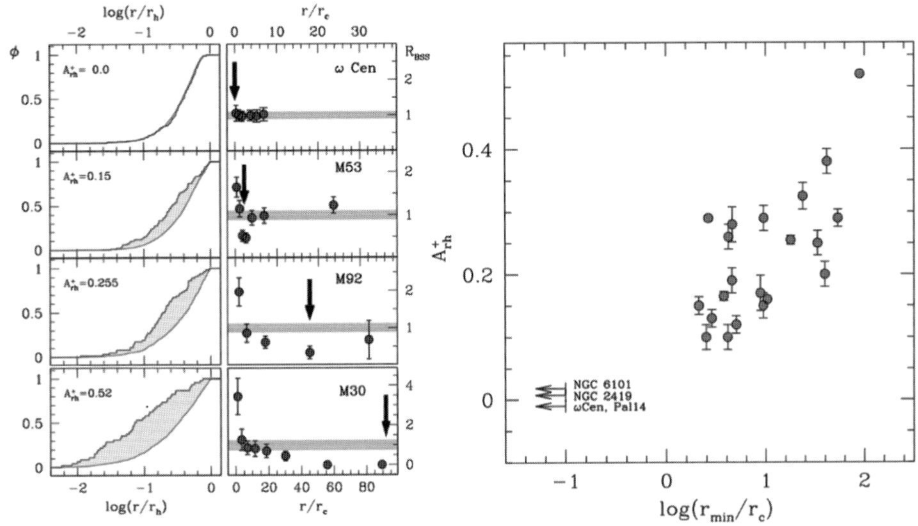

Figure 19.5 Left panels: On the left are cumulative radial distributions of BSSs (dark-grey line) and reference stars (grey line) observed within one half-mass radius (r_h) in four GCs (from top to bottom: ω Centauri, M53, M92 and M30). The size of the area between the two curves (shaded in grey) corresponds to the labelled value of A^+_{rh}. On the right are BSS normalised radial distributions ($R_{\rm BSS}$, dark-grey circles) for the same clusters. The black arrows mark the position of the BSS-nRD minimum ($r_{\rm min}/r_c$). The grey strips schematically show the distribution measured for the reference population. Right panel: Parameter A^+_{rh} as a function of the observed position of the BSS-nRD minimum normalised to the core radius ($r_{\rm min}/r_c$) and expressed in logarithmic units. An upper limit value $r_{\rm min}/r_c = 0.1$ has been assumed for Family I clusters (see arrows). From L16. Reproduced by permission of the AAS.

Lanzoni et al. (2016, hereafter L16) measured the new parameter out to one half-mass radius (A^+_{rh}) for the same set of 25 clusters discussed in F12, finding a tight connection with $r_{\rm min}$. Figure 19.5 (left panels) shows the cumulative radial distributions of BSSs and reference stars for four GCs in the L16/F12 sample. The labelled values of A^+_{rh} correspond to the size of the shaded areas: formally zero for ω Centauri, which has a nonsegregated BSS population (Family I cluster), and up to 0.52 for M30, which is one of the dynamically oldest system (Family III) in the sample, with BSSs much more centrally concentrated than the reference population. In the corresponding middle panels, we plot the BSS-nRD for the same four clusters, with the black arrows indicating the position of the minimum $r_{\rm min}/r_c$. The figure clearly shows that the increases of A^+_{rh} seen from top to bottom in the left panels are accompanied by a systematic increase of $r_{\rm min}/r_c$ in the corresponding BSS-nRD.

To better show the deep link between these two indicators, the right-hand panel of Figure 19.5 shows the relation between A^+_{rh} and the parameter $\log(r_{\rm min}/r_c)$ for the entire sample of 25 GCs. Indeed, the Spearman rank correlation coefficient between A^+_{rh} and $r_{\rm min}/r_c$ is $\rho = 0.77$, corresponding to a probability $P > 99.99\%$ that the two parameters are correlated. The significance remains very high ($P > 99.7\%$) even if one (arbitrarily) excludes from the sample the most extreme points (namely, M30 and the four Family I GCs), or if limiting the analysis to the 13 clusters with the smallest errors on A^+_{rh} ($\epsilon < 0.02$).

It should be noted that the definitions of these two parameters are completely independent and, in principle, they could be completely uncorrelated. Figure 19.5 shows, instead, that the two parameters are linked through a quite tight and direct correlation. This indicates that they describe the same phenomenon. They actually are different ways of measuring the progressive central segregation of BSSs: as discussed in F12, the underlying process is dynamical friction, which, as clusters get dynamically older, progressively removes BSSs at increasingly larger distances from the centre (thus generating a minimum at increasingly larger values of $r_{\rm min}/r_c$) and accumulates them towards the cluster centre (thus increasing A^+). This relation represents the strongest evidence of the deep connection between the two parameters and unequivocally links the minimum in the BSS-nRD with the effect of dynamical friction. This is at odds with the conclusion of Hypki and Giersz (2017), who confused the difficulty of their simulations in pinpointing the minimum of the BSS-nRD (which is intrinsically difficult for the reasons discussed earlier) with the effectiveness of this feature to trace the dynamical evolution of clusters.

The results obtained suggest that A^+ could be used as an alternative indicator of the level of dynamical evolution experienced by star clusters since their formation, and we should expect that it is related to other parameters measuring the dynamical evolution timescale. Indeed, F12 found a nice correlation between the clustercentric distance of the minimum of the BSS-nRD ($r_{\rm min}/r_c$) and the central relaxation time of the cluster (t_{rc}). Figure 19.6 shows that a tight relation exists also between A^+ (as measured in L16) and the number of current relaxation times since cluster formation $N_{relax} = t_{GC}/t_{rc}$, where $t_{GC} = 12$ Gyr is the average age of the GCs in the sample (from Forbes and Bridges, 2010).[2] The strong relation shown in Figure 19.6 fully confirms that A^+ is a direct and powerful indicator of cluster dynamical evolution.

In the framework of the dynamical clock, the new parameter A^+ corresponds to a new clock hand, while the engine of the clock remains the same (the dynamical friction process). From the observational point of view, the advantage of this new clock hand is that it is easier to measure and somewhat less 'fragile' than the position of the BSS-nRD minimum. In fact, for determining $r_{\rm min}$ it is necessary to sample the entire cluster radial extension, while only the central regions ($r < r_h$), possibly probed just by HST, are sufficient to measure A^+_{rh}. Moreover, it does not require any (somehow arbitrary) assumption on the radial binning (which is instead needed to build the BSS-nRD). Finally, from its own definition, $r_{\rm min}$ is located in a region of 'low signal', where dynamical friction is removing heavy stars and the number of BSSs reaches its minimum (the so-called 'zone of avoidance'; Mapelli et al., 2004, 2006); instead, A^+_{rh} is measured in a 'high signal' region, where dynamical friction is accumulating heavy stars and the number of BSSs reaches its maximum. Thus, the proposed change of clock hand makes the reading of the dynamical clock easier, less prone

[2] Note that the result remains unchanged if the individual cluster ages are assumed.

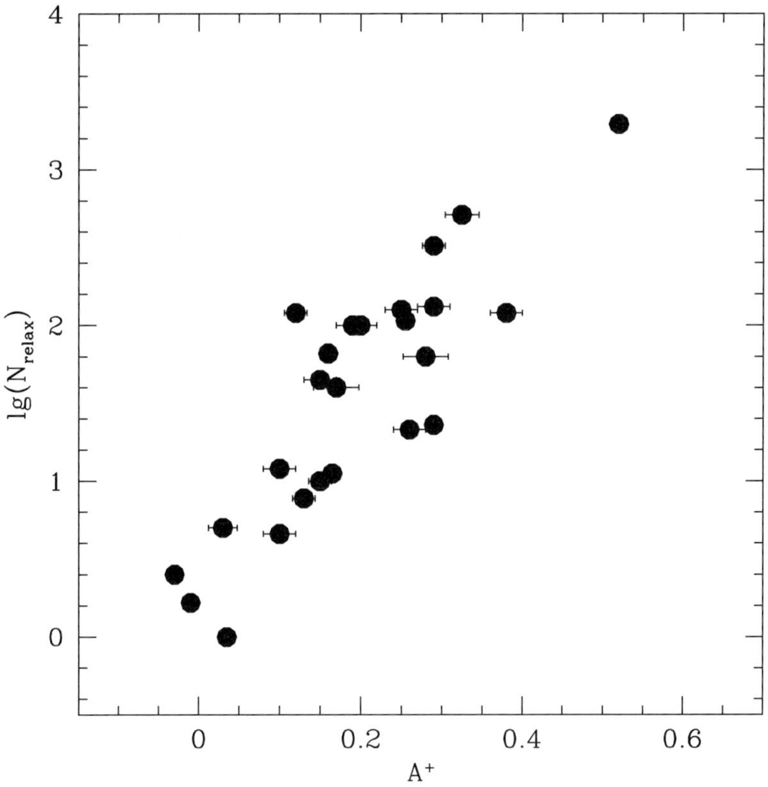

Figure 19.6 The strong correlation between A^+ (from L16) and the parameter $N_{relax} = t_{\rm GC}/t_{\rm rc}$ that measures the number of current central relaxation times occurred since cluster formation (we assumed an average GC age $t_{\rm GC} = 12$ Gyr).

to low statistics uncertainties and possibly reproducible even by the current generation of N-body and Monte Carlo simulations. Clearly, a detailed decription of how A^+ changes as a function of the cluster dynamical age requires simulations including virtually all the known ingredients (as massive stellar remnants, primordial binaries, etc.).

19.3 Scales: Searching for Evolved BSSs

Although BSSs have been routinely observed for 60 years now, only little information about their physical properties has been collected so far. Moreover, a crucial link of their evolutionary chain is entirely missing: while these stars certainly evolve along the red giant, horizontal and asymptotic giant branches (RGB, HB and AGB), only scant (mostly indirect) indications of the existence of evolved BSSs (E-BSSs) have been obtained so far. This is because, while BSSs can be easily identified during their hydrogen-burning phase in the CMD, E-BSSs are photometrically indistinguishable from their low-mass sisters. Three candidates have been recently identified in two open clusters (NGC 6791 and NGC 6819; Brogaard et al., 2012; Corsaro et al., 2012) from asteroseismology studies. Only a few candidate E-BSSs are known in GCs (see, e.g., the anomalous cepheid V19 in NGC 5466; Zinn and King, 1982). Based on photometric criteria suggested to optimise the search for E-BSSs in the CMD (Renzini and Fusi Pecci, 1988; Fusi Pecci et al., 1992), samples of candidate E-BSSs have been identified between the HB level and the base of the AGB in M3 (Ferraro et al., 1993), M80 (Ferraro et al., 1999), and 47 Tucanae (Beccari et al., 2006; Parada et al., 2016). The first spectroscopic confirmation of one of these E-BSS candidates has been recently provided (Ferraro et al., 2016) thanks to the definition of a spectroscopic *stellar scale* able to efficiently distinguish massive objects among normal low-mass stars.

19.3.1 The Stellar Scale

The basic concept of the stellar scale is quite simple: the abundance of a given chemical element obtained from the absorption lines of ionised atoms is sensitively dependent on the stellar gravity, while such a dependence is negligible for neutral spectral lines of the same element. Since the abundances obtained from the two measurements must agree, the difference between the two values can be used to derive the stellar mass. Indeed, the difference between the two derived abundances (from neutral and from ionised species) can be used as a sort of pointer of a weighting device: if it is zero, then the correct mass of the star is univocally determined. This is indeed a powerful approach since it allows one to perform a self-consistent and differential comparison in terms of stellar mass among photometrically indistinguishable objects (i.e., stars located in the same region of the CMD): this is quite important since it gets rid of any possible zero-point offset among different methods, making particularly solid and convincing any conclusion about a significant mass excess and the identification of anomalously massive stars (as evolved BSSs) into the dominant population of genuine (low-mass) stars in a stellar cluster.

19.3.2 The Discovery of the E-BSS1 in 47 Tucanae

By using a set of high-resolution spectra obtained with the Ultraviolet and Visual Echelle Spectrograph (UVES) at the Very Large Telescope (VLT), Ferraro et al. (2016) measured the iron abundance of three stars (bHB1, bHB2 and E-BSS1) in 47 Tucanae, from neutral and ionised iron lines separately (FeI and FeII, respectively). All the targets turned out to be members of the cluster (in terms of radial velocity), and they lie in the CMD region where E-BSSs experiencing the core helium-burning phase are expected to be observed, slightly brighter (by $\sim 0.2 - 0.5$ mag) than the HB level (see Figure 19.7, left panel). Thus atmospheric parameters appropriate for genuine cluster stars evolving off the HB have been adopted. For all the targets, the values of [FeI/H] turned out to be well in agreement with the mean metallicity of 47 Tuc ([Fe/H]$= -0.83$; Lapenna et al., 2014). However, the value of the iron abundance obtained from FeII lines agrees with that from FeI only for two objects, while it is ~ 0.2 dex lower for the other star. Clearly, this cannot be due to departures from local thermal equilibrium conditions, since these affect the neutral species and leave unaltered the abundances obtained from single ionised lines (e.g., Mashonkina et al., 2011). The only way to get the FeII abundance consistent with the cluster metallicity is to increase the adopted surface gravity, hence the stellar mass (instead [FeI/H] is insensible to these changes). Figure 19.7 (right panel) shows the difference between the abundances of FeII and FeI as a function of the stellar mass for the three targets: as apparent it systematically increases for increasing mass (gravity), and this is due to the increase of the FeII abundance while [FeI/H] remains constant (at the cluster metallicity value). Most notably, *while for two stars (bHB1 and bHB2) a good agreement between the FeI and FeII abundances is reached at* 0.6 $M_\odot$, *for E-BSS1 a significantly larger stellar mass* (1.4 $M_\odot$) *is required.* The same conclusions are reached by comparing the titanium abundances derived from TiI and TiII lines (see Figure 19.7).

The mass estimated for bHB1 and bHB2 (0.6–0.7 $M_\odot$) is exactly the mass expected for genuine HB stars evolving toward the AGB in 47 Tucanae (Gratton et al., 2010), where the main-sequence turnoff mass is $\sim$0.8 $M_\odot$. Instead, a mass of 1.4 $M_\odot$ is by far too large for any 'normal' star (apart from ultrafaint neutron stars), and BSSs are the only known stellar population in the cluster that can have masses as large as 1.6 $M_\odot$ (Shara et al., 1997; Gilliland et al., 1998; Fiorentino et al., 2014).

If E-BSS1 was the progeny of a 'normal' single star coeval to the other low-mass sisters populating the cluster, such a massive object would have already evolved into a white dwarf several billion years ago. The only possibility is that it formed more recently, through a mass-enhancement process: it could therefore be the descendant of a BSS. As any other

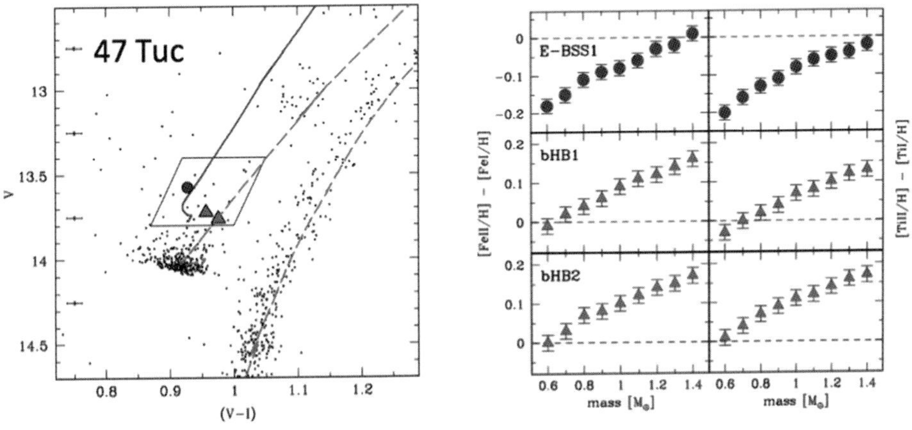

Figure 19.7 Left panel: CMD of 47 Tucanae zoomed around the HB. The position of E-BSS1 (dark-grey circle) and of the two genuine HB cluster stars (bHB1 and bHB2, grey triangles) is marked. The track of a 1.5 $M_\odot$ star (solid line) and the best-fitting isochrone to the cluster CMD (dashed line) are shown. Right panel: Difference between the iron abundance derived from ionised lines and that obtained from neutral lines, as a function of the assumed stellar mass for the three stars observed in 47 Tucanae. The same, but for the titanium abundances, is shown in the right-hand panels.

star, BSSs are expected to evolve along the various post-MS phases. Indeed, E-BSS1 is located in the region of the CMD where evolved BSSs experiencing the core helium-burning process are expected to lie. In fact, the collisional models of Sills et al. (2009) show that, during the core helium-burning stage, the progeny of collisional BSSs should populate the CMD slightly blueward of the red giant branch, between 0.2 and 1 mag brighter than the 'canonical' HB level of the host cluster.[3] Overall, the post-MS evolution of a collisional product in the CMD is very similar to that of a normal star of the same mass, the former being just a few tens of a degree hotter (see figure 6 of Sills et al., 2009, for the model of a $\sim 1.4\,M_{\odot}$ star). In Figure 19.7, we have superimposed to the CMD of 47 Tucanae the evolutionary track (from Pietrinferni et al., 2006) of a star with initial mass equal to $1.5\,M_{\odot}$ (reaching the HB phase with a mass of $1.4\,M_{\odot}$). It is interesting to note that EBSS-1 lies in a region very close to the red clump level of this track, corresponding to an effective temperature $T_{\rm eff} \simeq 5011$ K and a surface gravity $\log g \sim 2.47$ dex. These values are in very good agreement with those derived from the chemical analysis of this object. It seems therefore perfectly reasonable to identify E-BSS1 with an evolved BSS that is currently experiencing its red clump (core helium-burning) phase before ascending the AGB.

The proposed stellar scale efficiently pinpoints (heavy) evolved BSSs into the dominant and photometrically indistinguishable population of genuine (low-mass) stars. Hence, it finally enables systematic searches for evolved BSSs in GCs. Moreover, any detailed spectroscopic follow-up of these objects can provide the concrete possibility to even go back to their formation channel. In fact, a few characterising features impressed by the formation process could be still observable in such advanced stages of the evolution. One of the most solid predictions of the mass-transfer scenario is that MT-BSSs should be bound in a binary system with a compact (degenerate) companion star (the peeled core of the donor, probably a helium white dwarf). This was recently confirmed in the open cluster NGC 188 (Gosnell et al., 2015). Thanks to the high luminosity of evolved BSSs, spectroscopic follow-up observations would make such a prediction easily testable through the measurement of periodic radial velocity variations. Since no companion is expected in the collisional scenario (which ends up with the merger of the two progenitors), this kind of study is particularly important in GCs, where both formation channels are expected to be active but their relative efficiency is still unknown. Moreover, detailed spectroscopic follow-ups providing the entire chemical pattern of evolved BSSs represent an additional and highly fruitful route towards the full characterisation of their evolution. In fact, significant depletion of chemical species such as carbon and oxygen has been observed in a subsample of BSSs in 47 Tucanae, and it has been interpreted as the chemical signature of the MT origin of these objects (Ferraro et al., 2006b). However it is still unknown whether this signature is transient and, if so, on which timescales. Hence, any additional information obtained from more advanced stages of the evolution can provide new clues about the

[3] Unfortunately no specific tracks for the post-MS evolution of mass-transfer BSSs in GCs are available at the moment. However, the models specifically built for the open cluster M67 (Tian et al., 2006) and the globular cluster M30 (Xin et al., 2015) suggest that, after mass transfer, BSSs behave largely as normal single stars of comparable mass. Hence they are also expected to populate the same region of the CMD.

degree of mixing experienced by these stars. Indeed, after the detection of E-BSS1, the proposed stellar scale promises to boost the identification of evolved BSSs, thus providing unprecedented constraints to the theoretical modelling of these exotica and opening a new perspective on the comprehension of their evolutionary paths and formation processes.

19.4 What Next?

Because of their high surface temperatures ($T_{\rm eff} \sim$ 6500-9000 K), BSSs are among the brightest objects at ultraviolet (UV) wavelengths in Galactic GCs. Their systematic study thus strongly benefitted from the advent of UV space facilities. Approximately 20 years ago, our group first promoted the so-called *UV route to the study of BSSs in GCs* (see Ferraro et al. 1997a,b, 1999, 2001, 2003). This approach allowed, over the years, to derive complete samples of BSSs in several Galactic GCs, including systems of very high central density (see Lanzoni et al., 2007a,b,c; Dalessandro et al., 2008a, 2009; Contreras Ramos et al., 2012; Sanna et al., 2012, 2014; Dalessandro et al., 2013a,b; see also the recent work by Parada et al., 2016). The dataset recently acquired within the HST UV Legacy Survey of Galactic Globular Clusters (Piotto et al., 2015) allows to extend this approach to a significant number of additional clusters. As a first step, Raso et al. (2017), by studying four clusters (namely NGC 2808, NGC 6388, NGC 6541 and NGC 7078), demonstrated the clear advantages of the *UV-guided search*, with respect to the *optical-guided search*, for deriving complete samples of BSSs. The adopted procedure consists in using the F275W images to construct the master list of detected sources and then force it to the images acquired in the other filters. Such an approach optimises the detection of relatively hot stars and allows the collection of complete sample of BSSs even in the central region of high-density clusters. In fact, the light from the bright cool giants, which dominates the optical emission in old stellar systems, is sensibly reduced at UV wavelengths. The comparison between the BSS samples obtained with the UV-guided approach and those obtained in previous, optical-driven surveys clearly demonstrates the benefits and the efficiency of the former.

Raso et al. (2017) identified the $(m_{\rm F275W}, m_{\rm F275W} - m_{\rm F336W})$ CMD as the ideal diagram where the BSS population appears clearly distinguished from the other evolutionary sequences and can be safely selected (see Figure 19.1). In particular, (1) together with the hottest HB stars, BSSs appear to be the brightest objects in the CMD, while most of the RGB stars are significantly fainter; (2) BSSs define a clean sequence populating a $\sim$0.5 mag-wide strip spanning approximately three magnitudes; and (3) BSSs are clearly distinguishable from MS-TO objects. In order to perform a fully homogeneous selection of BSSs in clusters with different values of distance, reddening and metallicity, Raso et al. (2017) introduced the definition of a 'normalised' CMD, where the magnitudes and colours of all the measured stars in a given cluster are arbitrarily shifted in order to locate the cluster MS-TO at $m^*_{\rm F275W} = 0$ and $(m_{\rm F275W} - m_{\rm F336W})^* = 0$ (see, e.g., Figure 19.1). Figure 19.8 shows the normalised CMD location of more than 4,000 BSSs selected in 48 GCs observed in the HST UV Legacy Survey of Galactic Globular Clusters: this is the dataset

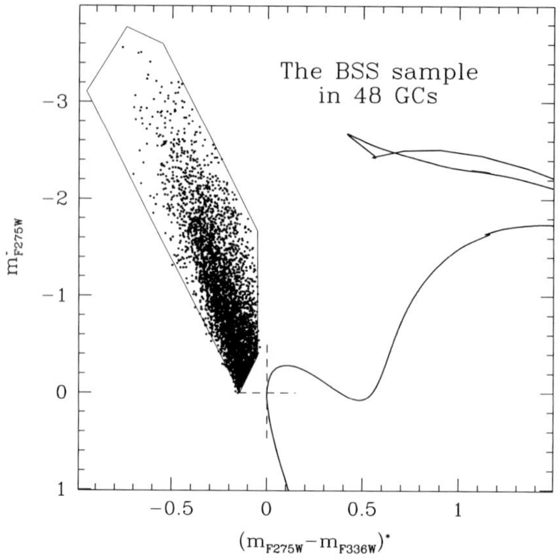

Figure 19.8 Global population of (more than 4,000) BSSs identified in the Galactic GCs observed within the HST UV Legacy Survey of Galactic Globular Clusters. All stars are shown in a 'normalised' UV CMD, accounting for the differences in distance, reddening and metallicity of the 48 surveyed GCs. The solid line corresponds to a 12 Gyr isochrone (from Pietrinferni et al., 2006) illustrative of the typical Galactic GC stellar population. The dashed cross marks the MS-TO point.

on which future BSS studies will be performed, possibly providing the final answers to the long-standing open issues about their origin, and likely opening many additional fascinating questions.

A first application has been presented in Ferraro et al. (2018), where the parameter A^+ has been measured in the 27 clusters observed in the HST UV Legacy Survey for which the HST-WFC3 observations sampled the cluster half-mass radius. In combination with the 21 clusters discussed by L17, this provides a global sample of 48 GCs, corresponding to about 32% of the Milky Way GC population.

Figure 19.9 shows the strong correlation emerging between A^+ and the number of current central relaxation times that occurred since cluster formation, considering this sample. The best-fit relation to the observed points turns out to be the following:

$$\log N_{\rm relax} = 5.1(\pm 0.5) \times A^+ + 0.79(\pm 0.12) \qquad (19.2)$$

This result provides further confirmation that BSSs are invaluable probes of GC internal dynamics and that the parameter A^+ is a powerful dynamical chronometer.

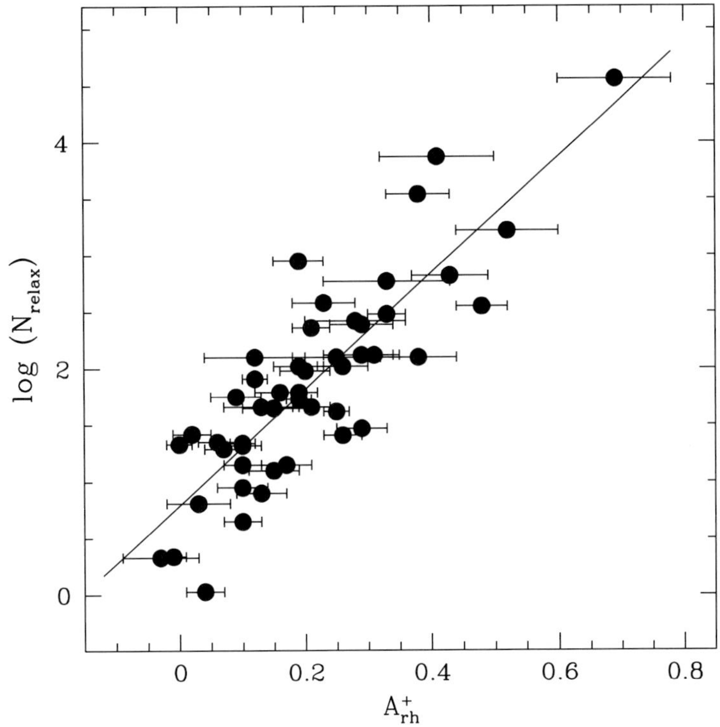

Figure 19.9 Relation between A^+ and $\log(N_{relax})$ for the sample of 48 GCs discussed in Ferraro et al. (2018). The parameter N_{relax} quantifies the number of current central relaxation times that occurred since cluster formation. The tight best-fit relation between these two parameters is shown.

References

Alessandrini, E., Lanzoni, B., Miocchi, Ferraro, F. R. and Vesperini, E. 2016. Investigating the Mass Segregation Process in Globular Clusters with Blue Straggler Stars: The Impact of Dark Remnants. *ApJ*, **833**, 252–262.

Bailyn, C. D. 1995. Blue Stragglers and Other Stellar Anomalies: Implications for the Dynamics of Globular Clusters. *ARA&A*, **33**, 133–162.

Beccari, G., Dalessandro, E., Lanzoni, B., et al. 2013. Deep Multi-Telescope Photometry of NGC 5466. I. Blue Stragglers and Binary Systems. *ApJ*, **776**, 60–69.

Beccari, G., Ferraro, F. R., Dalessandro, E., et al. 2018. Discovery of a Double Blue Straggler Sequence in M15: New Insight into the Core-Collapse Process. Submitted to ApJ.

Beccari, G., Ferraro, F. R., Lanzoni, B. and Bellazzini, M. 2006. A Population of Binaries in the Asymptotic Giant Branch of 47 Tucanae? *ApJ*, **652**, L121–L124.

Beccari, G., Sollima, A., Ferraro, F. R., et al. 2011. The Non-Segregated Population of Blue Straggler Stars in the Remote Globular Cluster Palomar 14. *ApJ*, **737**, L3–L8.

Brogaard, K., VandenBerg, D. A., Bruntt, H., et al. 2012. Age and Helium Content of the Open Cluster NGC 6791 from Multiple Eclipsing Binary Members. II. Age Dependencies and New Insights. *A&A*, **543**, A106.

Chatterjee, S., Rasio, F. A., Sills, A. and Glebbeek, E. 2013. Stellar Collisions and Blue Straggler Stars in Dense Globular Clusters. *ApJ*, **777**, 106–119.

Chen, X. and Han, Z. 2009. Primordial Binary Evolution and Blue Stragglers. *MNRAS*, **395**, 1822–1836

Contreras Ramos, R., Ferraro, F. R., Dalessandro, E., Lanzoni, B. and Rood, R. T. 2012. The Unimodal Distribution of Blue Straggler Stars in M75 (NGC 6864). *ApJ*, **748**, 91–99.

Corsaro, E., Stello, D., Huber, D., et al. 2012, Asteroseismology of the Open Clusters NGC 6791, NGC 6811, and NGC 6819 from 19 Months of Kepler Photometry. *ApJ*, **757**, 190–202.

Dalessandro, E., Beccari, G., Lanzoni, B., et al. 2009. Multiwavelength Photometry in the Globular Cluster M2. *ApJ S*, **182**, 509–518.

Dalessandro, E., Ferraro, F. R., Lanzoni, B., et al. 2013b. Ultraviolet Observations of the Globular Cluster M10 from HST and GALEX: The BSS Population. *ApJ*, **770**, 45–52.

Dalessandro, E., Ferraro, F. R., Massari, D., et al. 2013a. Double Blue Straggler Sequences in Globular Clusters: The Case of NGC 362. *ApJ*, **778**, 135–149.

Dalessandro, E., Lanzoni, B., Ferraro, F. R., et al. 2008a. Blue Straggler Stars in the Unusual Globular Cluster NGC 6388. *ApJ*, **677**, 1069–1079.

Dalessandro, E., Lanzoni, B., Ferraro, F.R., et al. 2008b. Another Nonsegregated Blue Straggler Population in a Globular Cluster: The Case of NGC 2419. *ApJ*, **681**, 311–319.

Davies, M. B., Piotto, G. and de Angeli, F. 2004. Blue Straggler Production in Globular Clusters. *MNRAS*, **349**, 129–134.

Djorgovski, S. and King, I. R. 1986. A Preliminary Survey of Collapsed Cores in Globular Clusters. *ApJ*, **305**, L61–L65.

Ferraro, F. R., Beccari, G., Dalessandro, E., et al. 2009. Two Distinct Sequences of Blue Straggler Stars in the Globular Cluster M 30. *Nature*, **462**, 1028–1031.

Ferraro F. R., Beccari G., Rood, R. T., Bellazzini M., Sills A. and Sabbi E. 2004. Discovery of Another Peculiar Radial Distribution of Blue Stragglers in Globular Clusters: The Case of 47 Tucanae. *ApJ*, **603**, 127–134.

Ferraro, F. R., D'Amico, N., Possenti, A. Mignani, R. P., and Paltrinieri, B. 2001. Blue Stragglers, Young White Dwarfs, and UV-Excess Stars in the Core of 47 Tucanae *ApJ*, **561**, 337–345.

Ferraro, F. R., Fusi Pecci, F. and Buonanno, R. 1992. The Galactic Globular Cluster NGC 5897 and Its Population of Blue Stragglers. *MNRAS*, **256**, 376–390.

Ferraro, F. R., Fusi Pecci, F., Cacciari, C., et al. 1993. Blue Stragglers in the Galactic Globular Clusters M3: Evidence for Two Populations *AJ*, **106**, 2324–2334.

Ferraro, F. R., Lanzoni, B., Dalessandro, E., et al. 2012. Dynamical Age Differences among Coeval Star Clusters as Revealed by Blue Stragglers. *Nature*, **492**, 393–395.

Ferraro, F. R., Lanzoni, B., Dalessandro, E., Mucciarelli, A. and Lovisi, L. 2015. Blue Straggler Stars in Globular Clusters: A Powerful Tool to Probe the Internal Dynamical Evolution of Stellar Systems. In Boffin, H. M. J., Carraro G. and Beccari, G. (eds.), *Ecology of Blue Straggler Stars*, Astrophysics and Space Science Library, **413**, 99–127. Springer

Ferraro, F. R., Lanzoni, B., Raso, S., et al. 2018. The Hubble Space Telescope UV Legacy Survey of Galactic Globular Clusters. XV. The Dynamical Clock: Reading Cluster Dynamical Evolution from the Segregation Level of Blue Straggler stars. *ApJ*, **860**, 36.

Ferraro, F. R., Lapenna, E., Mucciarelli, A., et al. 2016. Weighing Stars: The Identification of an Evolved Blue Straggler Star in the Globular Cluster 47 Tucanae. *ApJ*, **816**, 70–77.

Ferraro, F. R., Paltrinieri, B., Fusi Pecci, F., et al. 1997a. HST Observations of Blue Straggler Stars in the Core of the Globular Cluster M 3. *A&A*, **324**, 915–928.

Ferraro, F. R., Paltrinieri, B., Fusi Pecci, F., et al. 1997b. Hubble Space Telescope Ultraviolet Observations of the Cores of M3 and M13. *ApJ L*, **484**, L145–L148.

Ferraro, F. R., Paltrinieri, B., Rood, R.T. and Dorman, B. 1999. Blue Straggler Stars: The Spectacular Population in M80. *ApJ*, **522**, 983–990.

Ferraro, F.R., Sabbi, E., Gratton R., et al. 2006b. Discovery of Carbon/Oxygen-Depleted Blue Straggler Stars in 47 Tucanae: The Chemical Signature of a Mass Transfer Formation Process. *ApJ*, **647**, L53–L56.

Ferraro, F. R., Sills, A., Rood, R.T., Paltrinieri, B. and Buonanno, R. 2003. Blue Straggler Stars: A Direct Comparison of Star Counts and Population Ratios in Six Galactic Globular Clusters. *ApJ*, **588**, 464–477.

Ferraro, F. R., Sollima, A., Rood, R. T., Origlia, L., Pancino, E. and Bellazzini, M. 2006a. The Pure Noncollisional Blue Straggler Population in the Giant Stellar System ω Centauri. *ApJ*, **638**, 433–439.

Fiorentino, G., Lanzoni, B., Dalessandro, E., et al. 2014. Blue Straggler Masses from Pulsation Properties. I. The Case of NGC 6541. *ApJ*, **783**, 34–41.

Forbes, D. and Bridhes, T. 2010. Accreted versus In Situ Milky Way Globular Clusters. *MNRAS*, **404**, 1203–1214.

Fusi Pecci, F., Ferraro, F. R., Corsi, C. E., Cacciari, C. and Buonanno, R. 1992. On the Blue Stragglers and Horizontal Branch Morphology in Galactic Globular Clusters: Some Speculations and a New Working Scenario. *AJ*, **104**, 1831–1849.

Gilliland, R. L., Bono, G., Edmonds, P. D., et al. 1998. Oscillating Blue Stragglers in the Core of 47 Tucanae. *ApJ*, **507**, 818–845.

Gosnell, N. M., Mathieu, R. D., Geller, A. M., et al. 2015. Implications for the Formation of Blue Straggler Stars from HST Ultraviolet Observations of NGC 188. *ApJ*, **814**, 163–174.

Gratton, R. G., Carretta, E., Bragaglia, A., Lucatello, S. and D'Orazi, V. 2010. The Second and Third Parameters of the Horizontal Branch in Globular Clusters. *A&A*, **517**, A81.

Harris, W. E. 1996. A Catalog of Parameters for Globular Clusters in the Milky Way. *AJ*, **112**, 1487.

Hills, J. G. and Day, C. A. 1976. Stellar Collisions in Globular Clusters. *ApJ*, **17**, L87–L93.

Hypki, A. and Giersz, M. 2017. MOCCA Code for Star Cluster Simulations. VI. Bimodal spatial distribution of blue stragglers. *MNRAS*, **471**, 2537-2552.

Jiang, D., Chen, X., Li, L. and Han, Z. 2017. Contribution of Primordial Binary Evolution to the Two Blue-Straggler Sequences in Globular Cluster M30. *ApJ*, **849**, 100–109.

Knigge, C., Leigh, N. and Sills, A. 2009. A Binary Origin for 'Blue Stragglers' in Globular clusters. *Nature*, **457**, 288–290.

Lanzoni, B., Dalessandro, E., Ferraro, F. R., et al. 2007b. The Blue Straggler Population of the Globular Cluster M5. *ApJ*, **663**, 267–276.

Lanzoni, B., Dalessandro, E., Ferraro, F. R., et al. 2007c. The Surface Density Profile of NGC 6388: A Good Candidate for Harboring an Intermediate-Mass Black Hole. *ApJ*, **668**, L139–L142.

Lanzoni, B., Ferraro, F. R., Alessandrini, E., et al. 2016. Refining the Dynamical Clock for Star Clusters. *ApJ*, **833**, L29–L34.

Lanzoni, B., Sanna, N., Ferraro, F. R., et al. 2007a. A Panchromatic Study of the Globular Cluster NGC 1904. I. The Blue Straggler Population. *ApJ*, **663**, 1040–1048.

Lapenna, E., Mucciarelli, A., Lanzoni, B., et al. 2014. Non-Local Thermodynamical Equilibrium Effects on the Iron Abundance of Asymptotic Giant Branch Stars in 47 Tucanae. *ApJ*, **797**, 124–132.

Leigh, N., Sills, A. and Knigge, C. 2007. Where the Blue Stragglers Roam: Searching for a Link between Formation and Environment. *ApJ*, **661**, 210–221.

Leigh, N., Knigge, C., Sills, A., et al. 2013. The Origins of Blue Stragglers and Binarity in Globular Clusters. *MNRAS*, **428**, 897–905.

Li, C., de Grijs, R., Deng, L. and Liu, X. 2013. Blue Straggler Evolution Caught in the Act in the Large Magellanic Cloud Globular Cluster Hodge 11. *ApJ*, **770**, L7–L12.

Mapelli, M., Sigurdsson, S., Colpi, M., et al. 2004. The Contribution of Primordial Binaries to the Blue Straggler Population in 47 Tucanae. *ApJ*, **605**, L29–L32.

Mapelli, M., Sigurdsson, S., Ferraro, F. R., et al. 2006. The Radial Distribution of Blue Straggler Stars and the Nature of Their Progenitors. *MNRAS*, **373**, 361–368.

Mashonkina, L., Gehren, T., Shi, J.-R., Korn, A. J. and Grupp, F. 2011. A Non-LTE Study of Neutral and Singly-Ionized Iron Line Spectra in 1D Models of the Sun and Selected Late-Type Stars. *A&A*, **528**, A87.

Mathieu, R. D. and Geller, A. M. 2009. A Binary Star Fraction of 76 Per Cent and Unusual Orbit Parameters for the Blue Stragglers of NGC 188. *Nature*, **462**, 1032–1035.

McCrea, W. H. 1964. Extended Main-Sequence of Some Stellar Clusters. *MNRAS***128**, 147–155.

Meylan, G. and Heggie, D. C. 1997. Internal Dynamics of Globular Clusters *A&A Rev.*, **8**, 1–143.

Milone, A. P., Piotto, G., Bedin, L. R., et al. 2012. The ACS Survey of Galactic Globular Clusters. XII. Photometric Binaries along the Main Sequence. *A&A*, **540**, A16.

Miocchi, P., Pasquato, M., Lanzoni, B., et al. 2015. Probing the Role of Dynamical Friction in Shaping the BSS Radial Distribution. I. Semi-Analytical Models and Preliminary N-Body Simulations. *ApJ*, **799**, 44–53.

Moretti, A., de Angeli, F. and Piotto, G. 2008. Environmental Effects on the Globular Cluster Blue Straggler Population: A Statistical Approach. *A&A*, **483**, 183–197.

Parada, J., Richer, H., Heyl, J., et al. 2016. Formation and Evolution of Blue Stragglers in 47 Tucanae. *ApJ*, **830**, 139–148.

Pietrinferni, A., Cassisi, S., Salaris, M., and Castelli, F. 2006. A Large Stellar Evolution Database for Population Synthesis Studies. II. Stellar Models and Isochrones for an α-Enhanced Metal Distribution. *ApJ*, **642**, 797–812.

Piotto, G., De Angeli, F., King, I.R., et al. 2004. Relative Frequencies of Blue Stragglers in Galactic Globular Clusters: Constraints for the Formation Mechanisms. *ApJ*, **604**, L109–L112.

Piotto, G., Milone, A. P., Bedin, L. R., et al. 2015. The Hubble Space Telescope UV Legacy Survey of Galactic Globular Clusters. I. Overview of the Project and Detection of Multiple Stellar Populations. *AJ*, **149**, 91–117.

Raso, S., Ferraro, F. R., Dalessandro, E., et al. 2017. The UV-Route to Search for Blue Straggler Stars in Globular Clusters: First Results from the HST UV Legacy Survey. *ApJ*, **839**, 64–74.

Renzini, A. and Fusi Pecci, F. 1988. Tests of Evolutionary Sequences Using Color-Magnitude Diagrams of Globular Clusters. *A&A Rev.*, **26**, 199–244.

Sandage, A. R. 1953. The Color-Magnitude Diagram for the Globular Cluster M 3. *AJ*, **58**, 61–75.

Sanna, N., Dalessandro, E., Lanzoni, B., et al. 2012. The blue Straggler Star Population in NGC 6229. *MNRAS*, **422**, 1171–1177.

Sanna, N., Dalessandro, E., Ferraro, F. R., et al. 2014. The WFPC2 Ultraviolet Survey: The Blue Straggler Population in NGC 5824. *ApJ*, **780**, 90–98.

Shara, M. M., Saffer, R. A. and Livio, M. 1997. The First Direct Measurement of the Mass of a Blue Straggler in the Core of a Globular Cluster: BSS 19 in 47 Tucanae. *ApJ*, **489**, L59–L62.

Sills, A., Glebbeek, E., Chatterjee, S., and Rasio, F. A. 2013. Mock Observations of Blue Stragglers in Globular Cluster Models. *ApJ*, **777**, 105–111.

Sills, A., Karakas, A. and Lattanzio, J. 2009. Blue Stragglers after the Main Sequence. *ApJ*, **692**, 1411–1420.

Simunovic, M. and Puzia, T. H. 2016. Blue Straggler Star Populations in Globular Clusters. II. Proper-Motion Cleaned HST Catalogues of BSSs in 38 Galactic GCs. *MNRAS*, **462**, 3401–3418.

Simunovic, M., Puzia, T. H. and Sills, A. 2014. The Blue Straggler Star Population in NGC 1261: Evidence for a Post-Core-Collapse Bounce State. *ApJ*, **795**, L10–L14.

Sollima, A., Lanzoni, B., Beccari, G., Ferraro, F. R. and Fusi Pecci, F. 2008. The Correlation between Blue Straggler and Binary Fractions in the Core of Galactic Globular clusters. *A&A*, **481**, 701–704.

Tian, B., Deng, L.-C., Han, Z.-W. and Zhang, X.-B. 2006. The Blue Stragglers Formed via Mass Transfer in Old Open Clusters. *A&A*, **455**, 247–254.

Zinn, R. and King, C. R. 1982. The Mass of the Anomalous Cepheid in the Globular Cluster NGC 5466. *ApJ*, **262**, 700–708.

Xin, Y., Ferraro, F. R., Lu, P., et al. 2015. The Binary Mass Transfer Origin of the Red Blue Straggler Sequence in M30. *ApJ*, **801**, 67–74.

20

Binaries at Very Low Metallicity

SARA LUCATELLO

20.1 Background

Our knowledge of binarity (and multiplicity), its incidence, properties and characteristics for different stellar types and environments, has no doubt widened dramatically in the last decade or two. Improved instrumentation and large-scale stellar surveys have afforded samples of unprecedented sizes of stars across the Hertzsprung–Russell (HR) diagram, leading to considerable advancements. There are now studies reporting on multiplicity and characteristics of multiple stars ranging from massive objects (larger than $M \sim 20\ M_{\odot}$) all the way to – also thanks to the push from extrasolar planetary science – very low-mass stars (see, e.g., Luhman, 2012). There is also a growing body of work reporting on binaries in open and globular clusters (see, e.g., Mermilliod et al., 2008; Lucatello et al., 2015). Very little is known, however, about stellar multiplicity at very low metallicity.

Binaries play a key role in a number of astrophysical issues, ranging from supernovae to the dynamical evolution of stellar clusters, as it is discussed extensively in the previous chapters. Two aspects are of particular interest in the very metal-poor regime: binaries as tracers of very low metallicity star formation and as crucial element in the explanations for the high incidence of chemically peculiar objects found in this metal range, chemically enhanced metal-poor (CEMP) stars.

Throughout the chapter, the description extremely metal-poor (EMP hereafter) stars will indicate objects with $[Fe/H] < -2.5$ dex.[1] While Fe does not necessarily scale with overall metal content of stars, and even less so at low metallicity, because of the abundance of strong atomic transitions in the optical spectrum, it is generally one of the easiest abundances to measure and will be used in the chapter, just as it is in the literature, as a proxy for stellar metallicity.

20.1.1 *Binaries as Probes of Star Formation*

The first aspect of interest is that binaries provide indications of the conditions of star formation, as they shed light on the environment they were born in. The binary fraction,

[1] $[A/B] = \log_{10}(\frac{N_A}{N_B})_* - \log_{10}(\frac{N_A}{N_B})_\odot$ where N_A indicates the number density of the atomic species A in the atmosphere, $*$ indicates the star and $\odot$ indicates the Sun.

orbital period (P) and mass-ratio (q) distributions are constraints for star-forming models. Even if details are still not well established, star formation is expected to be affected by metallicity. Some models for the formation of metal-free stars report that, because of the lack of metals that act as coolants, Population III contained only massive stars (see, e.g., Hirano et al. 2014), that have now long evolved and that no member of such populations is observable in its stellar form at the present time. On the other hand, there are works that argue that, because the disk that builds up around protostellar Pop III objects is highly susceptible to fragmentation, low-mass star formation should be possible (Clark et al., 2011) even in such regime.

The former idea is supported by the fact that, in spite of extensive searches past and still ongoing, no metal-free star has been found. In fact, even the current record-holding star as the most Fe poor (Keller et al., 2014) has small but measurable amounts of Mg and Ca and a C content not very different from that of many CEMPs that will be discussed later in this and in the following chapters. Moreover, the star with the lowest overall metallicity (Caffau et al., 2011) known to date has measurable Fe, Mg, Si, Ca, Ti and Ni.

The compositions of these objects require that the medium from which they formed was enriched from some previous generations of star, making them by definition extreme members of stellar Population II. Nevertheless, it must be kept in mind that it cannot be excluded, on the basis of observations, that metal-free low-mass stars did form, as they might just be exceedingly rare or even located in areas of the Galaxy (like the Bulge) where searches for them have just started (see, e.g., Howes et al., 2016).

The existence of EMP stars with metallicities as low as [Fe/H] ~ -5 (and below) does definitely, however, show that low-mass star formation does happen even at extremely low metallicities. Simulations show that in fact at $Z \sim 10^{-5}$ low-mass stars are formed, and in proportions not so different from what is expected to happen at solar metallicity (Clark et al., 2008).

Moreover, models of low-mass star formation with different metallicities by Machida (2008) suggest that, because in the evolution of the clouds adiabatic cores form at later phases at low metallicity, fragmentation takes place even with smaller rotational energy. Since fragmentation results in multiple stellar systems, binary (and multiple) systems are formed with higher frequency, and with shorter periods at very low metallicities with respect to solar or moderately low metallicities. This prediction is shared by several other authors (Clark et al., 2008, 2011; Greif et al., 2011). However, observational constraints on the binary fraction and period distribution among EMP stars and as a function of [Fe/H] are needed to test these predictions and constrain a new generations of models.

20.1.2 Binaries and CEMP Stars

The low metallicity regime is characterised by a very large number of C-rich stars, the so-called CEMP stars.[2] Different authors have adopted different criteria to select them,

[2] This section provides a short summary of the issue; the reader is referred to Chapter 4 for a thorough discussion of the issue of the origin of CEMP stars.

using -2 or -2.5 dex as the upper cutoff metallicities and various thresholds for C enhancements, with [C/Fe]> 0.5, 0.7 or 1.0 dex, some even advocated using a luminosity-dependent cutoff, to account for the reduction of [C/Fe] due to evolution along the giant branch.

Early reported fraction of CEMP stars vary considerably in the literature, ranging from overall rates of 9% of Frebel et al. (2006) to a lower limit of 21% reported by Lucatello et al. (2006). More recent work, based on Sloan Digital Sky Surveys (SDSSs) SEGUE and SEGUE-2, indicate that such fraction varies with metallicity (from $\sim$14% at [Fe/H] ~ -2.5 to 39% at ~ -3.5 dex; Lee et al., 2013) and that increases with distance from the Galactic plane (Carollo et al., 2012). While the individual figures on incidence of CEMP stars are quite different (which is perhaps to be expected given the different definitions adopted by different groups), all works agree on the fact that the fraction of such objects is much higher than at solar metallicity.

When examined closely, CEMP stars show different chemical signatures. The most numerous class, which makes up the majority of CEMP stars in the $-2 \leq$[Fe/H]≤ -3.0 range, is that of the so-called CEMP-*s* stars, objects that exhibit enhanced C, N and *s*-process elements abundances. In some cases, these objects also show enhanced r-process elements abundances and are labelled by some authors as CEMP-r/s.[3]

CEMP-*s* stars have been shown to be the metal-poor analogy of classic Ba and CH stars (McClure, 1984), but to be far more common in the EMP regime than their solar metallicity and mildly metal-poor counterparts. Their incidence increases strongly below ~ -1.8 dex, where all the CEMP stars show overabundance of *s*-process elements, to then decrease rather sharply around -3 dex, where CEMP stars without *s*-process overabundance become more common. Radial velocity monitoring studies of CEMP-*s* objects established that they are compatible with all of them being in binary systems (Lucatello et al., 2005). The more massive companion ($\sim$1.5-4 $M_{\odot}$), now a faint, undetectable white dwarf, evolved up the Asymptotic Giant Branch (AGB) transferring material enriched by shell nucleosynthesis and hence imprinting a characteristic chemical signature on the star we observe today as CEMP-*s*.

Another group of CEMP stars show very low n-capture element abundances and are hence labelled CEMP-no stars. These objects, which are relatively rare around [Fe/H]~ -2.5 dex, become more and more common with decreasing Fe content, to the point that eight out of the nine stars known to date with [Fe/H] < -4.5 belong to this class. Radial velocity monitoring of these objects is still rather sparse; however, at the present time studies seem to suggest that their binary fraction is not significantly different from that found among solar metallicity stars (Starkenburg et al., 2014). The origin for the peculiar C abundance is then likely due to a different mechanism than the mass transfer in a binary that explains CEMP-*s*.

[3] For completeness sake, the reader should be aware that C and r-process enrichment (but no *s*-process enrichment) exists (Sneden et al., 2003; Hansen et al., 2015). Only very few stars are known to date to belong to the CEMP-r class, and while these objects are very interesting from a nucleosynthetic point of view, their small number prevents us from any statistically meaningful discussion on their connection to binarity. They will therefore not be discussed any further in this text.

The explanation for the strong increase in the overall incidence of CEMP stars, and of the variation of the relative proportions of them, is still not clear. Models can now account rather satisfactorily for the observed abundance patterns in CEMP-*s* stars (see, e.g., Hampel et al., 2016), it is however still challenging to account for the large observed fraction of them. Several uncertainties affect the modelling of the formation of these objects.

The shape of the initial mass function of the early Galaxy could affect the incidence of CEMP-*s* stars, and a small shift towards intermediate mass stars could account for the observed fractions, as discussed in Lucatello et al. (2005).

On the other hand, the amount of carbon and *s*-process elements produced by AGB nucleosynthesis as a function of the stellar mass and the efficiency of mass transfer at low metallicity are still not well established. The effect of variations of these parameters on the expected fraction of CEMP-*s* has been expored by Abate et al. (2013). The binary fraction and the distributions of periods and mass ratios also have an effect, and the models described in Abate et al. (2015) discuss the effect of binary characteristics on the expected CEMP-*s* fraction. All these parameters are, however, poorly constrained observationally. Moreover, no convincing attempt has been made to explain the sudden drop in incidence of these objects below ~ -3.0 dex. It is conceivable that a change in binarity properties as a function of metallicity is connected to the raise and then fall of CEMP-*s* fraction with decreasing metallicity.

There is currently no widely accepted explanation for the origin of the abundance patterns observed in CEMP-no stars, nor for their high frequency among EMP and for the lack of analogous solar metallicity. Even if observations seem to suggest that binarity is not responsible for their composition, a comparison of the binary fraction of CEMP-no to that of non-C-rich EMP stars, rather than to solar metallicity, remains a key piece of information to disentangle the origin of these objects. Observational constraints on the binary fraction and characteristics of EMP stars are hence crucial to shed light on this phenomenona.

20.2 Literature Studies of Binarity at Low Metallicity

In the last few years, a handful of studies have focused on determining the binary fraction at low metallicity. Data from large-scale Galactic surveys as well as smaller, but still extensive, targeted surveys have been used to probe the binary populations in Pop II stars.

Gao et al. (2014) using combined SDSS and Large Sky Area Multi-Object Fibre Spectroscopic Telescope (LAMOST) data find that data are consistent with binary fraction increasing from 30%±6% to 56%±5% among Galactic F, G, K stars with [Fe/H]> -0.6 and [Fe/H]< -1.1 dex respectively. This trend is confirmed by Gao et al. (2017), who studied Galactic dwarfs using LAMOST data. Badenes et al. (2018) also finds that the binary fraction below [Fe/H]< -0.5 is a factor of two to three larger than that found above such a threshold.

On the other hand, the frequency of visual binaries seems to be lower than at solar metallicity (see, e.g., Zapatero Osorio and Martín, 2004) even if perhaps not for K-M dwarfs (Jao et al., 2009).

Other studies, specifically designed to explore binarity also in very low metallicity stars, seem to not confirm the increase in binary fraction detected with lowering metallicity among moderately metal-poor stars. Carney et al. (2005) reported a mildly declining detected spectroscopic binary fraction between [Fe/H] −0.5 and −2.5. A similar finding is described by Rastegaev (2010), who also reports that data are consistent with the orbital period distribution at low metallicity being peaked around logP=2—3, a shorter value than that found among solar metallicity stars. There are still, however, too few well-studied Population II binaries to make a reliable estimate of the orbital period distribution and the mass-ratio distribution of metal-poor binaries. Note that the regime below −2.5 dex is completely unexplored in these studies, and that the determination of the binary fraction at −2.5 is based on just ∼20 stars.

More recently, Aoki et al. (2015) presented the results from a 1.5-year-long Subaru radial velocity monitoring dedicated to turnoff EMP ([Fe/H]< −3 dex) stars. They concentrate on double-lined binaries and on CEMP-*s* stars, and considering a sample of 109 turnoff stars, they report an estimated incidence of short-period binaries (P <1,000 d) in the ∼10% to 20% range. On the basis of this result, they argue that the binary fraction of EMP should be much higher than at higher metallicity or/and that the period distribution might be skewed towards lower values.

20.3 EMP Binarity

The dearth of published studies for binary fraction of EMP is not so surprising. They are rare, rather distant and faint objects, with low spacial density and magnitudes often beyond the limit of large-scale surveys. Moreover, their low metal content means their spectra have fewer lines than their solar or moderately metal-poor counterparts, which leads to less accurate radial velocity measurements and makes monitoring programs based on this quantity less effective in detecting binaries.

Presently, besides the previously cited Aoki et al. study of double-lined turnoff binaries, there is no published work on the overall EMP stars fraction. Radial velocity monitoring of EMP stars have generally concentrated on CEMP (mostly CEMP-*s*, but lately also CEMP-no; see Chapters 3 and 4 in this book and Section 20.1.2), because of interest in understanding their origin. In order to get a grasp of the overall EMP binarity fraction, it is obviously necessary to systematically probe binarity also in non-C rich stars. This is a key piece of information in the understanding of the origin of CEMP stars themselves, as any scenario for their formation has to take into account the binary fraction of the parent population.

The Stellar Abundances for Galactic Archaeology (SAGA) Database[4] (Suda et al., 2008) collects measurements based on high-quality data of atmospheric parameters, abundance ratios and radial velocity published in the literature for EMP stars. As of its latest update (September 2017), it contains a total of 1,053 stars with [Fe/H]≤ −2.5. Of these, 620

[4] http://sagadatabase.jp/.

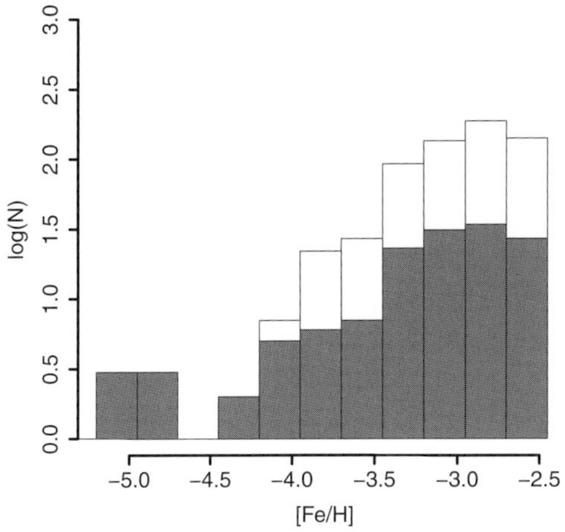

Figure 20.1 The metallicity distribution function of the stars in the SAGA database, in grey the distribution of stars with [C/Fe]>0.7

have reported C-abundance measurements (analyses that report upper only limits to C abundances are not included in this count), with 143 having [C/Fe]>0.7 dex, the threshold used in this discussion to consider a star CEMP. Therefore, the CEMP star fraction in the database is 23.1%±2.7%. Figure 20.1 shows the metallicity distribution of the SAGA sample, including that of CEMP stars.

For a subsection of these objects, multiple radial velocity measurements are reported and a derivation of the binary fraction from these data can be done. In many cases, these are measurements taken from contiguous spectra taken during the same night (or consecutive nights), which have very little significance in the detection of binaries, except for the very close pairs. Sixteen of these objects have reported orbital parameters, ranging from three to ~4,300 days. C abundance is reported only for seven of these objects, and in all cases it is [C/Fe>0.7 dex.

The database contains independent radial velocities with a baseline of at least one year for 57 EMP stars, with metallicity ranging from −2.5 to −4.1 dex. Twenty-eight of these objects are C-enhanced. The incidence in this sample is about a factor of 2 higher than when considering the fraction among all stars below −2.5. The reason behind this large discrepancy is the interest in establishing the binary status of CEMP stars, which mean those objects are much more likely to have multiple observed and published radial velocities, which bias considerably the statistics of the sample in favour of these objects.

Applying the same procedure as outlined in Lucatello et al. (2005), one can compute the probability P(χ^2—f) (where f are the degrees of freedom, i.e., f = 1–n with n number of measurements) that the radial velocity values measured for a given stars are compatible with being different measurements of the same values (in other words, the probability that the observed scatter is due to observational errors and not the actual variation of the measured radial velocity).

Adopting as a positive identification for radial velocity variation a value of P $\leq$0.01, the result on the SAGA sample is that 11 stars are radial velocity variables, for a fraction of 19%±5%. The fraction is consistent with that reported by Carney et al. for stars at ~ -2; however, the baseline of the observations for the SAGA-selected sample is much shorter than the Carney et al. one. When considering just non-C-enhanced star, five objects out of 26 meet the binary criteria, for a binary fraction of 17%±7%. The two fractions are virtually identical within the errors, probably due to the fact that the number of CEMP-*s* (likely all binaries) does not make up the majority of the CEMP objects sample. In spite of this fact, the sample containing a much larger fraction of CEMP stars of the actual EMP population cannot be considered representative of the overall population.

Monte Carlo simulations can be performed to explore the binary fraction of the parent population. Because of lack of information on the characteristics of EMP (both C and non-C-rich) binaries, uniform mass ratio, uniform eccentricity and Raghavan et al. (2010) period distribution are assumed for the sake of the present simulation.

Under these assumptions and using the selected sample observing patterns and observational errors, the derived overall and non-C-enhanced binary incidence is consistent with a ~45% to 70% parent population fraction. This seems to support the Aoki et al. (2015) findings of a fraction increasing at low metallicity.

This result is, however, obtained adopting a period distribution derived for solar-type stars, which might not be appropriate for the EMP binaries, and the detection of such a large number of binary stars might actually indicate that the period distribution is peaked at smaller values than assumed. Moreover, the reader must keep in mind that SAGA is a database made up of published literature measurements. It gathers data from different sources, collected with different purposes and using different instruments and strategies, analysed with different methods. It is therefore affected by a number of selection and observational biases as well as possible zero-point offsets in the measurements that are exceedingly difficult to account for. Therefore, extreme caution must be exerted when considering statistical properties of the sample, including the binary fraction.

20.4 Future Prospectives

Specific long-term radial velocity surveys targeted at extremely metal-poor stars are necessary to get a reliable estimate of the binary fraction and period distribution of EMP stars. The Aoki et al. program is still ongoing and monitoring the radial velocity of a sample of ~100 EMP stars. Because it focuses on turnoff stars for which measurement of C abundance is possible only in case of very high enhancements (over [C/Fe]~1.5 dex), for a considerable fraction of the objects it will be impossible to establish whether they are CEMP stars or not, hence hampering the possibility of deriving the binary fraction separately for non-C-rich stars, CEMP-*s* and CEMP-no stars. The same considerations apply to period distribution.

In order to determine the overall EMP and non-C-rich EMP binary fraction and period distribution, a radial velocity monitoring survey was started at Lick Observatory using the Hamilton Spectrograph. The survey plans also to start using the Lick Automatic Planet Finder in the near future. Targets have been selected among rather bright stars with [Fe/H] < -2.5 dex and currently includes $\sim$45 non-C-rich and $\sim$20 CEMP stars that either lack radial velocity monitoring or require more data to determine the orbital parameters. The data, combined with previous literature information, and data from ongoing and planned large-scale Galactic stellar surveys (e.g., APOGEE-2, GALAH, WEAVE, 4-MOST, etc.) are expected to lead to long-term radial velocity monitoring for at least $\sim$80 objects, of which $\sim$60 will be non-C-rich EMP stars. The aim is to derive the binary fraction for non-C-rich stars and place some constraints on their orbital period distribution. For what concerns CEMP stars, the data will be combined with the literature and other ongoing or planned observational efforts that aim at establishing the binary fraction of CEMP-no stars and derive the orbital period distribution of CEMP-*s* stars.

References

Abate, C., Pols, O. R., Izzard, R. G., Mohamed, S. S. and de Mink, S. E. 2013. Wind Roche-Lobe Overflow: Application to Carbon-Enhanced Metal-Poor Stars. *A&A*, **552**, A26.

Abate, C., Pols, O. R., Stancliffe, R. J., et al. 2015. Modelling the Observed Properties of Carbon-Enhanced Metal-Poor Stars Using Binary Population Synthesis. *A&A*, **581**, A62.

Aoki, W., Suda, T., Beers, T. C. and Honda, S. 2015. High-Resolution Spectroscopy of Extremely Metal-Poor Stars from SDSS/SEGUE. II. Binary Fraction. *AJ*, **149**, 39.

Badenes, C., Mazzola, C., Thompson, T. A., et al. 2018. Stellar Multiplicity Meets Stellar Evolution and Metallicity: The APOGEE View. *ApJ*, **854**, 147.

Caffau, E., Bonifacio, P., François, P., et al. 2011. An Extremely Primitive Star in the Galactic Halo. *Nature*, **477**, 67.

Carney, B. W., Aguilar, L. A., Latham, D. W. and Laird, J. B. 2005. A Survey of Proper-Motion Stars. XVII. A Deficiency of Binary Stars on Retrograde Galactic Orbits and the Possibility that Ω Centauri is Related to the Effect. *AJ*, **129**, 1886.

Carney, B. W., Latham, D. W., Laird, J. B. and Aguilar, L. A. 1994. A Survey of Proper Motion Stars. 12: An Expanded Sample. *AJ*, **107**, 2240.

Carney, B. W., Latham, D. W., Stefanik, R. P., Laird, J. B. and Morse, J. A. 2003. Spectroscopic Binaries, Velocity Jitter, and Rotation in Field Metal-Poor Red Giant and Red Horizontal-Branch Stars. *AJ*, **125**, 293.

Carollo, D., Beers, T. C., Bovy, J., et al. 2012. Carbon-Enhanced Metal-Poor Stars in the Inner and Outer Halo Components of the Milky Way. *ApJ*, **744**, 195.

Chini, R., Hoffmeister, V. H., Nasseri, A., Stahl, O. and Zinnecker, H. 2012. A Spectroscopic Survey on the Multiplicity of High-Mass Stars. *MNRAS*, **424**, 1925.

Clark, P. C., Glover, S. C. O. and Klessen, R. S. 2008. The First Stellar Cluster. *APJ*, **672**, 757–764.

Clark, P. C., Glover, S. C. O., Smith, R. J., et al. 2011. The Formation and Fragmentation of Disks around Primordial Protostars. *Science*, **331**, 1040.

Frebel, A., Christlieb, N., Norris, J. E., et al. 2006. Bright Metal-Poor Stars from the Hamburg/ESO Survey. I. Selection and Follow-up Observations from 329 Fields. *ApJ*, **652**, 1585.

Gao, S., Liu, C., Zhang, X., et al. 2014. The Binarity of Milky Way F,G,K Stars as a Function of Effective Temperature and Metallicity. *ApJL*, **788**, L37.

Gao, S., Zhao, H., Yang, H. and Gao, R. 2017. The Binarity of Galactic Dwarf Stars along with Effective Temperature and Metallicity. *MNRAS*, **469**, L68.

Greif, T. H., Springel, V., White, S. D. M., et al. 2011. Simulations on a Moving Mesh: The Clustered Formation of Population III Protostars. *ApJ*, **737**, 75.

Hampel, M., Stancliffe, R. J., Lugaro, M. and Meyer, B. S. 2016. The Intermediate Neutron-Capture Process and Carbon-Enhanced Metal-Poor Stars. *ApJ*, **831**, 171.

Hansen, T., Hansen, C. J., Christlieb, N., et al. 2015. An Elemental Assay of Very, Extremely, and Ultra-Metal-Poor Stars. *ApJ*, **807**, 173.

Hirano, S., Hosokawa, T., Yoshida, N., et al. 2014. One Hundred First Stars: Protostellar Evolution and the Final Masses. *APJ*, **781**, 60.

Howes, L. M., Asplund, M., Keller, S. C., et al. 2016. The EMBLA Survey: Metal-poor Stars in the Galactic Bulge. *MNRAS*, **460**, 884.

Jao, W.-C., Mason, B. D., Hartkopf, W. I., Henry, T. J., and Ramos, S. N. 2009. Cool Subdwarf Investigations. II. Multiplicity. *AJ*, **137**, 3800.

Keller, S. C., Bessell, M. S., Frebel, A., et al. 2014. A Single Low-Energy, Iron-Poor Supernova as the Source of Metals in the Star SMSS J031300.36-670839.3. *Nature*, **506**, 463.

Lee, Y. S., Beers, T. C., Masseron, T., et al. 2013. Carbon-Enhanced Metal-Poor Stars in SDSS/SEGUE. I. Carbon Abundance Estimation and Frequency of CEMP Stars. *AJ*, **146**, 132.

Lucatello, S., Beers, T. C., Christlieb, N., et al. 2006. The Frequency of Carbon-Enhanced Metal-Poor Stars in the Galaxy from the HERES Sample. *ApJL*, **652**, L37.

Lucatello, S., Sollima, A., Gratton, R., et al. 2015. The Incidence of Binaries in Globular Cluster Stellar Populations. *A&A*, **584**, A52.

Lucatello, S., Gratton, R. G., Beers, T. C. and Carretta, E. 2005. Observational Evidence for a Different Initial Mass Function in the Early Galaxy. *ApJ*, **625**, 833.

Luhman, K. L. 2012. The Formation and Early Evolution of Low-Mass Stars and Brown Dwarfs. *Annual Review of Astronomy and Astrophysics*, **50**, 65.

Machida, M. N. 2008. Binary Formation in Star-Forming Clouds with Various Metallicities. *ApJL*, **682**, L1.

McClure, R. D. 1984. The Binary Nature of CH-Stars. *ApJL*, **280**, L31.

Mermilliod, J.-C., Grenon, M. and Mayor, M. 2008. Membership, Binarity, and Rotation of F-G-K Stars in the Open Cluster Blanco 1. *A&A*, **491**, 951.

Raghavan, D., McAlister, H. A., Henry, T. J., et al. 2010. A Survey of Stellar Families: Multiplicity of Solar-Type Stars. *ApJS*, **190**, 1.

Rastegaev, D. A. 2010. Multiplicity and Period Distribution of Population II Field Stars in Solar Vicinity. *AJ*, **140**, 2013.

Sneden, C., Cowan, J. J., Lawler, J. E., et al. 2003. The Extremely Metal-Poor, Neutron Capture-Rich Star CS 22892-052: A Comprehensive Abundance Analysis. *ApJ*, **591**, 936.

Starkenburg, E., Shetrone, M. D., McConnachie, A. W. and Venn, K. A. 2014. Binarity in Carbon-Enhanced Metal-Poor Stars. *MNRAS*, **441**, 1217.

Suda, T., Katsuta, Y., Yamada, S., et al. 2008. Stellar Abundances for the Galactic Archeology (SAGA) Database: Compilation of the Characteristics of Known Extremely Metal-Poor Stars. *PASJ*, **60**, 1159.

Zapatero Osorio, M. R. and Martín, E. L. 2004. A CCD Imaging Search for Wide Metal-Poor Binaries. *A&A*, **419**, 167.

21

Population and Spectral Synthesis: It Doesn't Work without Binaries

J. J. ELDRIDGE[1] AND ELIZABETH R. STANWAY

21.1 What Is Population and Spectral Synthesis?

Population synthesis involves predicting the parameters of a population of stars through a combination of individual stellar models or empirical templates. Examples of stellar populations include those in individual star clusters and galaxies or those that give rise to a sample of transient events such as core-collapse supernovae or gravitational wave merger 'chirps'. To make such predictions through population synthesis, a number of stellar models, each predicting the properties and evolution of one star, are combined together with an initial parameter distribution to calculate a synthetic population that can be compared to observational constraints including stellar type ratios and transient event rates.

Spectral synthesis is the term for the combination of such a synthetic population with matched stellar atmosphere models that predict how each star of known surface temperature, gravity and composition would appear if observed across all wavelengths. The resulting composite spectral energy distribution can be convolved with the known parameters of individual detectors and telescopes. This step effectively observes the model population so it can be directly compared to the observational data. This technique is especially important when attempting to interpret unresolved stellar populations where individual stars cannot be studied and classified individually.

The first population synthesis models can be traced back to the first attempts to understand the stellar content of galaxies and star clusters (e.g., Tinsley, 1968; Tinsley and Gunn, 1976). There are now many codes that perform population and spectral synthesis. The evolutionary models most widely used to date carry the implicit assumption that most stars evolve as single stars, without any significant binary interaction, and thus relatively few evolutionary avenues are available (e.g., Leitherer et al., 1999; Bruzual and Charlot, 2003; Choi et al., 2016). While these codes have allowed us to deepen our knowledge of stellar systems, they are limited by their single-star evolutionary paradigm. This is fundamentally an unphysical assumption. It is clear from other chapters in this book that the fraction of binary stars in stellar populations range from 20% for the lowest-mass stars to 100% for more massive stars. As this volume also indicates, in many binaries at least one of the stars will have a substantially different evolutionary path from that of a single star. As a result, a failure to account for binary effects would be expected to compromise interpretation of stellar system parameters in any regime where these are significant for the dominant population.

Over the past few decades, a number of groups have been attempting to account for interacting binaries in their population synthesis, and more rarely spectral synthesis. Groups computing population synthesis models include, for example, Tutukov and Yungelson (1996); Hurley et al. (2002); Willems and Kolb (2004); De Donder and Vanbeveren (2004); Zhang et al. (2005); Belczynski et al. (2007); Lipunov et al. (2009); Izzard et al. (2009); Toonen and Nelemans (2013); and Eldridge et al. (2017). However, from this list only De Donder and Vanbeveren (2004), Zhang et al. (2005) and Eldridge et al. (2017) also make spectral synthesis predictions. Inclusion of spectral synthesis within binary population synthesis models has confirmed that interacting binaries make a substantial change to the predictions of spectral synthesis and that the use of single-star models can leads to errors in interpreting certain aspects of stellar populations.

In this chapter, we first give a brief overview of how population and spectral synthesis is performed and highlight some of the key uncertainties. We then discuss selected example observations where interacting binaries make a substantial difference to how we understand the underlying stellar systems. These include the HR diagram of resolved stellar clusters, number counts of massive stars in galaxies, core-collapse supernovae, gravitational wave sources, the stellar populations of distant galaxies and the epoch of reionisation.

21.2 How Do You Do It?

To perform a population synthesis, one first has to gather or create a set of stellar evolution models from which to construct the synthetic population. It is possible either to use extant stellar models (e.g., Ekström et al., 2012) or to calculate your own (e.g., Choi et al., 2016; Eldridge et al., 2017). It should be noted that these models themselves incorporate substantial uncertainties. There are many known problems in stellar evolution such as determination of mass-loss rates, rotation rates and how to implement convection that all impact on the certainty with which predictions of stellar properties at a given mass and age can be made. We can, however, test these models by comparing their properties to observed individual stars such as the Sun or the physically well-constrained main-sequence stars in eclipsing binaries.

When computing models of interacting binary stars, the situation is more complicated due to the extra physics that needs to be included in evolution codes. Prescriptions for Roche-lobe overflow (RLOF) and common-envelope evolution (CEE) must be included. An extensive body of literature exists exploring the uncertainties on each of these and the tunable parameters in any given prescription. Nonetheless, these are vital to account for the mass transfer from one star to the other in a binary, and to shrink the orbit to produce the tight post-mass-transfer binaries we see in the Galaxy. The effects of any mass transfer on the subsequent evolution must then be considered.

There are two different approaches to modelling the evolution of binary stars. One of these is to use a detailed stellar evolution code (e.g., De Donder and Vanbeveren, 2004;

Eldridge et al., 2017); however, this has the drawback that each model can take several minutes to calculate. This means that to calculate a large grid of models in order to fully investigate the parameter space can take years on a single computer. With universities investing in computational resources and large clusters, it is beginning to become feasible to use detailed models to create the tens of thousands of models required for population synthesis.

Alternately, rapid codes, for example Hurley et al. (2002), approximate the evolution of binary stars by fitting functions to a smaller precalculated grid of detailed models and interpolating between them. While the evolution is only followed in an approximate manner, this method does allow multiple large grids to be rapidly calculated so that how uncertainties in the input physics affect the results of the population synthesis can be investigated (e.g., de Mink et al., 2013). Applications include exploring the effects of stellar rotation on evolutionary stages. Such a study would be challenging in a detailed code as each model run would need to follow the transport of angular momentum through the star in detail, which makes the models difficult (and also time consuming) to calculate (e.g., Cantiello et al., 2007).

Once the individual stellar models are calculated, they must be combined into a synthetic population. This is done by weighing each model by a factor that represents how likely it is to exist in a given scenario. For all populations, we use an initial mass function that determines how likely it is that a star of a given stellar mass exists at a zero main-sequence age after the initial star formation episode. Examples include Kroupa et al. (1993) and Chabrier (2003). These all are more modern versions of the model established by Salpeter (1955), where a power-law representation was used to capture the observed fact that there are many more low-mass than high-mass stars.

In a binary population, we must also describe the distribution of initial periods, eccentricities and secondary masses. Some of the most recent can be found in Moe and Di Stefano (2017). The binary fraction is around 100% of stars being in binaries for stars $\geq 10\ M_{\odot}$, with 70% of them close enough to have their evolution affected by binary interactions during their evolution (Sana et al., 2012). For less massive stars, less massive than $<1\ M_{\odot}$, the binary fraction drops to 20 to 40%.

The period distributions also vary with stellar mass. While Opik's law is a good first estimate, studies tend to indicate that there may be more close binaries. Nonetheless, a flat distribution in the mass ratio for the secondary seems a good approximation. More data are still required to gain a firm understanding, but the initial parameters of binaries are beginning to become well constrained.

To summarise, we now know how to weight each of our binary models and have a model that follows the evolution of the stars, but there is one final complication: the first supernova. When a supernova occurs in a binary, there are one of two possible outcomes: either the binary is unbound and the two stars go on to evolve as single stars, or they remain bound and become a binary with a compact companion. These may evolve into a X-ray binary.

Determining whether the system is unbound or not depends on a few factors. First is the amount of ejecta in the supernova (SN). It can be shown, assuming a circular pre-SN orbit, that if the ejecta mass is more than half the mass of the binary, then the system is unbound. Therefore, estimating how much mass goes into the remnant and how much is ejected is a key problem. Work such as Sukhbold et al. (2016) is showing that it may not be easy to determine this accurately. The final outcome of evolution appears to be chaotic over some mass ranges and depends on all the uncertainties put into the stellar model. However, some simpler schemes exist to estimate the remnant mass, and future work will allow us to understand this more.

The other factor is that in formation neutron stars and black holes appear to be given a momentum kick during core collapse that can range from a few 10s km/s up to 1,000 km/s! (e.g., Hobbs et al., 2005). The source of the kick is currently being investigated by many groups (e.g., Bray and Eldridge, 2016; Janka, 2017). It is important to consider this, as kicks can change the orbital velocity of the compact remnant in the SN, and this can lead to unbinding binaries that would have remained bound as well as keeping binaries bound that would otherwise have been unbound. Further complicating the issue is whether black hole have kicks or not; most prescriptions seem to indicate they must be weaker, but there is limited observational evidence (e.g., Mandel, 2016).

In our population synthesis, the first supernova therefore creates multiple possible future evolutionary pathways for each binary. These have to be calculated and are usually accounted for by Monte Carlo methods, making models with a large number of different random kicks in direction and magnitude and weighting the results by a probability distribution. This is again where the rapid models have an advantage, as many different models can be calculated quickly, while for detailed models some approximations must be made to calculate such stellar models in a reasonable timeframe.

Once a synthetic population is created, it can be compared to observed stellar populations. For observations such as number counts, distributions of luminosities or orbital period distributions, this can be fairly straightforward. The stumbling block is that typically the observations might only be in a few observed photometric filters or comprise only the optical spectrum of a star. Most stellar models only give a bolometric luminosity and effective temperature for the stars, and so linking this to a spectrum or photometric V band magnitude requires an additional step.

It is possible to take the observations and process them, accounting for their selection functions, to get the required numbers to compare to the models. This is somewhat risky when we have an incomplete dataset where there are large gaps in our knowledge. A more rigorous method is to take the stellar models and observe the models in a similar way. This can be done by attaching stellar atmosphere models to the stellar evolution models and creating a synthetic spectrum. Doing so requires bringing together a large number of different spectra from different sources, as the physics of cool and hot stellar atmospheres can be very different, and the best models for each regime may be produced by different teams and codes (e.g., Hamann et al., 2006; Smith et al., 2002; Westera et al., 2002; Gustafsson et al., 2008; Sander et al., 2012; Massey et al., 2013).

Matching the stellar model to the atmosphere model typically requires us to know the effective temperature, gravity and surface composition of our stellar model, as these are the primary factors that determine the stellar spectra. Once this match is done, it is possible to create either spectra for individual stars or for the entire population, allowing us to closely match the observations to the models and gain greater constraints even if the observational dataset is limited.

All these complications and uncertainties have in the past contributed to an air of distrust regarding binary population synthesis. Today, however, the number of firm constraints on the uncertain parameters is increasing. Single-model populations are being compared to a large number of varied observables, as performed by the Binary Population and Spectral Synthesis (BPASS) team in Eldridge et al. (2017), giving additional confidence in their application. Other groups, such as the Brussels group (e.g., Van Bever et al., 1999; Van Bever and Vanbeveren, 2000; Vanbeveren et al., 2007), the binary_c community (e.g., Izzard et al., 2009; Claeys et al., 2014; Abate et al., 2015) and Yunnan groups (e.g., Zhang et al., 2005; Han et al., 2007; Han and Han, 2014; Zhang et al., 2015), are also doing very similar work.

The key result emerging from all these studies, however, is that we *must* include interacting binaries in our population and spectral synthesis. If we do not, then we run the risk of drawing incorrect conclusions when studying systems using stellar population models as our tool.

21.3 Why Are Binaries Important?

Binaries are important as they provide evolutionary pathways that are simply not accessible from single-star evolution. Single-star models have for some time done a very good job of allowing us to understand stellar populations that, in the local Universe, are often old and metal rich and thus relatively unaffected by binary interactions. Recent observational evidence shows that stars are born with many of them in binaries, especially the most massive (e.g., Sana et al., 2012; Moe and Di Stefano, 2017), and so the effects on young stellar populations is likely to be more pronounced. As well as that, many stars stay in a binary system right to the last possible event in a binary's life, when the two remnants of the stars inspiral and merge due to the loss of orbital energy through gravitational radiation (e.g., Abbott et al., 2016, 2017).

This suggests we need to take a holistic view when attempting to model stellar populations. The effect of binaries will not always be obvious and may only become apparent in certain phases of evolution. For example, the main-sequence evolution is not significantly affected by binary interactions (although mergers do occur and there are a few exceptional cases) when looking at a population. Most differences between predictions from single stars and binaries become apparent in post–main-sequence phases of evolution and the eventual death throes of massive stars in core-collapse supernovae.

By not concentrating on one detail of population and spectral synthesis, we predict that we will introduce a varied number of observations and show that in each case interacting binaries are required to explain the observed systems, using our own BPASS models (Eldridge et al., 2017) as a demonstration. We will first consider stellar populations, then consider spectral synthesis before finally speculating about the future directions of what population synthesis will be able to predict in future.

21.3.1 Blue Stragglers on the HR Diagram

For resolved stellar populations in clusters, the effects of binary evolution are twofold. There are two main binary interactions that affect the position of stars on a Hertzsprung–Russell diagram, for example: mergers and mass transfer. These make more massive stars available at ages for a cluster well beyond their expected main-sequence turnoff. They are well known in globular clusters, for example as blue stragglers; for younger H II regions or open clusters, identifying such stars is more difficult.

For clusters, the age is frequently estimated by using single-star isochrones, primarily from attempting to fit the main-sequence turnoff. The existence of blue stragglers can directly affect the accuracy of this method. We show in Figure 21.1 an example fit to a single cluster, with single-star isochrones and a population including interacting binaries, which yield an isochronal contour plot. We see that for the well-known cluster Cygnus OB, to fit the most luminous stars, a maximum age of 3 Myr is required from single stars, but with the binary population an older age of 5 Myr is qualitatively a better match to the data.

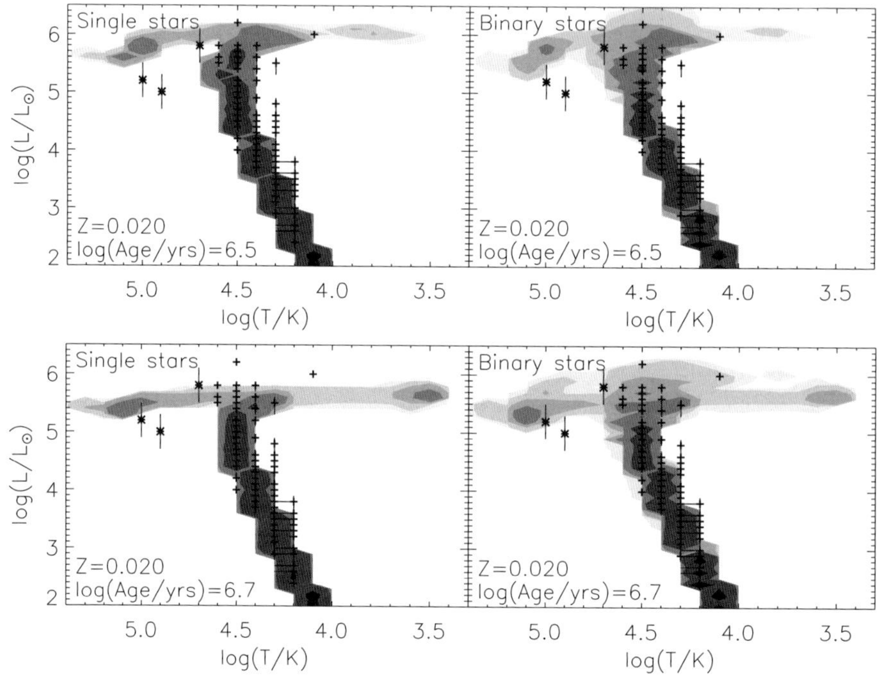

Figure 21.1 The HR diagram of Cygnus OB2 (Kiminki and Kobulnicky, 2012; Wright et al., 2015) compared to theoretical HR contours for single-star (left panels) and binary-star (right panels) populations. The models are for two different ages, 3 Myr in the upper panels and 5 Myr in the lower panels.

While this is an extreme example, it shows that there is something that must be taken into consideration when attempting to evaluate the age of resolved clusters.

21.3.2 Number Counts

Another basic observable of stellar populations are number counts. If stars can be typed then, say, the total number of O stars can be compared to the number of red supergiants (RSG) or Wolf–Rayet (WR) stars. This is a measurement of the number of main-sequence to post–main-sequence stars: something that depends on the nuclear evolution of a star's core as well as the mass loss from the surface. For example, in Figure 21.2 we show an example comparing the observed number of red supergiants to Wolf–Rayet stars, post–main-sequence stars that have retained or lost their hydrogen envelopes respectively. We can see here the dramatic change in the ratio of these two stellar types that arises due to the inclusion of binary interactions. The change is roughly around an order of magnitude due to the binary interactions, providing an extra avenue for mass loss rather than stellar winds alone. The binary models appear to match the observations slightly better, although the lowest metallicity point in the Small Magellanic Cloud (SMC) is closer to the single-star evolution line. However, the assumed scaling of mass-loss rates might be too strong or the assumption of constant star formation may also be wrong with the plot only based on ≈100 RSGs and 13 WRs.

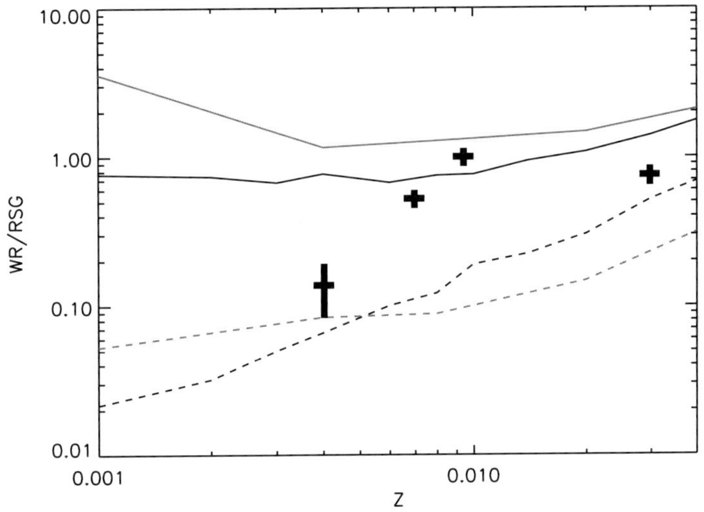

Figure 21.2 The ratio of the number of Wolf-Rayet stars to red supergiants as observed in galaxies at a single metallicity. Solid lines show BPASS binary models and dashed lines show single models. Points with error bars show observational constraints taken from the literature. All ratios include only stars with $\log(L/L_{\odot}) > 4.9$. The light grey lines show results for constant star formation from BPASS v1.0, while the black lines show similar results from BPASS v2.1 with IMF $M_{\max} = 100\,M_{\odot}$, respectively). The WR/RSG observed ratios come from P. Massey (private communication).

21.3.3 SN Progenitors

If the effect of interacting binaries is so dramatic that we can infer its presence from the number ratios of stars, we should also find this reflected in the number of stars that die with their hydrogen envelopes intact or removed. When core-collapse supernovae occur, they have historically been classified by their observational characteristics, primarily their spectra and light curves. The broadest classification is made on whether they are hydrogen-rich, type II, or hydrogen-free, type Ib/c (note type Ia supernovae arise from thermonuclear detonations of carbon-oxygen white dwarfs and are identified by strong silicon lines).

As we show in Figure 21.3, the type II supernovae progenitors are known to be mostly red supergiants with some blue and yellow supergiants (see Smartt, 2015). For type Ib/c supernovae, the progenitors were long expected to be Wolf–Rayet stars, but a problem with this was that if this was the case then they should have been directly detected in pre-explosion images of a sample of nearby explosions. This suggested that the progenitors were more likely to be lower-mass helium stars that were the result of binary interactions (Yoon et al., 2012; Eldridge et al., 2013). Confirmation that this is likely to be the case came from observations of the progenitor of type Ib supernova, iPTF13bvn. Detailed modelling by many groups have shown that the most likely progenitor of this event is a binary system (Bersten et al., 2014; Eldridge and Maund, 2016; Yoon et al., 2017). We can see this in Figure 21.3 as in single-star predictions that progenitors can only either be cool red supergiants or hot Wolf–Rayet stars. The binary predictions, however, fill a much greater space of the HR diagram, especially filling the region where most progenitors have been observed to date.

However, while progenitor detections provide direct evidence for the effects of binary evolution, the most compelling evidence comes from the relative rate of type Ib/c to type II progenitors. A consistent model should be able to predict both the WR/RSG ratio as well as the type Ib/c to II relative rate. The ratio is dependent on metallicity, with fewer type Ib/c supernovae at lower metallicities, but as shown by the recent study of Graur et al. (2017), the exact determination of the rate at a specific metallicity is difficult. If we restrict ourselves to nearby supernovae, where host galaxies have metallicities in the range from solar to maybe half-solar, we find a relative rate ratio of type Ib/c to type II supernovae or 0.29 ± 0.17 (Eldridge et al., 2013; Xiao and Eldridge, 2015), where again we are limited by large uncertainties. Despite these significant uncertainties, in any comparison single-star models predict a ratio below this value while binary models predict a ratio close to or above this value. An exact match is difficult because star-formation history and metallicities distributions must be modelled of the stellar population. But a stellar population with a high binary fraction generally provides a better match to the observed ratio. This is consistent with multiple equivalent studies over a significant period of time (e.g., Podsiadlowski et al., 1992; De Donder and Vanbeveren, 1998). Finally, recent work by Zapartas et al. (2017) shows that progenitors dominated by a population of interacting binary stars also reproduced the observed core-collapse supernova delay time distribution better than a single-star-only population.

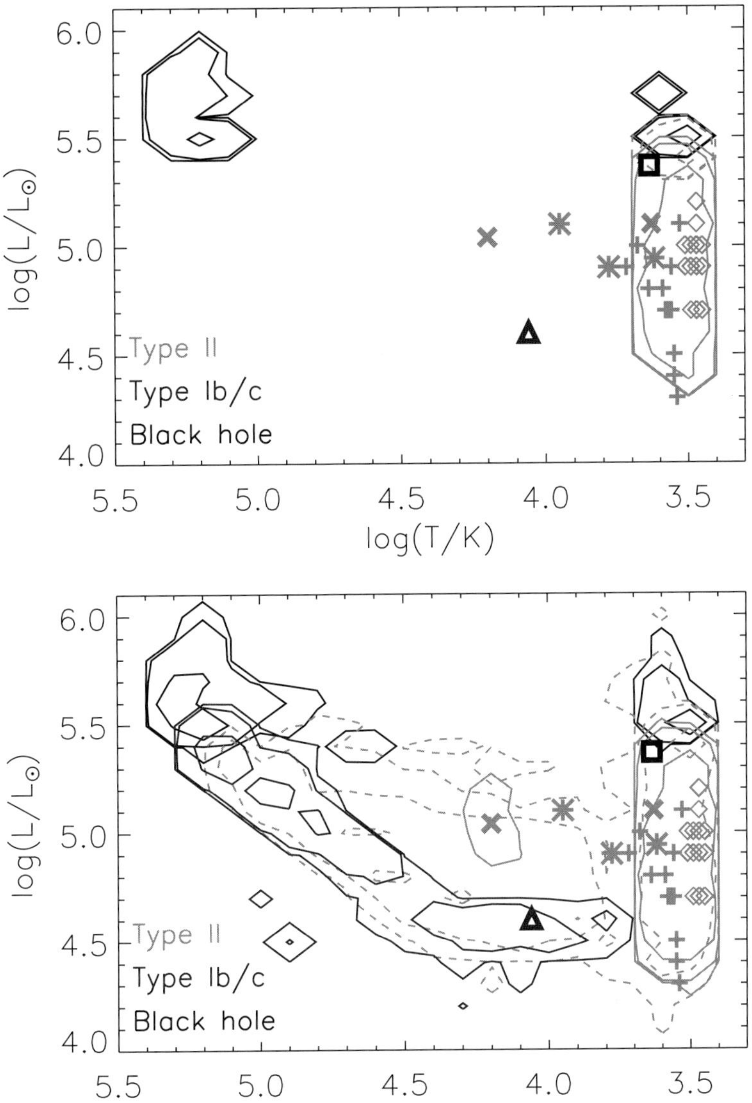

Figure 21.3 HR diagrams showing the predicted location of SN progenitors from BPASS v2.1 models compared to the position of observed SN progenitors. The upper panel is for single-star populations, and the lower panel is for binary star populations. The black contours are for progenitors expected to form black holes, the light grey contours are for type IIP (solid lines) and type IIL/b (dashed lines) progenitors and the dark grey contours are for type Ib/c progenitors. The points are observations taken from Smartt (2015), the pluses are observed type IIP progenitors, the diamonds are type IIP progenitors with upper limits on the luminosity, the asterisks are type IIb progenitors, the crosses are the progenitors of 1987A and 1993J, the triangle is for the type Ib progenitor of iPTF13bvn and the square is for the candidate black hole–forming event from Adams et al. (2017).

21.3.4 GW Mergers

In a massive binary system, if it can survive two supernovae and remain bound, then two remnants, either neutron stars or black holes, will slowly inspiral due to emission of gravitational radiation (e.g., Abbott et al., 2016, 2017). Since the detection of the first merging black hole binary, GW150917, there have been to date five pairs of merging black holes and one double neutron star merger detected. To round off our holistic study of population synthesis of massive stars, the same model that explains the observed post–main-sequence star ratios and supernova relative rates of different types also needs to predict the mass range and rate of these binaries. This is still work in progress, and there are many binary evolution codes now attempting to predict the rate of these mergers as well as the masses of the observed merging objects. While this is important, the same codes also need to make sure they correctly predict the other, more visible aspects of stellar evolution that lead up to formation of the remnants, rather than just the final ripples in space-time from the merger of those remnants.

21.4 Galaxies Near and Far

The populations and distribution of stars within the Milky Way and other galaxies in the local Universe reflects 13 billion years of evolution and complexity. While some of the stars, particularly the massive stars, likely formed in the locations where they are observed, others may have formed elsewhere within a gravitationally bound system, or even in satellite stellar populations or progenitor galaxies that merged over time into the current host. Most, if not all, incorporate metals and other material that have been processed by one or more earlier stellar populations and their resultant supernovae.

A full understanding of the stellar populations nearby thus requires an overview of processes that extend to cosmological scales in both time and distance, including (but not limited to) the thermal history of the Universe and variations in its chemical enrichment over cosmic time.

Our primary information on these processes is derived from starlight, sometimes reprocessed through dust or gas along the line of sight. However, for all but the most nearby galaxies, direct imaging and resolution of individual stars is technically impossible (although note the existence of a claimed single-star, caustic-crossing lensing event in a distant $z = 1.5$ galaxy; Kelly et al., 2016). The angular scales and sensitivity required exceed the limits of even an Extremely Large Telescope (ELT). In a best-case scenario, it may be possible to measure the integrated light from a single star-forming region (down to $\sim$30 pc in resolution), making use of the spatial magnification associated with gravitational lensing (e.g., Johnson et al., 2017). However, in a more normal case, resolving scales below 1 kpc in galaxies is difficult from space, and requires adaptive optics and/or exceptional conditions from the ground. As a result, the emission detected represents the integrated light of one or more stellar populations. In many cases, ultraviolet and optical light will be dominated by some combination of the youngest and most massive stars, simply because these outshine their more numerous but much fainter fellows. These can be fit with a simple stellar population, or with a straightforward model such as one continuously forming stars at a constant rate. At long wavelengths, moving towards the infrared, or for galaxies dominated by a massive underlying old stellar population, the contributions of different elements of the stellar population, or from different stages of the star-formation history, can be more equal (e.g., Kauffmann et al., 2003). In these cases, knowledge of both the stellar mass function and the star-formation history is required, or needs to be derived from the available data. In such instances, the outputs of a stellar population and spectral synthesis code are an essential tool for interpreting galaxy photometry and spectroscopy.

This has been recognised for many years, and a number of spectral synthesis models have been developed with the specific aim of modelling the complex stellar populations seen in galaxies. These include the very successful GALAXEV model set (Bruzual and Charlot, 2003, and later references), which combines theoretical stellar evolution tracks with simple star-formation histories and applies empirical models for nebula emission at young stellar ages, and the Multiwavelength Analysis of Galaxy Physical Properties (MAGPHYS) models (da Cunha et al., 2008), which aim primarily to explore the nonstellar components of galaxies, extending into the infrared and submillimetre, coupling these to the parent stellar population through an energy balance between dust absorption and reemission. These and similar models have been used with great success to explore the properties of galaxies in the nearby Universe, where empirical calibration and multiple analysis methods can be used to check their application and precision (e.g., Brinchmann et al., 2004; Tremonti et al., 2004; Taylor et al., 2011). However, as observational samples of galaxies push to earlier times, when the Universe was less metal-rich, star formation occurred with a higher specific rate density, and the stellar populations are significantly younger (see, e.g., Madau and Dickinson, 2014, for a recent review), we are moving into new regimes, where the efficacy of such models has not been tested. What is more, the identification of extreme classes of local galaxies that share similar physical conditions, and are thus classed as analogues to galaxies in the distant Universe (e.g., Heckman et al., 2005; Cardamone et al., 2009; Amorín et al., 2014; Stanway and Davies, 2014), has demonstrated that even nearby galaxy-scale stellar systems can lie well outside the normal range of properties predicted by the older generation of population synthesis models.

At the same time, the same distant or extreme galaxy spectra present an interesting opportunity to test and constrain stellar modelling. The conditions that prevail in different regions of the Universe, and at different epochs of cosmic history, are often rare in the local Universe. In particular, the distant Universe (where the galaxies we observe emitted their light within a few billion years after the Big Bang) has attracted interest in this respect. As mentioned previously, the stellar populations at those early times are typically young (<1 Gyr in age), and form from gas clouds with both a lower overall metal enrichment (well below half-solar) and potentially very different abundance ratios from those common in the local Universe (due to different enrichment processes). While old stellar populations such as Globular Clusters nearby tend to be enhanced in alpha-process elements, as the result of a higher ratio of core-collapse supernovae to other nucleosynthesis channels, at high redshifts we see the same alpha enhancement in much younger stellar populations, dominated by more massive stars. At the same time, observations of galaxies in the distant Universe are weighted towards the rest-frame ultraviolet spectral region, redshifted longwards of the atmospheric cutoff and into the observed frame optical or near-infrared.

Each of these conditions suggests that aspects of massive star evolution, including their spectral evolution, lifetimes and metal yields, are likely to be far more clearly manifest in distant galaxies (and their local analogues) than in typical galaxies in the local Universe. In this regime, binary interactions are expected to be important. Thus, while binary stellar population models are required to interpret the light of these integrated stellar populations, the same observations place a constraint on the models: any that fail to reproduce the observed properties of star-forming galaxies given plausible assumptions must be considered as suspect.

21.5 Ionising Radiation Fields and H II Regions

Early indications of the unusual stellar populations dominant in the distant Universe came in the detection of nebular line emission, particularly from lines requiring a hard ionising radiation spectrum such as the rest-frame ultraviolet He II 1640 Angstrom emission line from galaxies at $z > 2$. These appeared to include both broad (e.g., Shapley et al., 2003) and narrow (e.g., Erb et al., 2010) components, suggestive of strong stellar winds and intensely ionised nebular regions respectively, while exhibiting no evidence for an active galactic nucleus (AGN)-powered emission component. Studies of the He II feature have recently been complemented by observations of sources with C III] and C IV in emission again, with strong constraints indicating a likely stellar source for their photoionisation (e.g., Stark et al., 2015a,b; Maseda et al., 2017). At the same time, galaxies at still earlier times ($z > 5$) were exhibiting very blue ultraviolet colours (e.g., Stanway et al., 2005; Bouwens et al., 2010). While there is certainly evolution in the dust extinction in the galaxy population, the slopes observed were suggestive of a much hotter stellar spectrum than is typical at lower redshifts, and a resultant excess of hard ionising photons.

Early theoretical work suggesting that Population III (i.e., essentially metal-free, primordial stars) were required to explain these observations (e.g., Schaerer, 2003; Jimenez and Haiman, 2006) overlooked alternative sources of hard ionising photons, in particular the strongly metal-dependent contribution of binary evolution pathways and the hot stars they produce. Initial work on the spectral synthesis application of the BPASS binary stellar evolution models was inspired, at least in part, by the desire to confront these observations (see Eldridge and Stanway, 2009, 2012). It demonstrated that inclusion of binary evolution pathways was capable of generating stronger ionising photon radiation fields, and provided a good fit to a range of observations in both extreme local sources and to galaxies in the distant Universe.

More recently, both the weight of evidence for physical conditions in these systems and the models required to interpret them have developed significantly. The advent of multi-object near-infrared spectrographs on large telescopes (notably the Multi-Object

Spectrometer for Infrared Exploration [MOSFIRE] on Keck), have rendered the rest-frame optical accessible at $z > 2$, and allowed for direct comparison between the strong recombination line spectral diagnostics seen in large local surveys such as the Sloan Digital Sky Survey (SDSS) and those observed in distant galaxies. One of the most striking results has been the discovery of an offset in the Baldwin, Phillips and Terlevich (BPT, 1981) photoionisation-sensitive diagnostic diagrams, cementing the requirement for a hard ionising radiation spectrum (e.g., Masters et al., 2014; Holden et al., 2016). The same offset is seen in selected local analogue galaxies, which have similarly high star-formation densities (Stanway et al., 2014).

Crucially, a number of these galaxies, while clearly consistent with a star-forming galaxy locus, lie above the 'Maximal' starburst line defined by Kewley et al. (2001). This requires that their stellar radiation field is harder than that derived from a combination of single-star spectral synthesis and nebular emission models. As a number of studies have now shown (e.g., Stanway et al., 2014, 2016; Steidel et al., 2014, 2016; Götberg et al., 2017; Strom et al., 2017; Xiao et al., 2018) have now shown, binary spectral synthesis has two important effects that tend to produce harder galaxy spectral models. Firstly, binary interactions lead to the inclusion in synthetic populations of stripped helium stars, and also stars that undergo significant rotational mixing. The resultant higher surface temperatures result in atmosphere models that emit a significant fraction of their light shortwards of the Lyman limit, creating large and hot photoionisation regions. Secondly, the range of timescales for evolution of such stars extends the ionising lifetime of a stellar population beyond the few million years required for the most massive stars to live and die in a single-star model. This extended lifetime increases the fraction of galaxies and range of conditions in which large photoionised regions are expected, and also boosts the contribution of these to galaxies with continuous or ongoing starbursts.

Given the important role of metal opacities in driving stellar winds, and the effect these have on stellar evolution, such binary effects are strongly metallicity dependent. Consequently, the effects of binary evolution are relatively slight in the bulk of highly metal-enriched, slowly star-forming local galaxies, explaining the success of single-star models in fitting these. Nonetheless, an analysis of star-forming regions in local dwarf and spiral galaxies suggests that BPASS binary models perform at least as well as single-star models in fitting their recombination line ratios, while yielding slightly higher typical ages (Xiao et al, 2018).

The increasingly clear importance of binary population synthesis in interpreting distant galaxies can largely be attributed to the effects of cosmic metallicity evolution, together with a shift towards younger stellar populations.

21.6 Photon Production, Photon Escape and Reionisation

The hardness of the stellar radiation field is a key ingredient in understanding the role of galaxies in a key phase change in the thermal history of the Universe: the epoch of reionisation. During this period, the first luminous sources – most likely star-forming galaxies – gradually ionised their immediate surroundings for the first time since hydrogen atoms formed at $z \sim 1,100$. Individual galaxy-scale H II regions expanded and overlapped over an extended period, until the Universe was highly ionised. The topography of ionised gas in this epoch should thus reflect both the distribution of star-forming galaxies and the ionising photon output of their stellar populations. The Square Kilometer Array (SKA) will explore the power spectrum of the epoch of reionisation. Detailed mapping of the ionisation balance intergalactic medium at this time is still well beyond our capacity, and will likely remain so until the next generation of H I mapping telescopes is designed and constructed.

Nonetheless, strong observational constraints on the reionisation transition do already exist. Signals imprinted on the cosmic microwave background suggest that the Universe was 50% ionised somewhere around $z \sim 9$. Evolution in the characteristics of Lyman-alpha line emission and the detection of ionised troughs in the absorption spectra of distant sources (sometimes known as 'cosmic lighthouses') suggest that the ionised fraction was rapidly rising between $z \sim 6$ and $z \sim 10$ (see Planck Collaboration et al., 2016, and references therein), although line-of-sight variations make a definitive timeline challenging to establish.

However, reconciling this thermal history, observed in neutral hydrogen, with the star-formation history, observed in rest-frame ultraviolet stellar emission, has proved challenging. Considerable flexibility exists in estimates of parameters such as clumpiness (and hence self-shielding) in the intergalactic medium, its density and precise temperature. Nonetheless, most estimates have suggested that the observed galaxy population may struggle to produce sufficient photons to reionise the Universe (see, e.g., Robertson et al., 2015).

A key theoretical input into such estimates is the ionising photon production efficiency, $\xi_{\rm ion}$. This gives the rate of ionising photons (emitted shortwards of the Lyman limit) that will arise from a population with a given rest-1500 Angstrom continuum luminosity density. Since the former cannot be directly observed even in the local Universe, the latter is used as an observable proxy for ionisation, and $\xi_{\rm ion}$ is determined from theoretical arguments, indirect measurements or appropriate population synthesis models (or, more usually, a combination of all three). As Figure 21.4 demonstrates, the value of this parameter is sensitive to both the stellar metallicity and age of an ongoing continuous starburst. It is also sensitive to more complex star-formation history, and, needless to say, to the stellar population synthesis model being used. In particular, the presence of binary stars in a modelled population has a strong metallicity-dependent effect, suppressing $\xi_{\rm ion}$ at metallicities close to solar and strongly boosting it at low metallicities.

Both the requirements of reconciling reionisation timescales with galaxy observations (e.g., Ma et al., 2016; Stanway et al., 2016; Wilkins et al., 2016) and the indirect measurements that can be obtained by considering the reprocessed nebular emission spectrum in individual galaxies (e.g., Stark et al., 2015b; Shivaei et al., 2017) push the required values of $\xi_{\rm ion}$ well above those suggested by spectral synthesis models that neglect binary evolution effects, while favouring the BPASS models, with their harder ionising spectrum, and resultant higher ionising photon production efficiency.

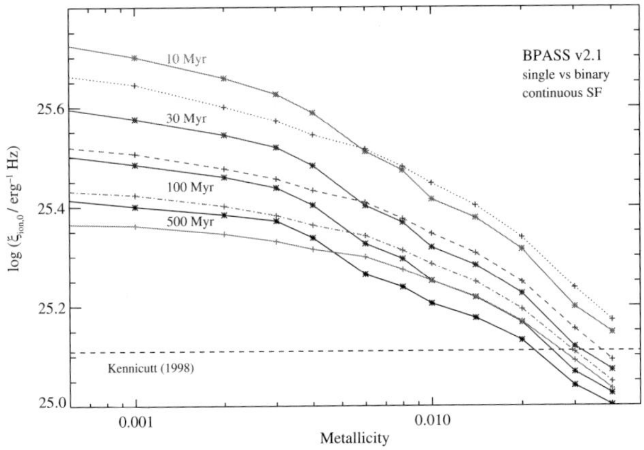

Figure 21.4 The metallicity and stellar population age dependence of the ionising photon production efficiency, $\xi_{\rm ion}$. This model-dependent parameter is crucial for placing observational constraints on the ionising photon production during the reionisation epoch. We show value for binary populations (solid lines) and single star populations (broken lines), at four different times after the onset of a constant, ongoing starburst event. Here we use the BPASS v2.1 models of Eldridge et al. (2017).

21.7 Looking Forward

While significant progress has been made on studying the effects of interacting binaries on the population and spectral synthesis, there is still much to do. Much of this revolves around extending the spectral synthesis to the supernovae created by a binary population.

As discussed previously, the relative rate of different SN types can only be reproduced by a stellar population including binaries, which increases the number of stars that lose their hydrogen envelope. However, an important question is, what will those supernovae actually look like? Do they match the observed lightcurves and spectra of the SNe? To date, SN lightcurve models consider a few different models, and only recently have large numbers of models been computed.

What is required is a supernova population and spectral synthesis, analogous to the stellar population and spectral synthesis. Groups are now these this models and using various open source supernova simulation codes to predict the observed light curves of SNe. The general result is that for single stars the light curves differ only slightly due to the different internal structure of the stellar models, as they are broadly the same. However, due to the variety of progenitor structures that arise from binary interactions, a much greater range of possible light curves are possible.

Future work and synthesis like this are key, especially studies and models that explore the full possible variety of binary evolution pathways. For the first time, we will have a prediction as to what the full variety of possible stellar explosions may be, not only so we can match models to observed explosions but also begin to understand what types of stellar deaths we may be missing due to their faintness of rapid evolution.

Further work is also required to improve the connection between modelling of simple stellar populations and that of entire galaxies. In the latter, the effects of dust, diffuse interstellar emission and complex star-formation histories are often important and difficult to disentangle or quantify. These can manifest in varied ways, affecting not only galaxy photometry and line emission but also less obvious properties, such as the chemical composition and abundance patterns resulting from previous generations of supernovae (commonly believed to lead to α-element enhancement in old stellar populations locally). Nonetheless, distant galaxies provide laboratories probing conditions atypical of the local Universe. There is increasing evidence from observations of such sources for specific phenomena that need to be explained. These appear common in the unusual, metal-poor and star-formation dense environments probed by young, distant galaxies, while remaining rare in the resolved stellar populations studied locally.

Recent observations of distant galaxies suggest that the photospheric iron abundance is significantly lower, relative to nebular oxygen abundance, than assumed in common binary (and most other) models (Steidel et al., 2016). With the imminent launch of the James Webb Space Telescope, observations of photospheric line blanketing in (rest-frame) far-ultraviolet spectra of distant galaxies will become more straightforward and may be complemented by improved measurements of both stellar and nebular abundance in the rest-frame optical. Improved modelling, with a better understanding of the abundance patterns and their effects on stellar evolution, will be required to fully interpret such data.

Another example of an ongoing challenge is the strength of the He II emission line observed in strongly star-forming galaxies. Observations suggest that this is underestimated in models, which implies that these underestimate the far-UV hard ionising radiation field. Either a change in the stellar population, for example a steeper initial mass function (IMF) or an improved treatment of components such as accreting compact binaries, may be required to resolve this challenge and must be accompanied by an understanding of why such changes are required in the conditions specific to the sources under observation.

21.8 Summary

In this chapter, we have discussed some of the current findings in population and spectral synthesis that show why including interacting binaries is important. Also we have shown, by discussing a broad range of observables, how by using a holistic view and considering varied and different observations it is becoming possible to firmly constrain the uncertainties of binary population synthesis and increase the predictive power and usefulness of the many binary pop synth codes.

References

Abate, C., Pols, O. R., Stancliffe, R. J., et al. 2015. Modelling the Observed Properties of Carbon-Enhanced Metal-Poor Stars Using Binary Population Synthesis. *A&A*, **581**, A62.

Abbott, B. P., Abbott, R. and Abbott, T. D., et al. 2016. Properties of the Binary Black Hole Merger GW150914. *Physical Review Letters*, **116**, 24.

Abbott, B. P., Abbott, R. and Abbott, T. D., et al. 2017. Multi-Messenger Observations of a Binary Neutron Star Merger. *ApJL*, **848**, L12.

Adams, S. M., Kochanek, C. S., Gerke, J. R., et al. 2017. The Search for Failed Supernovae with the Large Binocular Telescope: Constraints from 7 Yr of Data. *MNRAS*, **469**, 1445–1455.

Amorín, R., Sommariva, V., Castellano, M., et al. 2014. Discovering Extremely Compact and Metal-Poor, Star-Forming Dwarf Galaxies out to z ˜ 0.9 in the VIMOS Ultra-Deep Survey. *A&A*, **568**, L8.

Baldwin, J. A., Phillips, M. M. and Terlevich, R. 1981. Classification Parameters for the Emission-Line Spectra of Extragalactic Objects. *PASP*, **93**, 5–19.

Belczynski, K., Taam, R. E., Kalogera, V., et al. 2007. On the Rarity of Double Black Hole Binaries: Consequences for Gravitational Wave Detection. *ApJ*, **662**, 504–511.

Bersten, M. C., Benvenuto, O. G., Folatelli, G., et al. 2014. iPTF13bvn: The First Evidence of a Binary Progenitor for a Type Ib Supernova. *AJ*, **148**, 68.

Bouwens, R. J., Illingworth, G. D., Oesch, P. A., et al. 2010. Very Blue UV-Continuum Slope β of Low Luminosity $z \sim 7$ Galaxies from WFC3/IR: Evidence for Extremely Low Metallicities? *ApJL*, **708**, L69–L73.

Bray, J. C. and Eldridge, J. J. 2016. Neutron Star Kicks and Their Relationship to Supernovae Ejecta Mass. *MNRAS*, **461**, 3747–3759.

Brinchmann, J., Charlot, S., White, S. D. M., et al. 2004. The Physical Properties of Star-Forming Galaxies in the Low-Redshift Universe. *MNRAS*, **351**, 1151–1179.

Bruzual, G. and Charlot, S. 2003. Stellar Population Synthesis at the Resolution of 2003. *MNRAS*, **344**, 1000–1028.

Cantiello, M., Yoon, S.-C., Langer, N., et al. 2007. Binary Star Progenitors of Long Gamma-Ray Bursts. *A&A*, **465**, L29–L33.

Cardamone, C., Schawinski, K., Sarzi, M., et al. 2009. Galaxy Zoo Green Peas: Discovery of a Class of Compact Extremely Star-Forming Galaxies. *MNRAS*, **399**, 1191–1205.

Chabrier, G. 2003. Galactic Stellar and Substellar Initial Mass Function. *PASP*, **115**, 763–795.

Choi, J., Dotter, A., Conroy, C., et al. 2016. Mesa Isochrones and Stellar Tracks (MIST). I. Solar-Scaled Models. *ApJ*, **823**, 102.

Claeys, J. S. W., Pols, O. R., Izzard, R. G., et al. 2014. Theoretical Uncertainties of the Type Ia Supernova Rate. *A&A*, **563**, A83.

da Cunha, E., Charlot, S. and Elbaz, D. 2008. A Simple Model to Interpret the Ultraviolet, Optical and Infrared Emission from Galaxies. *MNRAS*, **388**, 1595–1617.

De Donder, E. and Vanbeveren, D. 1998. The Relative Frequency of Type II and I_(b,c) Supernovae and the Birth Rate of Double Compact Star Binaries. *A&A*, **333**, 557–564.

De Donder, E. and Vanbeveren, D. 2004. The Influence of Binaries on Galactic Chemical Evolution. *New Ast. Rev.*, **48**, 861–975.

de Mink, S. E., Langer, N., Izzard, R. G., et al. 2013. The Rotation Rates of Massive Stars: The Role of Binary Interaction through Tides, Mass Transfer, and Mergers. *ApJ*, **764**, 166.

Ekström, S., Georgy, C., Eggenberger, P., et al. 2012. Grids of Stellar Models with Rotation. I. Models from 0.8 to 120 $M_{\odot}$ at Solar Metallicity ($Z = 0.014$). *A&A*, **537**, A146.

Eldridge, J. J. and Maund, J. R. 2016. The Disappearance of the Helium-Giant Progenitor of the Type Ib Supernova iPTF13bvn and Constraints on Its Companion. *MNRAS*, **461**, L117–L121.

Eldridge, J. J. and Stanway, E. R. 2009. Spectral Population Synthesis Including Massive Binaries. *MNRAS*, **400**, 1019–1028.

Eldridge, J. J. and Stanway, E. R. 2012. The Effect of Stellar Evolution Uncertainties on the Rest-Frame Ultraviolet Stellar Lines of C IV and He II in High-Redshift Lyman-Break Galaxies. *MNRAS*, **419**, 479–489.

Eldridge, J. J., Fraser, M., Smartt, S. J., et al. 2013. The Death of Massive Stars. II. Observational Constraints on the Progenitors of Type Ibc Supernovae. *MNRAS*, **436**, 774–795.

Eldridge, J. J., Stanway, E. R., Xiao, L., et al. 2017. Binary Population and Spectral Synthesis Version 2.1: Construction, Observational Verification, and New Results. *PASA*, **34**, e058.

Erb, D. K., Pettini, M., Shapley, A. E., et al. 2010. Physical Conditions in a Young, Unreddened, Low-Metallicity Galaxy at High Redshift. *ApJ*, **719**, 1168–1190.

Götberg, Y., de Mink, S. E. and Groh, J. H. 2017. Ionizing Spectra of Stars that Lose Their Envelope through Interaction with a Binary Companion: Role of Metallicity. *A&A*, **608**, A11.

Graur, O., Bianco, F. B., Huang, S., et al. 2017. LOSS Revisited. I. Unraveling Correlations between Supernova Rates and Galaxy Properties, as Measured in a Reanalysis of the Lick Observatory Supernova Search. *ApJ*, **837**, 120.

Gustafsson, B., Edvardsson, B., Eriksson, K., et al. 2008. A Grid of MARCS Model Atmospheres for Late-Type Stars. I. Methods and General Properties. *A&A*, **486**, 951–970.

Hamann, W.-R., Gräfener, G. and Liermann, A. 2006. The Galactic WN Stars: Spectral Analyses with Line-Blanketed Model Atmospheres versus Stellar Evolution Models with and without Rotation. *A&A*, **457**, 1015–1031.

Han, Y. and Han, Z. 2014. BayeSED: A General Approach to Fitting the Spectral Energy Distribution of Galaxies. *ApJs*, **215**, 2.

Han, Z., Podsiadlowski, P. and Lynas-Gray, A. E. 2007. A Binary Model for the UV-Upturn of Elliptical Galaxies. *MNRAS*, **380**, 1098–1118.

Heckman, T. M., Hoopes, C. G., Seibert, M., et al. 2005. The Properties of Ultraviolet-Luminous Galaxies at the Current Epoch. *ApJL*, **619**, L35–L38.

Hobbs, G., Lorimer, D. R., Lyne, A. G., et al. 2005. A Statistical Study of 233 Pulsar Proper Motions. *MNRAS*, **360**, 974–992.

Holden, B. P., Oesch, P. A., Gonzlez, V. G., et al. 2016. Rest-Frame Optical Emission Lines in $z \sim 3.5$ Lyman-Break-Selected Galaxies: The Ubiquity of Unusually High [OIII]/Hβ Ratios at 2 Gyr. *ApJ*, **820**, 73.

Hurley, J. R., Tout, C. A. and Pols, O. R. 2002. Evolution of Binary Stars and the Effect of Tides on Binary Populations. *MNRAS*, **329**, 897–928.

Izzard, R. G., Glebbeek, E., Stancliffe, R. J., et al. 2009. Population Synthesis of Binary Carbon-Enhanced Metal-Poor Stars. *A&A*, **508**, 1359–1374.

Janka, H.-T. 2017. Neutron Star Kicks by the Gravitational Tug-Boat Mechanism in Asymmetric Supernova Explosions: Progenitor and Explosion Dependence. *ApJ*, **837**, 84.

Jimenez, R. and Haiman, Z. 2006. Significant Primordial Star Formation at Redshifts z ˜ 3 4. *Nature*, **440**, 501–504.

Johnson, T. L., Rigby, J. R., Sharon, K., et al. 2017. Star Formation at z = 2.481 in the Lensed Galaxy SDSS J1110+6459: Star Formation Down to 30 Pc Scales. *ApJL*, **843**, L21.

Kauffmann, G., Heckman, T. M., White, S. D. M., et al. 2003. Stellar Masses and Star Formation Histories for 10^5 Galaxies from the Sloan Digital Sky Survey. *MNRAS*, **341**, 33–53.

Kelly, P. L., Rodney, S., Diego, J. M., et al. 2016. Hubble Space Telescope Discovery of a Probable Caustic-Crossing Event in the MACS1149 Galaxy Cluster Field. *The Astronomer's Telegram*, 9097.

Kennicutt, Jr., R. C. 1998. The Global Schmidt Law in Star-Forming Galaxies, *ApJ*, **498**, 541–552.

Kewley, L. J., Dopita, M. A., Sutherland, R. S., et al. 2001. Theoretical Modeling of Starburst Galaxies. *ApJ*, **556**, 121–140.

Kiminki, D. C. and Kobulnicky, H. A. 2012. An Updated Look at Binary Characteristics of Massive Stars in the Cygnus OB2 Association. *ApJ*, 7 **51**, 4.

Kroupa, P., Tout, C. A. and Gilmore, G. 1993. The Distribution of Low-Mass Stars in the Galactic Disc. *MNRAS*, **262**, 545–587.

Leitherer, C., Schaerer, D., Goldader, J. D., et al. 1999. Starburst99: Synthesis Models for Galaxies with Active Star Formation. *ApJs*, **123**, 3–40.

Lipunov, V. M., Postnov, K. A., Prokhorov, M. E., et al. 2009. Description of the 'Scenario Machine'. *Astronomy Reports*, **53**, 915–940.

Ma, X., Hopkins, P. F., Kasen, D., et al. 2016. Binary Stars Can Provide the 'Missing Photons' Needed for Reionization. *MNRAS*, **459**, 3614–3619.

Madau, P. and Dickinson, M. 2014. Cosmic Star-Formation History. *ARA&A*, **52**, 415–486.

Mandel, I. 2016. Estimates of Black Hole Natal Kick Velocities from Observations of Low-Mass X-Ray Binaries. *MNRAS*, **456**, 578–581.

Maseda, M. V., Brinchmann, J., Franx, M., et al. 2017. The MUSE Hubble Ultra Deep Field Survey. IV. Global Properties of C III] Emitters. *A&A*, **608**, A4.

Massey, P., Neugent, K. F., Hillier, D. J., et al. 2013. A Bake-Off between CMFGEN and FASTWIND: Modeling the Physical Properties of SMC and LMC O-Type Stars. *ApJ*, **768**, 6.

Masters, D., McCarthy, P., Siana, B., et al. 2014. Physical Properties of Emission-Line Galaxies at z ˜ 2 from Near-Infrared Spectroscopy with Magellan FIRE. *ApJ*, **785**, 153.

Moe, M. and Di Stefano, R. 2017. Mind Your Ps and Qs: The Interrelation between Period (P) and Mass-Ratio (Q) Distributions of Binary Stars. *ApJS*, **230**, 15.

Planck Collaboration, Ade, P. A. R., Aghanim, N., Arnaud, M., et al. 2016. Planck 2015 Results. XIII. Cosmological Parameters. *A&A*, **594**, A13.

Podsiadlowski, P., Joss, P. C. and Hsu, J. J. L. 1992. Presupernova Evolution in Massive Interacting Binaries. *ApJ*, **391**, 246–264.

Robertson, B. E., Ellis, R. S., Furlanetto, S. R., et al. 2015. Cosmic Reionization and Early Star-Forming Galaxies: A Joint Analysis of New Constraints from Planck and the Hubble Space Telescope. *ApJL*, **802**, L19.

Salpeter, E. E. 1955. The Luminosity Function and Stellar Evolution. *ApJ*, **121**, 161.

Sana, H., de Mink, S. E., de Koter, A., et al. 2012. Binary Interaction Dominates the Evolution of Massive Stars. *Science*, **337**, 444.

Sander, A., Hamann, W.-R. and Todt, H. 2012. The Galactic WC Stars: Stellar Parameters from Spectral Analyses Indicate a New Evolutionary Sequence. *A&A*, **540**, A144.

Schaerer, D. 2003. The Transition from Population III to Normal Galaxies: Lyalpha and He II Emission and the Ionising Properties of High Redshift Starburst Galaxies. *A&A*, **397**, 527–538.

Shapley, A. E., Steidel, C. C., Pettini, M., et al. 2003. Rest-Frame Ultraviolet Spectra of z˜3 Lyman Break Galaxies. *ApJ*, **588**, 65–89.

Shivaei, I., Reddy, N. A., Siana, B., et al. 2017. The MOSDEF Survey: Direct Observational Constraints on the Ionizing Photon Production Efficiency, ξ_ion, at z˜2. *ApJ*, **855**, 42.

Smartt, S. J. 2015. Observational Constraints on the Progenitors of Core-Collapse Supernovae: The Case for Missing High-Mass Stars. *PASA*, **32**, e016.

Smith, L. J., Norris, R. P. F. and Crowther, P. A. 2002. Realistic Ionizing Fluxes for Young Stellar Populations from 0.05 to 2 Z_{solar}. *MNRAS*, **337**, 1309–1328.

Stanway, E. R. and Davies, L. J. M. 2014. Establishing an Analogue Population for the Most Distant Galaxies. *MNRAS*, **439**, 2474–2484.

Stanway, E. R., Eldridge, J. J. and Becker, G. D. 2016. Stellar Population Effects on the Inferred Photon Density at Reionization. *MNRAS*, **456**, 485–499.

Stanway, E. R., Eldridge, J. J., Greis, S. M. L., et al. 2014. Interpreting High [O III]/H β Ratios with Maturing Starbursts. *MNRAS*, **444**, 3466–3472.

Stanway, E. R., McMahon, R. G. and Bunker, A. J. 2005. Near-Infrared Properties of I-Drop Galaxies in the Hubble Ultra Deep Field. *MNRAS*, **359**, 1184–1192.

Stark, D. P., Richard, J., Charlot, S., et al. 2015a. Spectroscopic Detections of C III] λ1909 Å at $z \sim$ 6-7: A New Probe of Early Star-Forming Galaxies and Cosmic Reionization. *MNRAS*, **450**, 1846–1855.

Stark, D. P., Walth, G., Charlot, S., et al. 2015b. Spectroscopic Detection of C IV λ1548 in a Galaxy at z = 7.045: Implications for the Ionizing Spectra of Reionization-Era Galaxies. *MNRAS*, **454**, 1393–1403.

Steidel, C. C., Rudie, G. C., Strom, A. L., et al. 2014. Strong Nebular Line Ratios in the Spectra of z ˜ 2-3 Star Forming Galaxies: First Results from KBSS-MOSFIRE. *ApJ*, **795**, 165.

Steidel, C. C., Strom, A. L., Pettini, M., et al. 2016. Reconciling the Stellar and Nebular Spectra of High-Redshift Galaxies. *ApJ*, 8 **26**, 159.

Strom, A. L., Steidel, C. C., Rudie, G. C., et al. 2017. Nebular Emission Line Ratios in z $\sim$ 2-3 Star-Forming Galaxies with KBSS-MOSFIRE: Exploring the Impact of Ionization, Excitation, and Nitrogen-to-Oxygen Ratio. *ApJ*, **836**, 164.

Sukhbold, T., Ertl, T., Woosley, S. E., et al. 2016. Core-Collapse Supernovae from 9 to 120 Solar Masses Based on Neutrino-Powered Explosions. *ApJ*, **821**, 38.

Taylor, E. N., Hopkins, A. M., Baldry, I. K., et al. 2011. Galaxy and Mass Assembly (GAMA): Stellar Mass Estimates. *MNRAS*, **418**, 1587–1620.

Tinsley, B. M. 1968. Evolution of the Stars and Gas in Galaxies. *ApJ*, **151**, 547.

Tinsley, B. M. and Gunn, J. E. 1976. Evolutionary Synthesis of the Stellar Population in Elliptical Galaxies. I. Ingredients, Broad-Band Colors, and Infrared Features. *ApJ*, **203**, 52–62.

Toonen, S. and Nelemans, G. 2013. The Effect of Common-Envelope Evolution on the Visible population of Post-Common-Envelope Binaries. *A&A*, **557**, A87.

Tremonti, C. A., Heckman, T. M., Kauffmann, G., et al. 2004. The Origin of the Mass-Metallicity Relation: Insights from 53,000 Star-Forming Galaxies in the Sloan Digital Sky Survey. *ApJ*, **613**, 898–913.

Tutukov, A. and Yungelson, L. 1996. Double-Degenerate Semidetached Binaries with Helium Secondaries: Cataclysmic Variables, Supersoft X-Ray Sources, Supernovae and Accretion-Induced Collapses. *MNRAS*, **280**, 1035–1045.

Van Bever, J. and Vanbeveren, D. 2000. Hard X-Rays Emitted by Starbursts as Predicted by Population Synthesis Models Including a Realistic Fraction of Interacting Binaries. *A&A*, **358**, 462–470.

Van Bever, J., Belkus, H., Vanbeveren, D., et al. 1999. The Effect of Close Binary Evolution on Photoionization Models for Evolving Starbursts. *NewA*, **4**, 173–190.

Vanbeveren, D., Van Bever, J. & Belkus, H. 2007. The Wolf–Rayet Population Predicted by Massive Single Star and Massive Binary Evolution. *ApJL*, **662**, L107–L110.

Westera, P., Lejeune, T., Buser, R., et al. 2002. A Standard Stellar Library for Evolutionary Synthesis. III. Metallicity Calibration. *A&A*, **381**, 524–538.

Wilkins, S. M., Feng, Y., Di-Matteo, T., et al. 2016. The Lyman-Continuum Photon Production Efficiency in the High-Redshift Universe. *MNRAS*, **458**, L6–L9.

Willems, B. and Kolb, U. 2004. Detached White Dwarf Main-Sequence Star binaries. *A&A*, **419**, 1057–1076.

Wright, N. J., Drew, J. E. and Mohr-Smith, M. 2015. The Massive Star Population of Cygnus OB2. *MNRAS*, **449**, 741–760.

Xiao, L. and Eldridge, J. J. 2015. Core-Collapse Supernova Rate Synthesis within 11 Mpc. *MNRAS*, **452**, 2597–2605.

Xiao, L., Stanway, E. R. and Eldridge, J. J, 2018, Emission Line Diagnostics of HII Regions. *MNRAS*, **477**, 904–934.

Yoon, S.-C., Gräfener, G., Vink, J. S., et al. 2012. On the Nature and Detectability of Type Ib/c Supernova Progenitors. *A&A*, **544**, L11.

Yoon, S.-C., Dessart, L. and Clocchiatti, A. 2017. Type Ib and IIb Supernova Progenitors in Interacting Binary Systems. *ApJ*, **840**, 10.

Zapartas, E., de Mink, S. E., Izzard, R. G., et al. 2017. Delay-Time Distribution of Core-Collapse Supernovae with Late Events Resulting from Binary Interaction. *A&A*, **601**, A29.

Zhang, F., Li, L. and Han, Z. 2005. Evolutionary Population Synthesis for Binary Stellar Population at High Spectral Resolution: Integrated Spectral Energy Distributions and Absorption-Feature Indices. *MNRAS*, **364**, 503–514.

Zhang, F., Li, L., Cheng, L. and Wang, L. 2015. Radiation Fields of Intermediate-Age Stellar Populations with Binaries as Ionizing Sources of H II Regions. *MNRAS*, **447**, L21–L25.

Index